Civil and Structural Engineering

Civil and Structural Engineering

Editor: Armando Ruiz

NY RESEARCH PRESS

New York

Published by NY Research Press
118-35 Queens Blvd., Suite 400,
Forest Hills, NY 11375, USA
www.nyresearchpress.com

Civil and Structural Engineering
Edited by Armando Ruiz

International Standard Book Number: 978-1-63238-538-3 (Hardback)

Cataloging-in-Publication Data

Civil and structural engineering / edited by Armando Ruiz.
p. cm.
Includes bibliographical references and index.
ISBN 978-1-63238-538-3
1. Civil engineering. 2. Structural engineering. I. Ruiz, Armando.
TA145 .C58 2017
624--dc23

Printed in the United States of America.

Contents

Preface

This book on civil and structural engineering discusses the fundamental principles and practices of civil engineering. Civil engineering encompasses many different fields such as urban engineering, architectural engineering, water resources engineering and construction surveying. Structural engineering is a significant aspect of civil engineering as it focuses on the strength and stability of physical structures. The various advancements in this field are glanced at and their applications as well as ramifications are looked at in detail. This book, with its detailed analyses and data, will prove immensely beneficial to professionals and students involved in this area at various levels. Students, researchers, experts and all associated with civil engineering and structural engineering will benefit alike from this text.

After months of intensive research and writing, this book is the end result of all who devoted their time and efforts in the initiation and progress of this book. It will surely be a source of reference in enhancing the required knowledge of the new developments in the area. During the course of developing this book, certain measures such as accuracy, authenticity and research focused analytical studies were given preference in order to produce a comprehensive book in the area of study.

This book would not have been possible without the efforts of the authors and the publisher. I extend my sincere thanks to them. Secondly, I express my gratitude to my family and well-wishers. And most importantly, I thank my students for constantly expressing their willingness and curiosity in enhancing their knowledge in the field, which encourages me to take up further research projects for the advancement of the area.

Editor

Seismic Vulnerability Assessment of Existing Building Stocks at Chandgaon in Chittagong city, Bangladesh

Atik Sarraz[*], Md. Khorshed Ali and Debesh Chandra Das

Department of Civil Engineering, University of Information Technology and Sciences (UITS), Chittagong, Bangladesh

Email address:

sarraz.sust@gmail.com (A. Sarraz), khorshed.chem@gmail.com (M. K. Ali), debeshdas73@yahoo.com (D. C. Das)

Abstract: The draft of Bangladesh National Building Code (BNBC)-2012 has been updated the seismic coefficient of 0.28g (with Zone III) for Chittagong region, which is larger than the previous of 0.15g (with Zone II). Chittagong is the largest port city and commercial capital of Bangladesh, which has many development activities as like of planned residential areas. Although BNBC code is up-to-date with earthquake provisions since 1993 with interpreting several new clauses and provisions, but in case of pre-code revision structures it is quite unsafe. Thus it is quite impossible to reduce earthquake damage without considering the safety of pre-code revision structures. In this regards earthquake vulnerability of Chandgaon Residential Area(R/A) has been assessed on the basis of potential structural vulnerability of more than 300 buildings. Initial results reveal that there have large varieties of construction practices, however, predominantly RCC structures were found. RVS score of these structures reveal that in general buildings are of minimum quality and further evaluation and strengthening of buildings is recommended. Walk down evaluation encountered several factors which were responsible for comparatively lower range of vulnerability scores.

Keywords: Chandgaon, BNBC, RVS, FEMA, Soft Story, Short Column

1. Introduction

For providing seismic safety in building structures need to ensure their conformance to the current seismic design codes which is a valid approach for new buildings. However majority of the existing buildings in seismic environments do not satisfy modern code requirements. The need to predict the seismic vulnerability of existing buildings has led to an increase interest on research dealing with the development of seismic vulnerability assessment of existing RC buildings. In case of Bangladesh which is possibly one of the country's most vulnerable to potential earthquake threat and damage. An earthquake of even medium magnitude on Richter scale can produce a mass damage without any previous notice in major cities of the country, particularly Dhaka and Chittagong.

Earthquake vulnerability of any place largely depends on its geology and topography, population density, building density and quality, and finally the coping strategy of its people and it shows clear spatial variations.

The location of Bangladesh close to the boundary of two active plates: the Indian plate in the west and the Eurasian plate in the east and north. As a result the country is always under a potential threat to earthquake at any magnitude at any time, which might cause catastrophic death tolls in less than a minute. In the basic seismic zoning map of Bangladesh Chittagong region has been shown under Zone II with basic seismic coefficient of 0.15 [1], but recent repeated shocking around this region indicating the possibilities of potential threat of even much higher intensity than projected.

According to report of Professor Roger Bilham and Peter Molnar of Colorado University huge amount of pressure is created under two kilometer wide Himalayan fault which can produce earthquake of magnitude 8.1 to 8.3 (in RS) at any time [5]. If this will happen, about two lack people will die and fifty million will be injured and affected.

According to Global Hazard Assessment Program (GSHAP), the most hazardous division in Bangladesh is the Port City Chittagong. About 80-90 percent of buildings and physical infrastructures in Chittagong are vulnerable to future massive earthquake measuring RS 6-7 magnitudes, as most of these were not designed to withstand against seismic load.

Hilly terrain of this city corporation area may create huge land slide during a heavy earthquake. As, most of the building contain sloppy ground around them. Asian Disaster

Preparation Center (ADPC) Seismic Hazards assessment has carried out at the Chittagong City Corporation Area of some buildings and found many vulnerable existing buildings. Now further evaluation of the seismic resistance and the assessment of possible damage are quite imperative in order to take preventive measures and reduce the potential damage to civil engineering structures and loss of human lives during possible future earthquakes. Several studies on seismic vulnerability assessment have been carried out at Chittagong City Corporation, but those performed by the consideration of showing the region at zone II of seismic zoning map of BNBC. But new update of national building code (BNBC) proposed this region at zone III with a seismic coefficient value of 0.28g [Fig. 2]. Thus a pilot application in a residential area named by Chandgaon Residential Area (R/A) of Chittagong City Corporation has been conducted which is situated on the banks of Karnaphuli River and is a most densely populated area of the city. Seismic risks of RC structures were evaluated and the concerned authority will be noticed of the probable disaster by providing these data.

1.1. Status of Earthquakes in and around Bangladesh

Bangladesh is surrounded by the regions of high seismicity which include the Himalayan Arc and Shillong Plateau in the north, the Burmese Arc, Arakan Yoma anticlinorium in the east and complex Naga-Disang-Jaflong thrust zones in the northeast. It is also the site of the Dauki Fault system along with numerous subsurface active faults and a flexure zone called Hinge Zone. These weak regions are believed to provide the necessary zones for movements within the basin area. In the generalized tectonic map of Bangladesh the distribution of epicenters is found to be linear along the Dauki Fault system and random in other regions of Bangladesh. The investigation of the map demonstrates that the epicenters are lying in the weak zones comprising surface or subsurface faults. Most of the events are of moderate rank (magnitude 4-6) and lie at a shallow depth, which suggests that the recent movements occurred in the sediments overlying the basement rocks. In the northeastern region (SURMA BASIN), major events are controlled by the Dauki Fault system. The events located in and around the ADHUPUR TRACT also indicate shallow displacement in the faults separating the block from the ALLUVIUM. The first seismic zoning map of the subcontinent was compiled by the Geological Survey of India in 1935.

The Bangladesh Meteorological Department adopted a seismic zoning map in 1972. In 1977, the Government of Bangladesh constituted a Committee of Experts to examine the seismic problem and make appropriate recommendations. The Committee proposed a zoning map of Bangladesh in the same year.

1.2. Geologic and Tectonic Set-Up

Tectonically, Bangladesh lies in the northeastern Indian plate near the edge of the Indian carton and at the junction of three tectonic plates – the Indian plate, the Eurasian plate and the Burmese micro plate. These form two boundaries where plates converge the India- Eurasia plate boundary to the north forming the Himalaya Arc and the India-Burma plate boundary to the east forming the Burma Arc (Fig. 1).

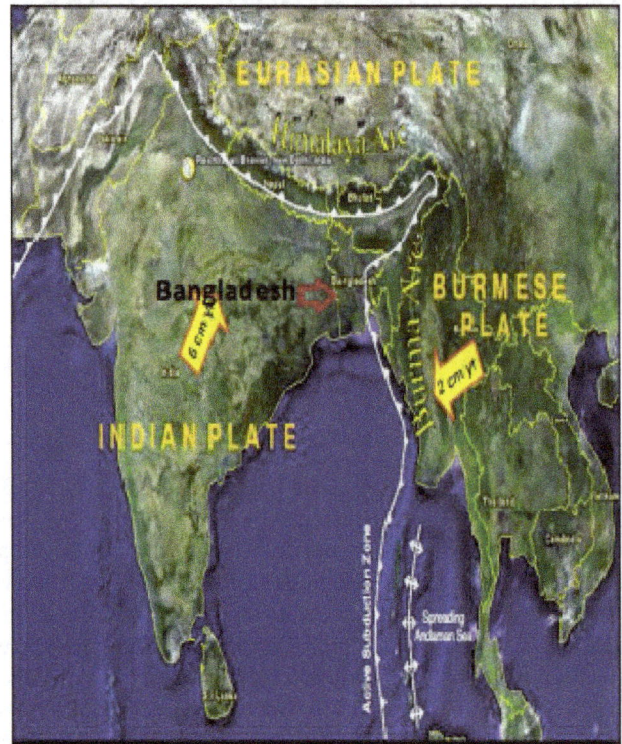

Figure 1. *Regional Tectonic Setup of Bangladesh with respect to Plate [20]*

The Indian plate is moving ~6 cm/yr in a northeast direction and sub ducting under the Eurasian (45 mm/yr) and the Burmese (35 mm/yr) plates in the north and east, respectively [2, 3]. This continuous motion is taken up by active faults. Active faults of regional scale capable of generating moderate to great earthquakes are present in and around Bangladesh. These include the Dauki fault, about 300 km long trending east-west and located along the southern edge of Shillong Plateau (Meghalaya- Bangladesh border), the 150 km long Madhupur fault trending north- south situated between Madhupur Tract and Jamuna flood plain, Assam-Sylhet fault, about 300 km long trending north-east-south-west located in the southern Surma basin and the Chittagong-Myanmar plate boundary fault, about 800 km long runs parallel to Chittagong-Myanmar coast (Fig. 2).

The Chittagong- Myanmar plate boundary continues south to Sumatra where it ruptured in the disastrous 26 December 2004 Mw 9.3 earthquake [4]. These faults are the surface expression of fault systems that underlie the northern and eastern parts of Bangladesh. Another tectonic element, the Himalayan Arc' is characterized by three well defined fault systems (HFT, MBT and MCT) that are 2500 km long stretching from northwest syntaxial bend in Pakistan in the west to northeast syntaxial bend in Assam in the east. It poses a great threat to Bangladesh as significant damaging historical earthquakes have occurred in this seismic belt [5, 6 & 7].

1.3. Seismic Sources of Bangladesh

Bolt (1987) analyzed the different seismic sources in and around Bangladesh and arrived at conclusion related to maximum likely earthquake magnitude [8]. The magnitudes of earthquake suggested by Bolt (1987) in Table 1 are the maximum magnitude generated in these tectonic blocks as recorded in the historical seismic catalogue. The historical seismic catalogue of the region covers approximately 250 years of recent seismicity of the region and such a meager data base does not provide a true picture of the seismicity of the tectonic provinces. For example, the Assam and the Tripura fault zones contain significant faults capable of producing magnitude 8.6 and 8.0 earthquakes, respectively, in the future. Similarly, earthquakes with maximum magnitude of 7.5 in Sub-Dauki fault zone and in Bogra fault zone are not unlikely events.

Table 1. Seismic Sources of Potential Earthquake [8]

Location	Maximum Likely Magnitude
(i)	(ii)
Assam fault	8.0
Tripura Fault	7.0
Sub-Dauki Fault	7.3
Bogra Fault	7.0

1.4. Earthquake Zone Co-Efficient

Fig. 2 presents the proposed seismic zoning map for Bangladesh based on PGA values for a return period of 2475 years. The country is divided into four seismic zones with zone coefficient Z equal to 0.12 (Zone 1), 0.2 (Zone 2), 0.28 (Zone 3) and 0.36 (Zone 4). In previous Bangladesh National Building Code (BNBC) of 1993 there were three (3) zones. As mention earlier in BNBC 1993 the zone co-efficient of Chittagong was 0.15 [1]. But in newer adopted code of BNBC 2010 it would be proposed to 0.28 which like to be nearly doubled of previous coefficient thus making the zone prone to be highly risk against Earthquake [9, 10]. The red circle zone is the zone of the study area which indicates the Zone III with a basic seismic co-efficient of 0.28 [Fig. 2].

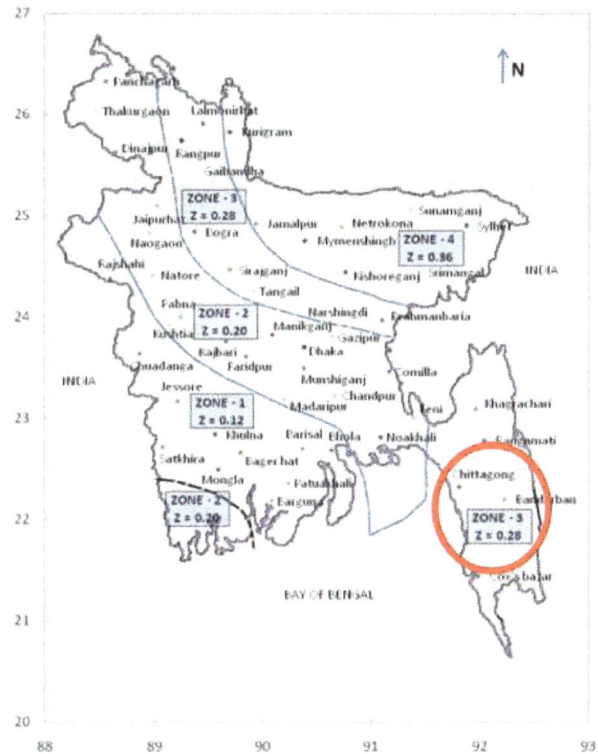

Figure 2. Update Earthquake Zoning Map of Bangladesh [9, 10]

1.5. Seismicity Records of Chittagong

Chittagong is the second largest city of Bangladesh which is located in a strategic geographic position at the south-eastern part of the country, contributes a lot in the national economy acting as a commercial hub being connected with the busiest sea- port. The city is exposed mostly tropical storm surges and earthquake. Moreover Chittagong City Corporation (CCC) area is situated approximately 70 km from the fault zone in Bangladesh-Myanmar Boarder. Historical information reveals that earthquakes of magnitude between 6 and 7 have been occurred around the city in the past [11].

For last two decades there were encountered about 8 major earthquake around the city as 1997- Bandarban (M 6.1), 1999- Moheshkhali (M 5.1), 2009- Chandanaish (M 5.2) earthquake etc. (Table 2). These earthquakes caused enormous amount of damages with causalities (Table 2)

Table 2. Lists of Recent Earthquakes around the Region

Date of Occurrence	Epicenter of Earthquake	Epicentral Distance, km	Magnitude (Mb)	Causalities
(i)	(ii)	(iii)	(iv)	(v)
21-11- 1997	Bandarban (Myanmar Border)	65	6.1	20 killed
22-07- 1999	Moheshkhali	184	5.1	6 Killed
19-12- 2001	Dhaka (Manikganj)	285	4.2	20 injured
22-07- 2005	Rangamati	37	5.5	2 killed
13-12- 2009	Chittagong	45	5.2	N/A
10-11- 2010	Chandpur	125	4.8	N/A
03-05- 2011	Comilla	115	4.6	N/A

2. Description of Study Area

The study area is Chandgaon R/A at ward no.4 of Chittagong City Corporation situated on the bank of Karnaphuli River. Ward no. 4 is the most densely populated (131,212) ward of Chittagong city and have a residential

building value of 129 million US dollar [12]. Chandgaon have a population near 30,000 and more than 400 buildings [13]. The global positioning of Chandgaon is around the 91° 52' 10" N and 22° 21' 40" S (Fig. 3).

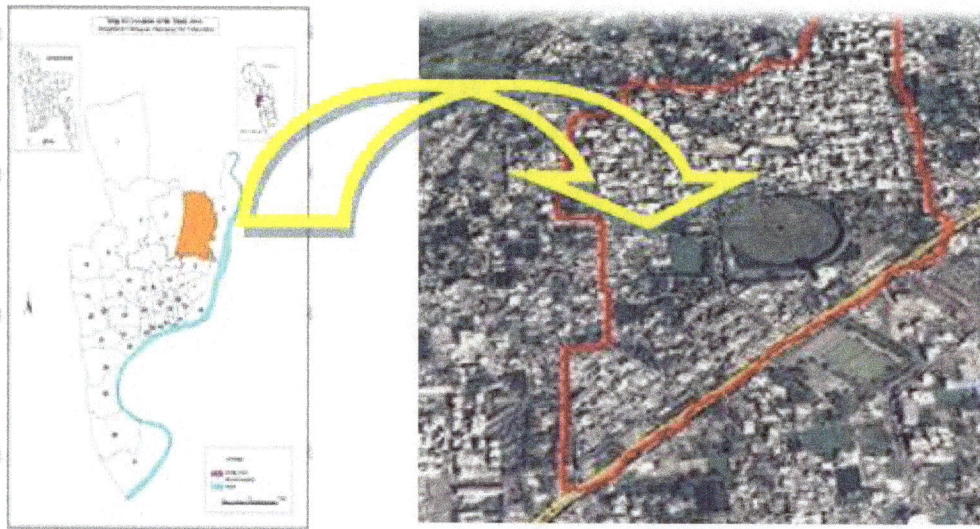

Figure 3. Satellite View of Chandgaon R/A; Source: CCC and Google map, 2014

3. Methodology

3.1. Methods of Seismic Vulnerability Assessment

Rapid Visual Screening (RVS) method was originally developed by the Applied Technology Council (ATC) in the late 1980's and published in 1988 in the FEMA 154 [14] report. It is a "sidewalk survey" approach that enabled users to classify surveyed buildings into two categories: those acceptable as to "risk to life safety" or those that may be seismically hazardous and should be evaluated in more detail. The Turkish Simple Survey procedure is a two level risk assessment procedure which has been proposed on the basis of statistical correlations obtained by Sucuoglu and Yazgan (2003) [15]. The first level incorporates recording of building parameters from the street side and in the second level, these are extended by structural parameters measured by entering into the ground storey.

Although there have no method developed in Bangladesh but some efforts found in India towards developing RVS methods. In this study the method of vulnerability assessment limited to the walk down survey (level-I). Two basic procedures employed in this study. One was the RVS of FEMA 154 [14] and another had done based on the RVS forms proposed by Sudhir K Jain et al. [16, 17] for Indian conditions.

3.2. Rapid Visual Screening (RVS)

Data collection form of RVS (FEMA 154) includes space for documenting building identification information,

including its use and size, a photograph of the building, sketches, and documentation of pertinent data related to seismic performance, including the development of a numeric seismic hazard score. There are three types of form in FEMA 154 [14]. The form for high seismicity region has been used to carry out the research work. Basic Structural Hazard Scores based on Lateral Force Resisting System for various building types are provided on the form, and the screener circles the appropriate one. The screener modifies the Basic Structural Hazard Score by identifying and circling Score Modifiers related to observed performance attributes, by adding (or subtracting) them a final Structural Score, 'S' is obtained.

3.3. Indian Method of Evaluation

Indian evaluation method is based on few parameters of RCC and Masonry building. The parameters of the RCC buildings are building height, frame action, pounding effect, structural irregularity etc. as shown in Table 3. The scoring system adopted in this research is for seismic zone V in Indian seismic zoning map as the value of zone coefficient is relevant to the updated zone coefficient of Chittagong region as proposed by Dr. T.M. Al-Hussaini [9]. A "cut-off" performance score of 50 has been suggested for the study.

On the basis of parameters shown in Table 3, Performance Score (PS) of the buildings has been calculated by using the values of Base Score (BS), Vulnerability Score (VS) and Vulnerability Score Modifiers (VSM). The formula of the performance score is given as:

$$PS = (BS) - \Sigma [(VSM) \times (VS)] \qquad (1)$$

Where, BS = Base Score; VSM = Vulnerability Score Modifiers; VS = Vulnerability Score

The corresponding storey wise value of Vulnerability Score (VS) for different parameters compiled at Table 3. The BS value of zone V (= 0.36) of Indian seismic zoning map was selected as the Base Score (BS) due to the consideration of maximum risk and the coefficient value (= 0.36) is quite nearer to the value of study area (= 0.28). Table 3 shows the storey wise distribution of Base Score (BS). The value of Vulnerability Score Modifiers (VSM) of equation 1 also described at Table 5.

Table 3. Vulnerability Scores (VS) for RCC Building [16, 17]

No of Stories	1or 2	3	4	5	>5
Vulnerability Factors	Vulnerability Scores				
(i)	(ii)	(iii)	(iv)	(v)	(vi)
Soft Stories	0	-15	-20	-25	-30
Vertical Irregularities	-10	-10	-10	-10	-10
Plan Irregularities	-5	-5	-5	-5	-5
Heavy Overhanging	-5	-10	-10	-15	-15
Apparent Quality	-5	-10	-10	-15	-15
Short Column	-5	-5	-5	-5	-5
Pounding Effects	0	-2	-3	-3	-3
Soil Condition	10	10	10	10	10
Frame Action	10	10	10	10	10
Water Tank at Roof	0	-3	-4	-5	-5
Location of Water Tank	0	-3	-4	-5	-5
Basement-Full/Partial	0	3	4	5	5

Table 4. Base Scores (BS) for RCC Building [16, 17]

Number of Stories	Base Scores
(i)	(ii)
1 or 2	100
3	90
4	75
5	65
More than 5	60

Table 5. Vulnerability Score Modifiers (VSM) for RCC Building [16, 17]

Vulnerability Factors	Vulnerability Score Modifiers (VSM)		
(i)	(ii)	(iii)	(iv)
Soft Stories	Absent =0	Present =1	---
Vertical Irregularities	Absent =0	Present =1	---
Plan Irregularities	None =0	Moderate =1	Extreme =2
Heavy Overhanging	Absent =0	Present =1	---
Apparent Quality	Good =0	Moderate =1	Poor =2
Short Column	Absent =0	Present =1	---
Pounding Effects	Absent =0	Unaligned floors =2	Poor apparent quality of adjacent building =2
Soil Condition	Medium =0	Hard =1	Soft = -1
Frame Action	Absent =0	Present =1	Not sure =0
Roof Water Tank Capacity	Absent =0	<5000 L =0.5	>5000L =1.0
Location of Water Tank	Symmetric =0	Asymmetric =1	---
Basement-Full/Partial	Absent =0	Present =1	---

Based on the scores of RVS, some percentage of structures will be selected for preliminary evaluation and further for detailed evaluation. RVS is useful when the number of buildings to be evaluated is large and even non-engineers may collect data and assign scores.

4. General Description of Building

The study comprised of a detailed survey on 310 buildings of Chandgaon R/A. Although largest percentage (95%) of buildings are of residential occupancy but other types of occupancy like mosque, school (3%) and offices (2%) also present here. The structural systems are mainly encompassed of two types, one is the Moment Resisting Frame (MRF) and another is the Unreinforced Masonry Wall Infill Frame (URM) from the categories provided at RVS form. It was found from the study that about 99% building belongs to the category of URM-infill and only 1% belongs to the category of MRF.

4.1. Age of Buildings

The age of the building is attributed at both of the methods used in the study. From this information the building can be classified as range of four significant stage of development of BNBC code according to Table 6.

Study found that about 68% buildings built at applicative stage of code and 14% built at legislative stage (Fig. 4). But the main vulnerable objectives were 10% buildings built at precode and 8% at primitive stage. According to this finding Chandgaon can be termed as a newly developed residential area.

Table 6. *Stages of Building Age based on the development period of BNBC*

Stage	Range of Year Built	Remarks
(i)	(ii)	(iii)
Precode	< 1993	Code was proposed
Primitive	1993 – 1996	Updated for 1st time
Applicative	1996 – 2006	Updated for 2nd time
Legislative	> 2006	After the year of 2006

4.2. Vertical Height of Buildings

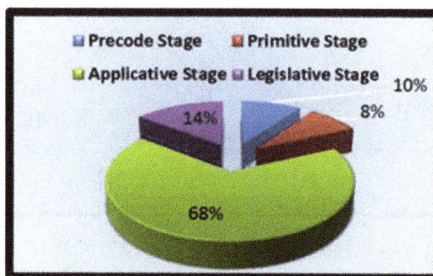

Figure 4. *Different Stages of Building Age*

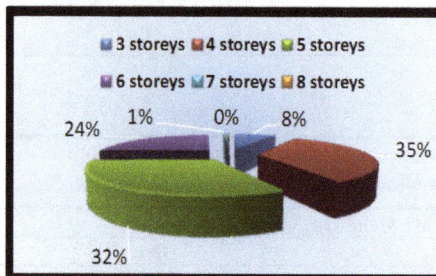

Figure 5. *Vertical Height of Buildings*

The vertical height of the building limited to 8 storeys. Figure 5 shows that majority of the buildings are 4storied (35%) and 5 storied (32%) where 24% buildings are 6 storied. A smallest percentage of 0 to 1% buildings are 7 to 8 storied (Fig. 5). Thus it can be said that Chandgaon area is occupied by low to medium rise building rather than the most vulnerable high rise structure.

5. Vulnerability Factors

5.1. Structural Irregularities

Structural irregularity is the combination of two vulnerability factors: plan irregularity and vertical irregularity. The criteria based definition of plan and vertical irregularities mentioned at both the FEMA and BNBC regulations [1, 14]. The Euro code 2008 showed a precise analysis of plan and vertical irregularities by showing the marginal values of setback in elevation and slope at contour [19].

The study found that the plan and vertical irregularities possess by only small percentages of building. These factors were encountered for a maximum of 7 and 6 numbers of four storied building of the Chandgaon area (Table 4).

5.2. Storey Drift Parameters

Major storey drift parameters are soft storey and pounding effects. A soft story usually exists in a building when one particular story, usually employed as a commercial space, has less stiffness and strength compared to the other stories [15]. In this study soft story problem encountered for 37, 55 and 64 buildings of 4, 5 and 6 storied respectively (Table 4). Soft story problem found as major vulnerability factor of the buildings (53.7%) of Chandgaon area.

When there is no sufficient clearance between adjacent buildings, they pound each other during an earthquake as a result of different vibration periods. Uneven floor levels aggravate the effect of pounding [15]. Pounding effects was found only for 2.2% buildings of Chandgaon area (Table 4). Short column is another factor which can be formed at frames with partial infill which sustain heavy damage since they are not designed for the high shear forces due to shortened heights that will result from a strong earthquake [15]. Here short column, heavy overhanging and poor apparent quality found for 2.6%, 0.97 and 1.94% building respectively (Table 7).

Table 7. *Vulnerability Factors Found in the Study*

	Vulnerability Parameters							Roof Water Tank Capacity		Position of Roof Water Tank	
	VI	PI	SC	HO	SS	PAQ	PE	≥ 5000 liter	< 5000 liter	Unsymmetrical	Symmetrical
	(i)	(ii)	(iii)	(iv)	(v)	(vi)	(vii)	(viii)	(ix)	(x)	(xi)
	Number of Buildings										
3 Storied	2	1	0	0	7	0	0	0	24	23	1
4 Storied	7	6	5	3	37	5	2	99	8	103	5
5 Storied	3	3	3	0	55	1	2	73	27	88	12
6 Storied	1	1	0	0	64	0	3	68	8	24	6
7 Storied	0	0	0	0	2	0	0	2	0	2	0
8 Storied	0	0	0	0	1	0	0	2	0	1	0
Total =	13	11	8	3	166	6	7	244	67	241	24
Percentage =	4.2	3.6	2.6	0.97	53.7	1.94	2.2	79	21.68	78	7.77

VI= Vertical Irregularities, PI= Plan Irregularities, SC= Short Column, HO= Heavy Overhanging, SS= Soft Storey, PAQ= Poor Apparent Quality, PE= Pounding Effects

In case of roof water tank, largest percentage (about 79%) of building possess a tank of more than 5000 liter capacity and 78% building have a tank located at an unsymmetrical position with respect to roof plan of the building (Table 4).

6. Results and Discussion

6.1. Assessment by RVS

As mentioned earlier, total number of 310 Buildings have been analyzed using Rapid Visual Screening (RVS) method. Considering Chittagong as a high Seismic Risk zone, the cut off value was determined as 2.0. The results show that no score for any building was found to touch the cut off value according to FEMA method and all of them require further detailed analysis for vulnerability to determine the level of actual risk. Table 8 shows summary of the RVS score for different storied buildings.

In fact the basic score for RC building in FEMA-RVS is only 1.6 (less than the cut off), which becomes smaller after being modified by the negative parameters. This is one of the reasons for the FEMA RVS score to be less.

Table 8. *Final Scores of RVS*

	Score Range			
	S≤0.4	0.4<S<0.7	0.7<S<1.2	1.2<S<2
	Number of Buildings			
3 Storied	4	0	14	6
4 Storied	9	0	78	21
5 Storied	3	0	79	18
6 Storied	0	0	59	17
7 Storied	0	0	1	1
Total =	16	0	231	63
Percentage =	5	0	75	20

The parameters contributing the scoring system are mainly, the height, irregularities of the buildings and type of the soil underneath. The parameters "Pre Code" and "Post Benchmark" remained inapplicable in the scoring. On a general view, the soil type of Chandgaon has been considered as Stiff, so this modifier also remains constant in the whole process.

6.2. Assessment by Indian Method

Here, the assessment has been performed considering the soil zonation of Chittagong in Zone-V of Indian method which PGA (Peak Ground Acceleration) value is analogous to Zone-III (Chittagong) of updated BNBC earthquake zoning map.

In this method much more variation in final scores has been observed as not only the basic score but also the influences of vulnerability parameters are very much dependent on the height of the building.

In fact the positive or negative score modifications due to vulnerability parameters are weighted multiplications based on their existence and number of stories of the buildings. As a result the score becomes high for low rise buildings in spite

of presence of negatively influential vulnerability parameters. Due to these dependent variations, it is comparatively tougher task here to classify the damage probabilities with minute specifications, as that of RVS-FEMA method, only from final scores. Rather it is easier to indicate an overall view on safety of the building comparing the final score with the cut off value and observing their relative difference.

Fig. 6 reflects that about 50% buildings from 6 storied and verily 5%, 1% from 5 and 4 storied scored less than 25 which less than the "cut-off" score, according to the survey. All of the 7 and 8 storied buildings scored less than 25. The building with a score of 0 to 25 might be possessed to the most critical state of vulnerability and must be investigated further. Fig. 6 also reveals that more than 50% of 5 storied and 80% of 6 storied building have a score less than 50 (the "cutoff score"). These buildings also should be assessed in details. Thus rest of the buildings does not need further detailing on vulnerability assessment, or they can be termed as safe.

When the performance scores of the buildings have been placed against building story, it was found that few 4 storied and more than 80% of 3 storied buildings scored above 75 (Fig. 6) and all the buildings scoring greater than 50 (Fig. 6) were low to midrise (3 to 5 storied). The observed general trend is that the taller the buildings the higher the presence of negative parameters and thus the lower becomes the scores.

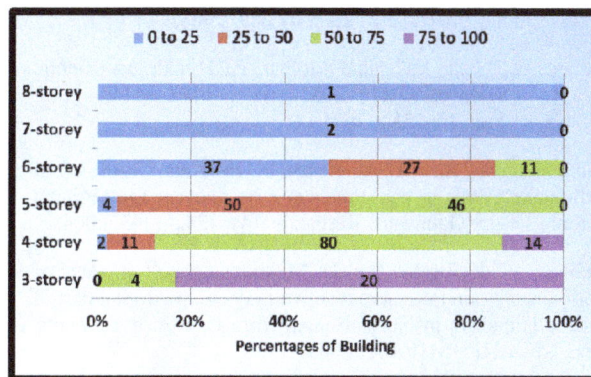

Figure 6. *Performance Score of Assessment by Indian Method*

7. Conclusions

The occurrence of earthquakes is a part of the natural process in the earth's geophysical system. The earthquake tremors cannot be stopped or reduced and the causalities and damages are caused mainly due to the collapse of the infrastructures. The infrastructures of different areas will not be equally vulnerable to any earthquake. In this research, an attempt has been made to develop firsthand vulnerability assessment for Chandgaon Residential Area (R/A).These works are expected to be useful for especially pre-disaster planning and capital investment planning.

In the present research, an inventory with huge information of the building features has been made for residential buildings of Chandgaon area. The vulnerability of the buildings has been assessed so that the buildings prone to earthquake can be identified and repair, restoration or evacuation plans can be prepared easily. The results of the vulnerability analyses can also be useful for providing guidance in the construction of seismic resistant buildings.

While, it is not possible and also not economic to abandon all the vulnerable structures, future buildings and structures in the area are recommended to be brought under strict building code as well as to be constructed according to the land use planning zoning ordinances so that earthquake vulnerability of the region can be mitigated.

References

[1] BNBC 2006, "Bangladesh National Building Code", House and Building Research Institute, Dhaka, Bangladesh (http://buildingcode.gov.bd/), 2006.

[2] Sella, G. F., T. H. Dixon, and A. Mao, "REVEL: A model for recent plate velocities from space geodesy, J. Geophys. Res., 107(B4), 2081, doi: 10.1029/2000JB000033, 2002.

[3] Bilham, R.,"Earthquakes in India and the Himalaya: tectonics, geodesy and history", Annals of Geophysics, Vol. 47, N. 2/3, April/June 2004.

[4] Steckler, M. S., Akhter, S. H., Seeber, L. Collision of the Ganges-Brahmaputra Delta with the Burma Arc: Implications for earthquake hazard. Earth and Planetary Science Letters, 273, 367-378. doi 10.1016/j.epsl.2008.07.009, 2008.

[5] Bilham, R., Gaur, V.K., and Molnar, P., Himalayan seismic hazard: Science, v. 293, p. 1442–1444, doi: 10.1126/science.1062584, 2001.

[6] Mukhopadhyay, B., S. Dasgupta and S. Dasgupta, Clustering of earthquake events in the Himalaya – Its relevance to regional tectonic set-up. Gondwana Research 7(4): 1242-1247, 2004.

[7] Mullick, M., F. Riguzzi and D. Mukhopadhyay, Estimates of motion and strain rates across active faults in the frontal part of eastern Himalays in North Bengal from GPS measurements. Terra Nova 21(5): 410-415, 2009.

[8] Bolt B.A., Fault Zone Analysis for Seismic Hazards, Journal of Seismology, Vol. 1, No. 3, 1987.

[9] Al-Hussaini, T.M., Hossain, T.R. and Al-Noman, M.N., "Proposed Changes to the Geotechnical Earthquake Engineering Provisions of the Bangladesh National Building Code", Geotechnical Engineering Journal of the SEAGS & AGSSEA, Vol. 43, No.2, 1-7 (www.seags.ait.ac.th), 2012.

[10] Ansary M.A. and Dhar A. S., "Earthquake Vulnerability Assessment of Cox's Bazar District", A report to CDMP, Department of Civil Engineering, BUET, Bangladesh, 2009.

[11] M. J. Alam, A.R. Khan and A. Paul, "Seismic Vulnerability Assessment of Existing RC Buildings in GIS Environment", Online Proc. of Fourteenth World Conference on Earthquake Engineering (WCEE), Beijing, China (http://www.nicee.org/wcee/index2.php), October 12-17, 2008.

[12] CDMP (Comprehensive Disaster Management Programme), "Earthquake Risk Assessment of Dhaka, Chittagong and Sylhet City Corporation Area", General report, Table A-8, Appendix B, pp. 113. Ministry of Food and Disaster Management, Govt. of Bangladesh, June 2009.

[13] BBS (Bangladesh Bureau of Statistics), "Bangladesh Population Census 2011", vol-3, Urban Area Report, Chittagong: Statistics Division, Ministry of Planning, Govt. of Bangladesh, 2011.

[14] FEMA 154, "A Hand Book on Rapid Visual Screening of Buildings for Potential Seismic Hazards", Handbook, Federal Emergency Management Agency, Washington DC, USA, 2002.

[15] Sucuoglu, H. and Yazgan, U., "Simple Survey Procedures for Seismic Risk Assessment in Urban Building Stocks", Seismic Assessment and Rehabilitation of Existing Buildings, NATO Science Series, IV/29, pp. 97-118, 2003.

[16] Sudhir K Jain, Keya Mitra, Praseeda KI, "A Proposed Rapid Visual Screening Procedure for Seismic Evaluation of Buildings in India", January, 2004.

[17] Srikanth T., Kumar R. P., Singh, A. P., Rastogi, B. and Kumar, S., "Earthquake Vulnerability Assessment of Existing Buildings in Gandhidham and Adipur Cities Kachchh, Gujarat (India)", European Journal of Scientific Research , Vol.41 No.3, pp. 336-353 (http://www.eurojournals.com/ejsr.html), 2010.

[18] Ozcebe, G.,Yucemen, M. S., Aydogan, V., and A. Yakut, "Preliminary Seismic Vulnerability Assessment of Existing Reinforced Concrete Buildings in Turkey- Part I: Statistical Model Based on Structural Characteristics", Seismic Assessment and Rehabilitation of Existing Buildings, NATO, Science Series IV/29, pp. 29-42, 2003.

[19] Euro Code 8, "Design of structures for earthquake resistance-Part-1.", European Standard PREN 1998–1. Draft no. 4. Brussels: European Committee for Standardization, 2001.

[20] www.cdmp.org.bd/modules.php?name=Publications&download=13_Time_Predictable_Fault_Modeling__of_BD.pdf

Thermo-Physical Properties of Local Materials Used in the Construction of Chad

Ahmat-Charfadine Mahamat[1], Mahamat Barka[1], Abakar Mahamat Tahir[1], Malloum Soultan[1], Salif Gaye[2], Aboubakar Cheikh Beye[2]

[1]Laboratoire des Energies Renouvelables et des Matériaux Locaux de Faculté des Sciences Exactes et Appliquées de l'Université de N'Djaména, Tchad

[2]Laboratoire de Matériaux, Mécanique et Hydraulique de la Faculté des Sciences et Techniques de l'Université de Thiès, Sénégal

Email address:

yakoussou@yahoo.fr (Ahmat-Charfadine M.), mahamat.barka@gmail.com (M. Barka), abakarmt@gmail.com (Abakar. M. T.), malloum.sultan@gmail.com (M. Soultan), sgaye@univ-thies.sn (S. Gaye), aboubakar.c.b@ucad.sn (A. C. Beye)

Abstract: As earth is the most commonly used construction material in Chad, the purpose of the present study is to characterize it as a whole, and to characterize its thermo physical properties. Various tests were conducted on it and mainly on its thermo-physical characteristics, as these significantly affect the comfort of a house. The study focused on earth taken from a climate zone of Chad, mixed with straw at a given proportion or used alone. Different values of thermal conductivity and diffusivity of these materials as well as their variations in relation with water continence are presented.

Keywords: Construction Materials, Earth/Clay, Thermo-Physical Characteristics, Thermal Conductivity, Thermal Diffusivity, Housing

1. Introduction

Nowadays in rural or urban Sahel which Chad is a part, various materials are used for the construction of housing. These materials range from earth mixed with straw commonly called "banco", to the mixture of cement and sand or even concrete. The purpose of this research is the study of thermo-physical properties of the material earth used in the construction of housing in Chad. Given the diversity of these materials in either parts of the country, it is important to model the thermo-physical behavior of this type of construction material. As for the materials considered as porous, there are many possible models at our disposal.

Assuming that earth shaped as bricks is treated as a porous material; we selected the parallel series configuration with a distribution θ when the pores are filled with air. We present the different testing methods used to study the thermo-physical properties of the material under consideration and the test results conducted on the different materials studied on the one hand, and we carried out the practical study of thermal modeling of these materials on the other.

2. Studied Materials

Raw earth, as used in building construction since the earliest times and evidenced by traditional housing in numerous parts of the planet, is the subject of our study. As it is a highly heterogeneous material with very different characteristics from one region to another, as well as being a very large resource, its precise knowledge before any use in construction is important. Numerous laboratory tests and in situ tests can be performed to identify and characterize the thermal conditions of a soil. In order to establish a reliable diagnosis for the use of earth in construction, a number of these tests are absolutely necessary.

2.1. Preparation of Samples

2.1.1. Choice of Site

Given the very large size of the country and taking into account the diversity of its climate zones and meteorological data, we limited our study to the earth materials used in the city of N'Djamena, located in the Sahelian zone. The earth studied is one taken from quarries closest to the abovementioned city and from extraction depths comprised between 1.50 and 2.50 meters. On that soil, we carried out a few determination tests on its thermo-physical properties and assessed its potential for use in construction as well as its contribution to thermal comfort.

2.1.2. Production of Samples

The making of earth samples derives from the traditional technique of making mud bricks [2], [3], [20]. A certain amount of earth is taken from the collected material, poured into a tank, and a certain amount of water is added. The whole is mixed until a consistency closer to plasticity is reached or even exceeded, generally. The paste is then packed into a mold designed for this purpose until completely filled. The molded mixture is left for 24 hours at room temperature in the laboratory.

After 24 hours of drying, that mixture is placed in an oven at a temperature of 105 °C ± 5 °C for 24 hours. After drying, the samples are left to cool for 15 to 20 minutes before the last weighing. In order to obtain baked bricks, the samples are made, dried, and left to bake in an oven at a temperature of 300 ° C to 400 ° C for twelve hours. To measure the thermo physical characteristics of the material by the method known as "method of boxes" presented in Chapter 2, we made samples of adobe (earth + straw) destined to be specifically placed in the measuring cell. Table 1show the dimensions and detailed compositions of samples made and dried in an oven for 24 hours.

Table 1. *Dimensions and compositions of the samples.*

Designation Composition	N'Djamena Earth alone	Earth and straw
Mass of earth (kg)	3,5	0,5
Mass of straw (kg)	0	0,05
Volume (cm3)	23,28 x 22,80 x 2,54	17,52 x 8,29 x 1,84
Mass (dry) in kg	2,786	0,409
a (kg/m3)	2068	1529

3. Measurement Methodology and Description of the Material

Conductivity, diffusivity, mass-related heat and thermal resistance are the most important features to consider when choosing a heat exchanger material. It is important to remember that conductivity is measured in steady state while diffusivity is involved in transient state and, in most cases these two quantities are the measured ones. The two others are derived by computation. Many methods are used to measure conductivity and thermal diffusivity. We used the *Boxes Method* developed in the Laboratory of Solar and Thermal Studies of Claude Bernard University. It is relatively fast, of a simple handling technique, and has a level of precision comparable to others. This method has been the subject of several publications [1], [5], [4], [17], [21]; It uses samples of significant sizes, and the associated measurements are made in the actual conditions of use of the materials. Figure 1 provides an overview of the experimental device the

main elements of which are described below:

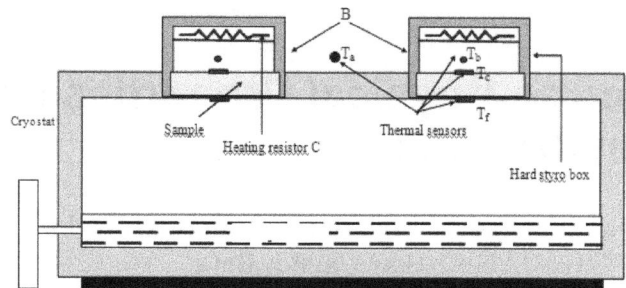

Figure 1. *Sectional view of the cold capacitor and hot boxes.*

The measuring cell consists of:

A lower chamber of size 2m x 1m x 0.45m that is highly insulated and kept at low temperature (up to-10 ° C). It is supplied with brine cooler. This cooler that we have named "exchanger" is powered by a HAAKE circulating bath cryostat of type DC1.

Two identical plywood boxes B isolated from the inside by the styrodur and each equipped with a heat emitter C allow two simultaneous measurements of conductivity and diffusivity.

- In order to measure the level of conductivity, the box is coated on the inside of its upper surface with a film of low power (<10 W) heating element in Zegadi [4], [13], [14]whose heat output is controlled with a rheostat.
- For the measurement of diffusivity, the box is equipped with a high power incandescent lamp of 500 to 1500 W.

All the wiring connections of the temperature sensors and film heating power source are connected to a box that can in turn be connected either to a measurement console or a type E 7100 hybrid central data acquisition recorder with 32 inlets and equipped with an RS232C communication interface. It allows the use of a 286 PC compatible remote computer to access the information programmed in Turbo Pascal 6.0 language for storage or modification, and obtains measurement results automatically, among others.

Temperature sensors that are surface and ambient atmosphere probes divided between the two faces of the sample that measures the temperature of the hot surface (T_c) and that of the cold surface (T_f) as well as within the box that measures the ambient temperature(T_b). Another sensor common to both boxes measures the ambient temperature of the room (T_a). The samples used have parallelepiped shapes and 3 to 7 cm thick. Their square sections have 27 cm sides.

3.1. Measurement of Thermal Conductivity

3.1.1. Evaluation of Thermal Conductivity

Conductivity is a function of temperature. The expression of conductivity averaged over the interval [T_c, T_f] is given by

$$\lambda_m = \frac{1}{T_c - T_f} \int_{T_c}^{T_f} \lambda(T) dT \qquad (1.1)$$

Thus, λ_m is the conductivity corresponding to the average

temperature

$T_m = (T_c + T_f) / 2$ for a temperature difference $\Delta T = T_c - T_f$.

This correspondence between λ_m and T_m requires linear dependence between the two measurements. In steady state, the heat flow that goes through the sample is written as

$$\lambda_{eff} = \left[\frac{\theta}{(1 - P_T).\lambda_s + \lambda_f.P_T} + (1 - \theta).\left(\frac{1 - P_T}{\lambda_s} + \frac{P_T}{\lambda_f} \right) \right]^{-1}$$

with:

$\dot{Q}$: Power emitted by the resistance formulated by $\frac{V^2}{R}$

(where V is the voltage across resistor R)

$\dot{Q}_1$: Power exchanged between the inside and outside of the box (it is formulated by $\dot{Q}_1 = C(T_b - T_a)$, C is the coefficient of heat loss through the box and is expressed in W / °C. It is either determined experimentally or calculated).

$\dot{Q}_2$: Thermal power passing through the sample. It is formulated by $\dot{Q}_2 = \frac{\lambda}{e}S(T_c - T_f)$ where e represents the thickness of the sample, S the useful area also known as measurement area and slightly less than (27 x 27) cm² in size. In our calculations, we used the area described as corrected

$S = \left(\frac{1 + \ln}{2} \right)^2$ instead, taking into account the very geometry of the box (Figure 1.2).

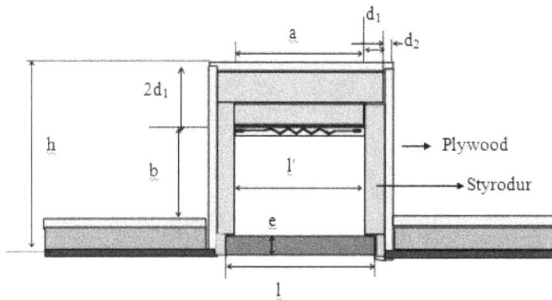

Figure 2. Cross section of a measuring box.

In order to calculate C, we can use the formula given by [5]

$$C = b \left(\frac{4a}{d} + 2.23 \right)\lambda + \left(\frac{a}{d'} + 2.57 \right)\lambda' + 0.6\left(\frac{d}{a} \right)^2 \quad (1.2)$$

With $d = d_1 + d_2$ and $d' = 2d_1 + d2$;

$$\frac{\lambda}{d} = \frac{1}{\frac{d_1}{\lambda_1} + \frac{d_2}{\lambda_2}} \quad \frac{\lambda n}{dn} = \frac{1}{\frac{2d_1}{\lambda_1} + \frac{d_2}{\lambda_2}}$$

λ_1: thermal conductivity of the hard styro that equals 0,033 W.m⁻¹ styrodur⁻¹.K⁻¹

λ_2: thermal conductivity of the plywood that equals 0,12 W.m⁻¹plated.K⁻¹.

In order to minimize the lateral losses and to determine

conductivity with good accuracy, the emission of heat inside the box is adjusted in such a way that T_b becomes slightly greater than T_a and $T_b - T_a < 1$ °C.

Using the heat balance expressed above, the value of the effective thermal conductivity is obtained in continuous state by:

$$\lambda = \frac{e}{S(T_c - T_f)}\left[\frac{V^2}{R} - C(T_b - T_a) \right] \quad (1.3)$$

Errors corresponding to the test results are systematic errors and random errors. The calculation of this error resulted in: $\frac{\Delta\lambda}{\lambda} \approx 5\%$. It can be said that the relative error in the determination of continuous thermal conductivity by the method of box is about 5%.

3.1.2. Measurements

a) Measurement of Tb - Ta

An ambient heat sensor is placed near the boxes and measures the room temperature T_a. Another one that measures T_b, is placed inside the box in a suitable position (middle), so that the temperature is averaged throughout the volume. In the determination of $T_b - T_a$, and up to the measurements checking, one must assure that the outer atmosphere had been stable for at least two hours. Indeed, because of the high inertia of the box, a change in temperature of the room is not transmitted immediately inside the box.

b) Measurement of Tc-Tf

Two contact probes are distributed on both sides of the sample. On the cold side, the temperature is strictly uniform (homogeneous). However, on the warm side, the edges of the sample are slightly colder than the center. The difference being constant, a systematic correction is made.

c) Measurement of V and of R

Heat is generated by Joule effect, at the beginning of the test, an approximate voltage has to be applied to the resistor. After an initial stabilization of temperatures, especially that of the room T_b, voltage adjustments can be made. When the steady state is reached, we record the precise value of the voltage reading and that of the resistance obtained with a DVM (Digital Voltmeter).

d) .Conductivity measurement by embedding

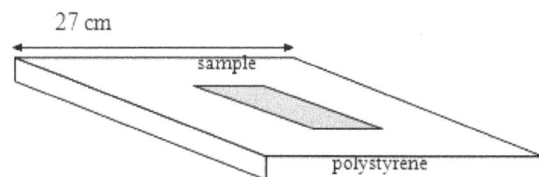

Figure 3. Diagram of an embedded sample.

This method is used if the sample under consideration is small. In order to be able to use samples with sections smaller than standard, and to determine their conductivity or diffusivity, they have to be placed in a parallelepipedic polystyrene matrix, since the box is designed to receive samples of size 27 cm x 27 cm. The sample is placed as show in Figure 3.

In steady state, a part of the flow passes through the sample and another through the matrix of polystyrene. Considering that the flow remains unidirectional, its preservation for a given medium is written as follows:

$$\dot{Q} = \dot{Q}_1 + \dot{Q}_2 + \dot{Q}_3$$

With:

$\dot{Q}$: Electrical power emitted by the resistance (film) shaped by $\dfrac{V^2}{R}$. $\dot{Q}_1$: Lateral losses toward the room atmosphere (expressed by: $\dot{Q}_1 = C(T_b - T_a)$). $\dot{Q}_2$: Thermal power through the polystyrene, expressed by: $\dot{Q}_2 = \dfrac{\lambda_p}{e_p} S_p (T_{p.sup} - T_{p.inf})$ where e_p is the thickness of the polystyrene, S_p, also called measurement area, is its area useful. $\dot{Q}_3$: Thermal power running through the sample and expressed as: $\dot{Q}_3 = \dfrac{\lambda_b}{e} S_b (T_{b.sup} - T_{b.inf})$.

The index b refers to the box. In steady state, we express:

$$\lambda_b = \frac{e}{S_b(T_{b.sup} - T_{b.inf})} \left[\frac{V^2}{R} - C(T_b - T_a) - \frac{\lambda_p}{e_p} S_p (T_{p.sup} - T_{p.inf}) \right] \quad (1.4)$$

Therefore, we conducted an experimental verification of the results given by this method and another on a sample of size 27 x 27 of same nature.

4. Introduction and Use of the Experimental Results Related to the Thermo Physical Characteristics

In the following paragraphs, we try to accurately follow the behavior of thermo physical parameters as a function of water content. These parameters are essentially the thermal conductivity (λ) and thermal diffusivity (a). We will study the influence of moisture on these two parameters.

4.1. Change as a Function of Water Content

4.1.1. Thermal Conductivity

Conductivity is the measurement that characterizes the ability of the materials to let heat through. We know that the more a material is a conductor of heat, the higher it conductivity is. The choice made on the dirt as a reference material is justified by the fact that among the construction materials used in Chad, it plays an important role. For the tests, we varied the water content of the soil in each composition, starting from a sample containing a certain amount of water that is gradually dried until completely dry.

The initial water content is obtained from an addition of a certain amount of water to the sample previously conditioned for 48 hours in a closed environment with less variable room temperature and humidity in order to allow the water to be distributed the most evenly possible. We simultaneously report the results of two selected soils.

a) Earth only

The results obtained are shown in the following figure.

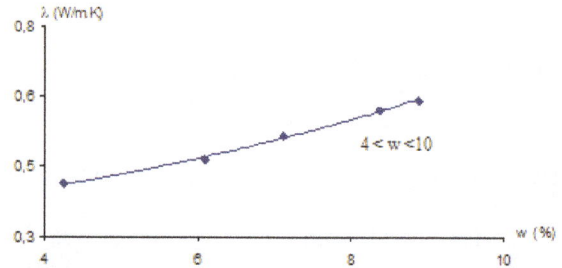

Figure 4. Variation of conductivity λ according to the water content w: dirt only (N'Djamena).

b) Earth and straw

The results obtained are shown in Figure 5. The moisture content w was calculated in relation to the mass of the completely dry sample obtained by baking it in an oven for 24 hours and at a temperature of 105 ° C.

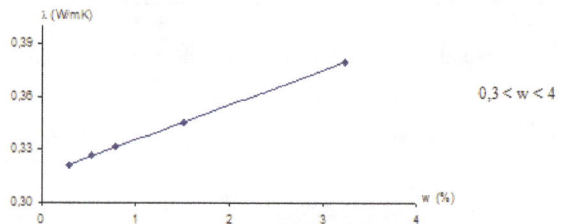

Figure 5. Variation of conductivity as a function of water content: dirt and straw (N'Djamena).

Interpretation of results:

With regards to the results shown in the previous figures, we notice that the thermal conductivity increases with water content, regardless of the composition of the samples. Since the samples are porous materials, continuous membranes of liquid water with higher thermal conductivity than the dry air contained in the pores are formed in the capillary spaces, consequently resulting in an increase of the apparent thermal conductivity. In fact, KALBOUSSI [10] had shown in his work that, for light materials, thermal conductivity increases in a roughly parabolic function of density. Thermal conductivity of the samples can be estimated in relation with their density by connecting them through a degree 2 polynomial. And as the density and water content change linearly, we connected the points of the thermal conductivity of our measurements in relation with the water content by a degree 2 polynomial.

4.1.2. Thermal Diffusivity

Thermal diffusivity is related to a propagation speed of the thermal wave. In presenting the results, we followed the previously adopted plan in the case of thermal conductivity. The results given below are obtained from the formulations derived from the expressions of the model of the flash pulse on the sample [3], [6], [13].

a) Earth only

The measurements yielded the following results summarized in Figure 6.

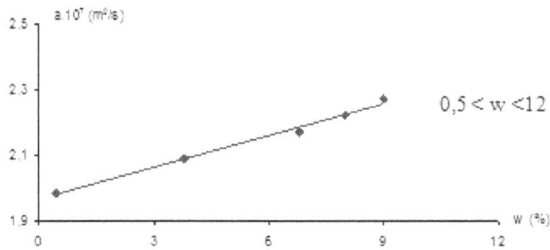

Figure 6. *Variation of the diffusivity as a function of water content: Earth only (N'Djamena).*

b) Earth and straw

In order to assess the amount of straw actually introduced, we can compare the densities of earth only and earth plus straw. For the form of use of dirt and straw combination, we obtained the following results:

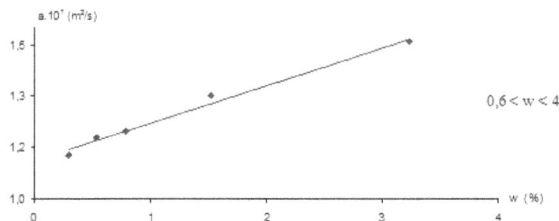

Figure 7. *Variation of the diffusivity as a function of water content: earth and straw (N'Djamena).*

Interpretation of results: For all the measurements done, we found that diffusivity increases slightly with water content, which shows that when dirt is wet, it transmits heat more than when it is dry.

4.1.3. Mass-Related Heat

Here we present the values of specific heat taken from the measured values obtained previously. Indeed, having obtained the thermal conductivity and diffusivity by direct measurements, we deduced the specific heat by the relation $Cp = \lambda/a\rho$.

a) Earth only

The calculation gave us the following results for earth used alone:

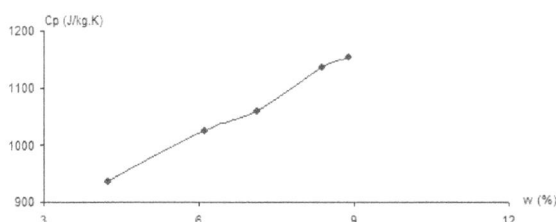

Figure 8. *Variation of mass-related heat as a function of water content: land only (N'Djamena).*

b) Earth and straw

For dirt mixed with straw, we obtained the results collected in Figure 9.

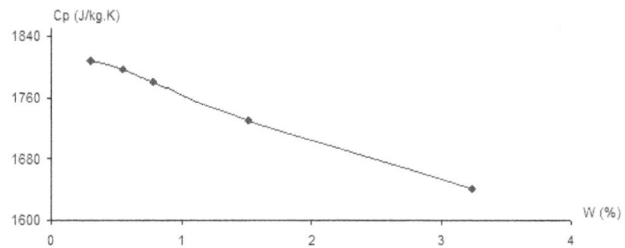

Figure 9. *Variation of mass-related heat as a function of water content: earth and straw (N'Djamena).*

4.1.4. Thermal Resistance

In all the tables that follow, the values for thermal resistance are obtained using the formula: $R_{th} = e/\lambda$ where e is the thickness of the sample under consideration [18].

a) Earth only

The calculation results of thermal resistance for earth alone without addition are shown in fig. 10.

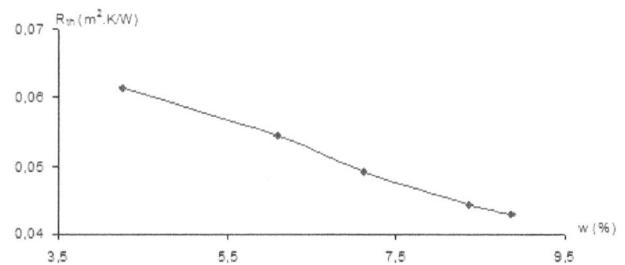

Figure 10. *Variation of thermal resistance as a function of water content: land only (N'Djamena).*

b) Earth and straw

The calculation results of thermal resistance for earth with added straw are shown in Fig.11.

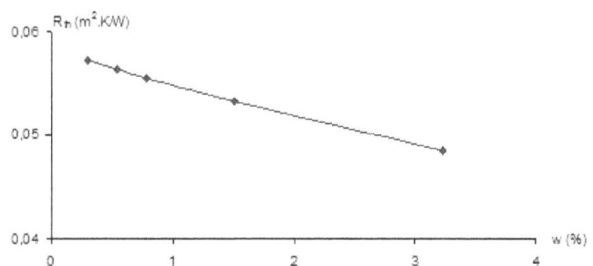

Figure 11. *Variation of thermal resistance as a function of water content: earth and straw (N'Djamena).*

4.2. Change in Thermal Conductivity as a Function of Temperature

For the influence of temperature, we only studied the case of the thermal conductivity λ we consider an important parameter of the out layer in the transfer of heat. For this, we varied the atmosphere of the measurement boxes by working on the supply voltage of the heating film. We performed the measurement of the parameter sought (λ) each time the steady state was established.

According to the theory of gases, the temperature of an element of matter is proportional to the kinetic energy of microscopic particles that constitute it. Thus, the higher the temperature is, the faster its particles move. The extension of these results to the solid and liquid media shows that the pure electrolytic conductivity is a function of average temperature. The theoretical approach to the study of the apparent conductivity of a porous medium according to the average temperature may be conducted based on parallel series model [11], [10], [15] [19]. This approach as we have just shown for porous composite materials, would fit better for the case of our samples made of earth. As for porous medium that are opaque to radiation, the apparent thermal conductivity is given by:

$$\lambda_{eff} = \left[\frac{\theta}{(1-P_T).\lambda_s + \lambda_f.P_T} + (1-\theta).\left(\frac{1-P_T}{\lambda_s} + \frac{P_T}{\lambda_f} \right) \right]^{-1} +$$

$$\frac{4.\sigma.e.T_m^3}{n.\left(\frac{2}{\varepsilon} - 1 \right)} \quad (3.1)$$

$$\lambda_s = \lambda_{os} (1 + \beta_s T_m) \quad (3.2)$$

$$\lambda_f = \lambda_{of} (1 + \beta_f T_m) \quad (3.3)$$

θ: Parallel conduction fraction

P_T: total porosity.

λ_s: thermal conductivity of the solid (earth and straw).

λ_f: thermal conductivity of the fluid part.

β_s: temperature coefficient of the solid medium.

β_f: temperature coefficient of the fluid medium.

λ_{os}: thermal conductivity of pure conduction at 0 °C, the solid part.

λ_{of}: thermal conductivity of pure conduction at 0 °C, the fluid part.

n: number of heat shields used in the model.

The combination of equations (3.1), (3.2) and (3.3) gives an expression of the form:

$$\lambda_{app} = \lambda_{c0} + \lambda_{ct} + \lambda_r \quad (3.4)$$

λ_{c0}: thermal conductivity of pure conduction at 0 °C

λ_{ct}: term reflecting the change in conduction pure in relation with temperature

λ_r: thermal conductivity of internal radiation in the pores.

The amount λ_r is negligible as compared to λ_c and λ_{ct} for the temperature range considered in our experiments (0 to 50 °C).

Thus theoretically, we obtain the following expression for the thermal conductivity:

$$\lambda_{app} = A + B.T_m \quad (3.5)$$

As A and B are constants, it can be assumed in first approximation that thermal conductivity is a linear function of the average temperature of the environment under consideration. In order to verify the theoretical reasoning outlined, we varied the voltage of the heating film. The variation of the supply voltage of the heating film increased the average temperature of the sample. Once the steady state was established, we proceeded to determine λ with varied the average temperature of materials in temperature ranges corresponding to the dry environment in the tropical zones. The results obtained are shown in the tables below. Then we plotted least-squares curve of fitness between the thermal conductivity and the average temperature of the sample.

a) Dirt only sample

We present the results of measurements of thermal conductivity for the composition of earth alone without addition in relation with temperature in Figure 12.

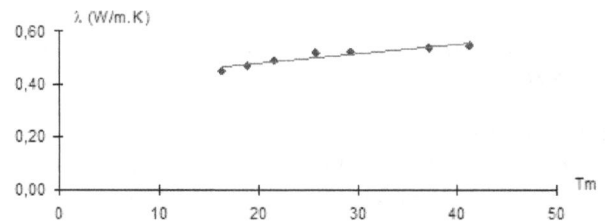

Figure 12. Variation of conductivity as a function of the average temperature: earth only (N'Djamena).

b) Dirt and straw sample

We present the results of measurements of thermal conductivity for the soil with added straw as a function of temperature in Figure 13.

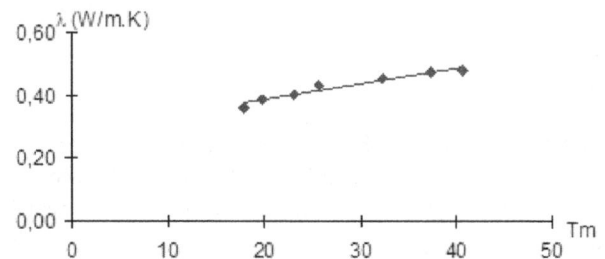

Figure 13. Variation of conductivity as a function of the average temperature: earth and straw (N'Djamena).

Interpretation of results:

For each sample of the material studied, the curves obtained show that the conductivity varies linearly with the average temperature. In our case, this variation is very small and it is confirmed by the narrow range of temperatures considered [0 - 50 °C]. We relied on these temperatures because they are broadly representative of ambient temperatures in the Sahel.

5. Testing of Thermal Modeling and Comparison with Experimental Results

This part of our work is a study approach designed to

present key information needed to assess the variation of the apparent thermal conductivity of materials previously studied as a function of water content.

5.1. Proposed Model

For modeling, [8], [9] we assume that the materials studied in relation to the water content are tri-phasic porous materials: a gas, liquid and a solid phase by:

$$\dot{Q}_r = \frac{\sigma(T_c^4 - T_f^4)}{\frac{3.a.e}{4} + n.\left(\frac{2}{\varepsilon} - 1\right)} \qquad (3.5)$$

For low temperature gradients $T_c - T_f$ below 30 K it can be assumed that:

$$T_c^4 - T_f^4 \cong 4.T_m^3 . (T_c - T_f) \qquad (3.6)$$

With $T_m = \dfrac{T_c + T_f}{2}$, the average temperature of the environment.

Assuming that the gas is non-absorbent ($a = 0$) the density of heat flux exchanged by radiation within the material is:

$$\dot{Q}_r = \frac{4.\sigma.T_m^3 (T_c - T_f)}{n.\left(\frac{2}{\varepsilon} - 1\right)} \qquad (3.7)$$

Thus, thermal conductivity induced by radiative exchange and by analogy with the thermal conductivity is:

$$\lambda_r = \frac{4.\sigma.T_m^3 . D}{\left(\frac{2}{\varepsilon} - 1\right)} \qquad (3.8)$$

With $D = \dfrac{e}{n}$, the average pore diameter.

Taking into account the experimental measuring conditions and the pore dimensions of the materials we studied, convection and radiation transfers are negligible.

To simplify the study, we adopt the model proposed by KRISCHER for two phases taking into account the above assumptions discussed above.

5.2. Validity of the Model

In order to calculate the thermal conductivity of the materials studied with the model we proposed, we had to experimentally determine two parameters, namely the thermal conductivity of the solid matrix λ_s and the volume fraction of the parallel phase θ. For us, the gas phase is the air the thermal conductivity of which we fix at 0.023 W/mK and the liquid phase is water with a conductivity estimated at 0.57 W/mK, assuming that on average, measurements were performed at 20 °C.

5.2.1. Determination of λ_s and θ

In order to model the heat level of these materials, we had to search for the conductivity of the solid structure λ_s of each of them. The structure of solid or solid matrix material is the dry part of material, excluding the pores.

Thus, modeling requires knowledge of λ_s that corresponds to the conductivity of the material compressed up to the total disappearance of the pores, which is even more difficult to achieve as this does not happen in a vacuum and air is incompressible. However, knowledge of ρ_s helps to obtain the conductivity of the solid structure by calculation. To determine ρ_s we used the simple method described below. We had a scaled cylinder in which we poured a quantity of water with known mass me and volume Ve. By adding an amount of crushed material and measuring the mass mT and the volume VT as a whole, we obtained the mass m and volume V of the added material. Density of the solid matrix is given by the relation:

$$\rho_s = \frac{m}{V} \qquad (3.9)$$

With: $m = m_T - m_e$ and $V = V_T - V_e$

Measurements of the density of different samples gave results that we recorded in Table 2.

***Table 2.** Values of the density of the solid structure of Samples.*

Name	N'Djamena	
Measure	earth only	earth and straw
mT (g)	82,531	90,747
VT (cm3)	76,000	82,000
me (g)	70,060	69,810
m (g)	15,471	20,937
V (cm3)	6,000	12,000
s (kg/m3)	2579	1745

After determining ρ_s we made the determination of the porosity P_T. The porosity of a sample is given by the relation:

$$P_T = \frac{V_P}{V_a} \qquad (3.10)$$

with: V_P, pores volume and V_a the total apparent volume. The porous material consisting of solid and pores, this relationship can be expressed as:

$$P_T = 1 - \frac{V_s}{V_a} \qquad (3.11)$$

Where V_s: the volume of the solid structure.

In representing the density of the solid structure by ρ_s and the apparent density of the dry material by ρ_a, we show that the above relation is marked as shown below:

$$P_T = 1 - \frac{\rho_a}{\rho_2} \qquad (3.12)$$

We use this last expression (4.12) to determine the porosity of the materials studied. The calculations provided the results shown below in Table 3.

Table 3. Calculated values of the porosity of different samples.

Name	N'Djamena	
Measure	earth only	earth and straw
a (kg/m3)	2068	1529
s (kg/m3)	2579	1745
PT	0,198	0,125

In order to determine λ_s we conducted two tests one of which on dry material and dry and material in saturated state in the other. In the dry state, the fluid phase was air, and in the saturated condition, the fluid phase was water. The apparent thermal conductivity depends on the thermal conductivity of the solid structure and the thermal conductivity of fluids of the phase or phase λ_f as well as on the degree of resistance on the part of the arrangement of phases. The distribution of each phase is random in the porous material.

It is believed that the porous material is replaced by a system of flat plates (Figure 6), each representing a phase. The plates are arranged in parallel, then in series so that the thermal conductivity reached by the model equals that of the material under consideration.

The thermal conductivity of a parallel configuration of volume fraction θ is given by:

$$\lambda_{eff} = \lambda_{//} = (1 - P_T)\lambda_s + P_T\lambda_f \qquad (3.13)$$

And the configuration of a series of volume fraction (1 - θ) is:

$$\lambda_{eff} = \lambda_{\perp} = \frac{1}{\dfrac{1-P_T}{\lambda_s} + \dfrac{P_T}{\lambda_f}} \qquad (3.14)$$

When the porous body is dry, its thermal conductivity depends on the thermal conductivity of the solid structure λ_s and that of the thermal conductivity of the air filling the volume of the pores. And when it is saturated, its thermal conductivity depends on that of the water included in the volume of the pores and of λ_s. By applying the relationship 3.1, we will have the following systems of equation:

$$\frac{1}{\lambda_{dry}} = \frac{\theta}{(1 - P_T).\lambda_s + \lambda_{air}.P_T} + (1 - \theta).\left(\frac{1-P_T}{\lambda_s} + \frac{P_T}{\lambda_{air}}\right)$$
$$\frac{1}{\lambda_{saturated}} = \frac{\theta}{(1 - P_T).\lambda_s + \lambda_{water}.P_T} + (1 - \theta).\left(\frac{1-P_T}{\lambda_s} + \frac{P_T}{\lambda_{water}}\right) \qquad (3.15)$$

Table 4. Calculated values of λ_s and θ for the different samples.

N'Djamena		
Measure	Earth only	Earthand straw
λ_s (W/m.K)	0,63	0,59
θ	0,88	0,75

To solve this system of nonlinear equations with two unknowns by the software Mathcad provided the results reported in Table 4.

5.2.2. Results of the Calculation of Thermal Conductivity of the Material from the Model

In order to calculate the apparent thermal conductivity of the materials studied from the model for given water content, we need to know the rate of water by any fraction of volume.

The rate of water volume σ_e is the ratio between the volume of water contained in the material and the apparent volume. When the material is dry, the pores are empty and σ_e equals zero. The value of σ_e in any state is given by:

$$\sigma_e = w.\frac{m_a}{\rho_e.V_a} \qquad (3.16)$$

With: w, water contents of the sample m_a, and V_a respectively the mass and the apparent volume of the sample and ρ_e the density of water.

From σe, we deduce the rate of air volume in $\sigma_a(\sigma_a = P_T - \sigma_e)$ for a given value of water content and we present the results by the study area as shown below. The following are samples made with the dirt of N'Djamena.

Table 5. Thermal conductivities Values calculated with the proposed model (N'Djamena).

Land only				Earth and straw			
w (%)	σ_e(%)	σ_a(%)	λ_{eff} (W/m.K)	w (%)	σ_e(%)	σ_a(%)	λ_{eff} (W/m.K)
0,0	0,0	19,8	0,34	0,0	0,0	12,5	0,32
4,3	8,9	10,9	0,43	0,3	0,5	12,0	0,32
6,1	12,6	7,2	0,48	0,5	0,8	11,7	0,33
7,1	14,7	5,1	0,52	0,8	1,2	11,3	0,33
8,4	17,4	2,4	0,57	1,5	2,3	10,2	0,35
8,9	18,4	1,4	0,59	3,2	4,9	7,6	0,39

5.2.3. Comparison of Theoretical Models with Experimental Results

To compare experimental results with theoretical models presented in the previous section of this study, we present the variations of thermal conductivity calculated and experimentally determined as figures for this area. The results for samples made with earth taken from N'Djamena are presented in the following figures:

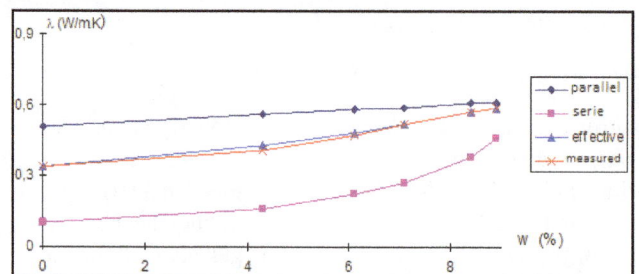

Figure 14. Calculated and measured thermal conductivity variations: earth only (N'Djamena).

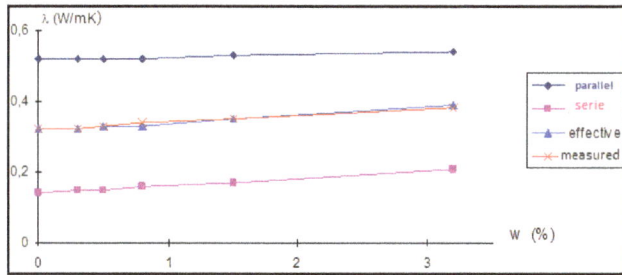

Figure 15. *Calculated and measured thermal conductivity variations: earthand straw (N'Djamena).*

5.3. Conclusion

The comparison of the values of the thermal conductivity of the material with the model for given water content and the experimental results shows very good correspondence of results and justifies the validity of this model. Indeed, with the error margins of about 5% on the values of λ_{mes} the figures show that the curves of the proposed model have behaviors essentially identical to the experimental results for the materials studied. As for the values given by the serial and parallel theoretical models, they are situated on either side of the measured thermal conductivity values.

6. General Conclusion

In the work presented above, we relied on experimental results. We would like to complete this work by analyzing how a thermal study can be done in practice as well as in theory, by applying it to concrete cases in order to identify the common measures that could be used in other studies.

In practical terms, we found interesting to look deeply and physically into the properties characterizing the thermal behavior of an envelope between two different ambient environments such as thermal conductivity, thermal diffusivity, the thermal resistance and thermal capacity. Just as in our so-called Sahel countries, the most widely used material for construction of habitat is earth. We chose this material as a reference material even though in the mechanical point of view it is not a durable one. We presented the techniques for measuring conductivity and thermal diffusivity of the material earth. Different values of thermal conductivity, thermal diffusivity, thermal resistance and capacity of this material were presented and their variation as a function of parameters such as moisture content and temperature (for a single case). A model of distribution of the structure of earth was selected (parallel series). A comparison of the measured values with those calculated by the model used showed good agreement. As a result, we believe to contribute to a better definition of the thermal parameters of this material in order to help designers to more precisely evaluate the exchanges established between the external environment and within an enclosure made of earth.

References

[1] S. GAYE. Contribution à l'étude de confort thermique en climat tropical: influence de la mouillure cutanée. Thèse de doctorat: UCAD. DAKAR (Sénégal). 1998.125p.

[2] M. AHMAT-CHARFADINE. Etude expérimentale des propriétés thermo physiques des matériaux locaux du Tchad. Application au confort thermique de l'habitat. Thèse de Doctorat: Energétique: Université Claude Bernard Lyon 1: 06 Mars 2002 - 187p.

[3] R. YEZOU. Contribution à l'étude des propriétés des matériaux de construction cohérentes et non cohérentes.- 221p. Thèse: INSA Lyon (FRANCE): 1978.

[4] S. ZEGADI. Etude théorique et expérimentale des transferts thermiques dans les milieux granulaires.- 195p. Thèse de Doctorat: Energétique: Université Claude Bernard Lyon 1: 22 Mai 1997; -97p.

[5] S. BOUSAD., EZBAKHE, A. EL BAKKOUR, T. AJZOUL, A. EL BOUARDI. Etude thermique de la terre stabilisée au ciment utilisée en construction au Nord du Maroc. Revue Energies Renouvelables: Journées de Thermique, 2001 pp. 69-72.

[6] A. DEGIOVANNI. Diffusivité et Méthode de Flash. Contribution à l'étude de la diffusivité thermique. Revue Générale de Thermique, 185, 1977 pp. 420-442.

[7] AHMED ALI Ep. AIT KADI, S. Performances thermiques du matériau terre pour un habitat durable des régions arides et semi-arides: cas de Timimoune. Mémoire de Magistère: Université Mouloud Mammeri. TIZI-OUZOU (Algérie). 2012 - 109p.

[8] R. SIEGEL, J. R. HOWELL. Thermal radiation heat transfer Mc GRAW HILL Book Company, N.Y 1972.

[9] G. MENGUY, H. EZBAKHE, J. LEVEAU. Influence de la porosité sur les caractéristiques thermiques des matériaux de construction. First international Congress Held by RILEM of AFREM, vol.1, pp.269-270, Versailles (France). Septembre 1987.

[10] A. KALBOUSSI. Transfert de chaleur en régime stationnaire et dynamique à travers les milieux poreux humides. - 238 p. Thèse de Doctorat.: Energétique: Université Claude Bernard Lyon 1: 1990; 40-90.

[11] Mourtada, A. (1993). Thermique du bâtiment en climat chaud: Actes de l'Atelier sur "la maîtrise de l'énergie dans les bâtiments".Yaoundé, Cameroun: PRISME / IEPF / ADEME. - pp.32-51.

[12] A. Mokhtari, K. Brahimi et R. Benziada, 'Architecture et Confort Thermique dansles Zones Arides, Application au Cas de la Ville de Béchar', Revue des Energies Renouvelables, Vol. 11, N°2, pp. 307 – 315, 2008.

[13] Philippi I., Batsale J. C., Maillet D., Degiovanni A., Measurement of thermal diffusivity through processing of infrared images, Rev. Sci. Instrum. 66(1) 1995, 182-192.

[14] Krapez JC, Spagnolo L., FrieB M., Maier H. P., Neuer, Measurement of in-plane diffusivity in nonhomogeneousslabs by applying flash thermography, International Journal of Thermal Sciences 43. (2004) 967-977.

[15] A. W. Aregba, C. Pradere, J.-C. Batsale. Measurements thermo physical properties of materials by infrared thermography. 2IE. Conférence Internationale: Eco matériaux de Construction. Du 10 au 12 Juin 2013 à Ouagadougou. 180p.

[16] V. K. Mathur. Composite materials from local resources Construction and Building Materials. Volume 20, Issue 7, September 2006, Pages 470–477

[17] Emmanuel Ouedraogo and al. Mechanical and Thermophysical Properties of Cement and/or Paper (Cellulose) Stabilized Compressed Clay Bricks. Journal of Materials and Engineering Structures 2 (2015) 68–76.

[18] Mustapha BOUMHAOUT and al. Mesure de la conductivité thermique des matériaux de construction de différentes tailles par la méthode des boites. 3ème Congres de l'Association Marocaine de Thermique. Agadir (Maroc) 21-22 Avril 2014. p1-6.

[19] K. GADRI & A. GUETTALA. Etude des caractéristiques physico-mécaniques des bétons de sable à base de fumée de silice. MATEC Web of Conferences. Volume 11, 2014. International Congress on Materials & Structural Stability.

[20] Chao-Lung Hwang, Trong-Phuoc Huynh. Investigation into the use of unground rice husk ash to produce eco-friendly construction bricks. Construction and Building Materials. Volume 93, September 2015, Pages 335–341.

3

Experimental Analysis of Seepage in Soil Beneath a Model of a Gravity Dam

Najm Obaid Salim Alghazali[1], Hala Kathem Taeh Alnealy[2]

[1]Corresponding author, Asst. Prof. Doctor, Civil Engineering Department, Babylon University, Iraq
[2]M. Sc. Student, Civil Engineering Department, Babylon University, Iraq

Email address:
drnajm59@gmail.com (N. O. S. Alghazali), Halhhalh300@yahoo.com (H. K. T. Alnealy)

Abstract: In this research the experimental method by using Hydraulic modeling used to determination the flow net in order to analyses seepage flow through single- layer soil foundation underneath hydraulic structure. As well as steady the consequence of the cut-off inclination angle on exit gradient, factor of safety, uplift pressure and quantity of seepage by using seepage tank were designed in the laboratory with proper dimensions with two cutoffs. The physical model (seepage tank) was designed in two downstream cutoff angles, which are (90, and 120°) and upstream cutoff angles (90, 45, 120°). After steady state flow the flow line is constructed by dye injection in the soil from the upstream side in front view of the seepage tank, and the equipotential line can be constructed by piezometer fixed to measure the total head. From the result It is concluded that using downstream cut-off inclined towards the downstream side with Θ equal 120° that given value of redaction (25%) is beneficial in increasing the safety factor against the piping phenomenon. Using upstream cut-off inclined towards the downstream side with Θ equal 45° that given value of redaction (52%) is beneficial in decreasing uplift pressure and quantity of seepage.

Keywords: Flow Net, Inclined Cutoff, Seepage Tank, Single Layer, Soil

1. Introduction

Hydraulic structures are a specific type of engineering structures designed and executed in such a way in order to control natural water or save industrial sources to guarantee optimum use of water. These structures are frequently build on soil materials and the foundation thickness must be thick so as to be safe against uplift pressure [1]. The differential head in water levels between the upstream and downstream acts on the foundation and causes seepage flow [2]. The Groundwater flow depends on the type of flow, the soil media, and the boundary conditions. Seepage of water is one of the main problems which effect on hydraulic structures [3]. There are Different methods solution can be used to analysis the seepage problem such as experimental works using physical model as well as numerical models electrical analog models[4]. A flow net is in fact a solution of Laplace's equation in two dimensions. The model of seepage tank (sand tank)is very useful in studying the conditions of fluid flow under the hydraulic structure [5]. The paths taken by moving particles of water as the flow through a permeable material may be represented by a series of flow lines. These flow lines are nearly parallel curved lines The hydraulic modeling method to determine the flow net in this study represented by a physical model was built to study the phenomenon' of seepage through soil.

2. Aim of the Study

The main objectives of this work can be summarized by the following points:

1. locate the equipotential lines and flow lines for single-layer soil used in research work.

2. Study the effect of inclined cutoffs at different angles of inclination on exit gradient, uplift pressure underneath the hydraulic structure quantity of seepage and find the best inclination angle of cut-off for upstream and downstream side of hydraulic structure for all types of soils placed in different position under the foundation of hydraulic structure used in research work.

3. Experimental Work

The results obtained by the present hydraulic model using the seepage tank that designed and carried out at the hydraulic laboratory of the Engineering College at Babylon University. The major purpose of the physical model adopted in the present research is to study the flow net and calculate the values of uplift pressure underneath the hydraulic structure, distribution of exit gradient, quantity of seepage for different types of soil at different position under the hydraulic structure foundation.

3.1. Model Description

Laboratory experiments have been conducted in a seepage

tank that has been designed with hypnotically dimensions of 1.6 m long, 0.5m width and 1.1m height The bottom and sidewalls of this tank were made of Acrylic of (10mm) thickness. Figure (1) shows the seepage tank used in present study (front view). The bottom of the tank was filled with this material of soil to a depth of 60 cm. Acrylic walls were used to build the body of the superstructure which consists of two parts. The first part simulated as foundation of a structure (40 cm long × 50 cm wide).This base is connected with the upstream and downstream cutoffs by gluing rubber strips. 10 piezometers were placed at the right side of the tank at different location. All piezometers are fixing to the board.

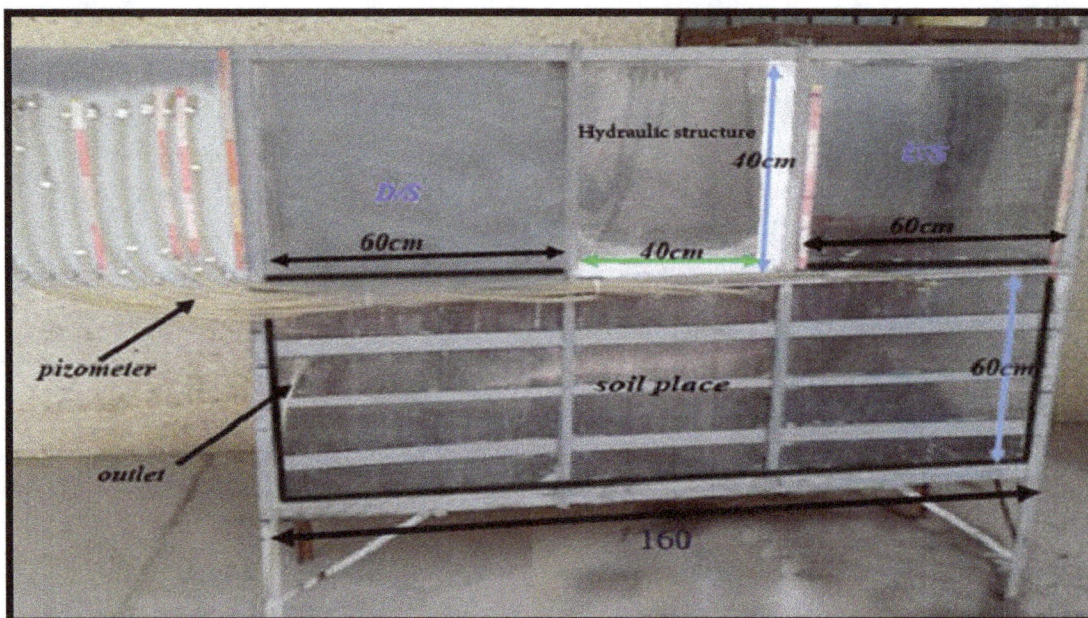

Figure (1). The seepage tank used in present study (front view).

3.2. Engineering Properties of Soil Used in Research

The experimental soil sample was taken from Hilla city region. The soil profile of any region contains many soil horizons, the difference between these horizons is marginal (no-homogenous soil). In the model tests, the profile is assumed to consist of one horizon (homogenous soil) [6]. The tests used here to analyze the soil specimen in order to determine soil distribution and other engineering properties

were conducted as per the Unified Soil Classification System, also the hydraulic conductivity values of the soil samples was measured in laboratory method. Table 1 shows a summary of the physical properties that measured of the soils which consist of three layers arranged in descending order. Used in the present study. Figure (1) shown the arranged for type of soil used in present study.

Table (1). Physical properties of used soils.

No of layer.	Type off soil Arrange in descending order	Void ratio	Unit weight KN/m3	Gs	Value of Hydraulic Conductivity m/sec
1	Sandy silty clay	0.52	20.65	2.75	$5.23*10^{-7}$

3.3. The Experimental Procedures

Taking the datum to be at the bottom of the tank, Install the soil in the form of three layers each layer thickness of 0.2m and it is monitoring the process of using. The metal cylinder.

Feeding the water to the seepage tank through the inlet hose until the water level in the upstream region reached the overflow hose level previously adjusted to meet the desired upstream water level.

After reaching steady-state flow dye is injected from dye

bottles which placed in the specific points , after a period of time flow lines were drawn Flux, which represents how the flow of water within the soil particles

After drawing flow lines, the vertical piezometers were installed transparent glass vertically into the soil to measure the total head in the points to draw the equipotential line.

Measure the discharge of drained water collected from the downstream funnel using the volumetric method by using jar.

Record the reading of the piezometric head of all installed piezometers under the base and downstream side.

Put the cutoff at upstream side with the angle of inclination for upstream($\Theta=45$, $\Theta=120$, $\Theta=90$) and repeat the step (7-8-9) to find the best angle to gave less value of uplift pressure and quantity of seepage.

Put the cutoff at downstream side with the angle of inclination ($\Theta=90$, $\Theta=120$) and repeat the step (7-8-9) to find the best angle to give max value factor of safety.

Table (2) shown the type of testing that made on multi layer soil

Table (2). The type of testing that made on soil.

General case (without any cutoff)

Cutoff at the upstream side $\Theta=90$

(cutoff at the downstream side $\Theta=90$)

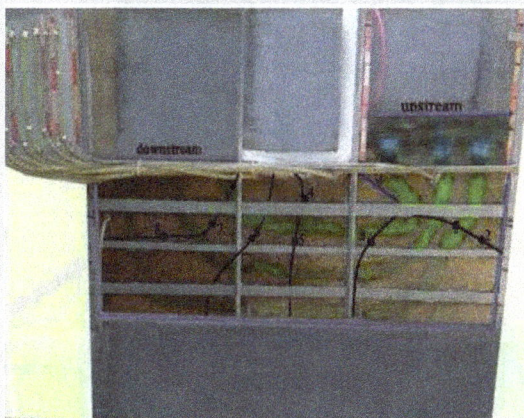

(cutoff at the upstream side $\Theta=45$)

(cutoff at the upstream side $\Theta=120$)

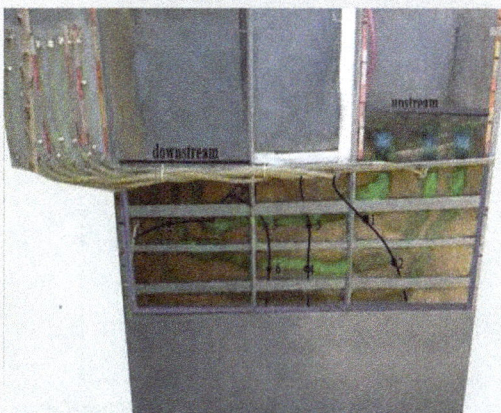

(cutoff at the downstream side $\Theta=120$)

4. The Results and Discussions

Herein, the discussions of the results for single layer (sandy silty clay) according the following parameters:

4.1. Effect of Inclination Cutoff and its Position on the Uplift Head

As shown in figure (2). When the Cutoff in upstream side of hydraulic structure was inclination with different angles, its noticed that the uplift pressure underneath the hydraulic structure decreases as(Θ) decreases toward U/S side for Θ (45° , 90°,120°) where the maximum redaction in uplift pressure according to the general case Θ=0(without any cutoff) was (57% , 45% ,22%) respectively so that the best angle is 45°.

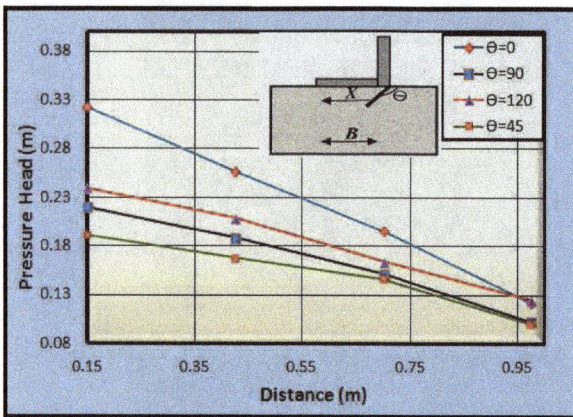

Figure (2). Uplift head ratio through foundation . A range of (Θ) values for cutoff in U/S.

From figure (3). When the Cutoff in downstream side of hydraulic structure is used, the uplift pressure obtained decreases as (Θ) decreases toward U/S side for Θ (90°and 120°) and the maximum redaction in uplift pressure according to the general case Θ=0 was (2.3% ,and2.74%). It is noticed that the redaction of the uplift pressure is small to the replacing of the cut-off ,therefore, it is not suggested at any angle of inclination.

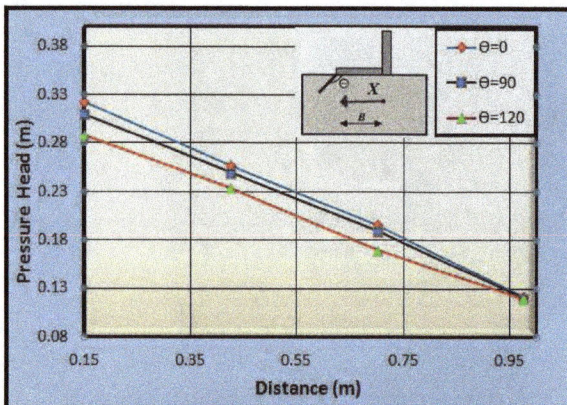

Figure (3). the value of uplift head for different values of (Θ) values for cutoff in D/S part of structure.

When the cutoff positioned in U/S and D/S part of structure as shown figure(4), the uplift reduced strongly and value of redaction is 38% as compared with general case (Θ=0) .

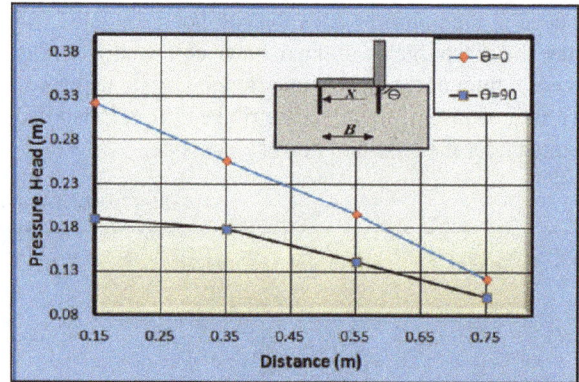

Figure (4). Uplift head ratio through foundation a range of (Θ) values for cutoff in U/S and D/S.

4.2. Effect of inclination Cutoff and its Position on the Exit Gradient

The exit gradient was studied at the end of the hydraulic structure for the cases that will be discussed herein and the results are represented graphically. The factor of safety for each angle of inclination must be calculated where is equal to the division of exit gradient on critical gradient (Icr) which is dependent on the specific gravity (Gs) and void ratio (e) of the soil particles [Icr =(Gs-1)/(1+e)]. In this study and for this type of soil (Gs =2.75 ,e =0.77, Icr =0.988).

In figure(5) when the cutoff put in upstream side of hydraulic structure it is found that the redaction in values of exit gradient were so small and as follows for Θ(45°,90°,120°) where the maximum redaction in according to the general case Θ=0(without any cutoff) was (1.2% , 1.87% ,2.47%) respectively

The result shows that the upstream cut-off inclination angle has no noticed effect on exit gradient. From the result the factor of safety against piping for this case were shown in table (3).

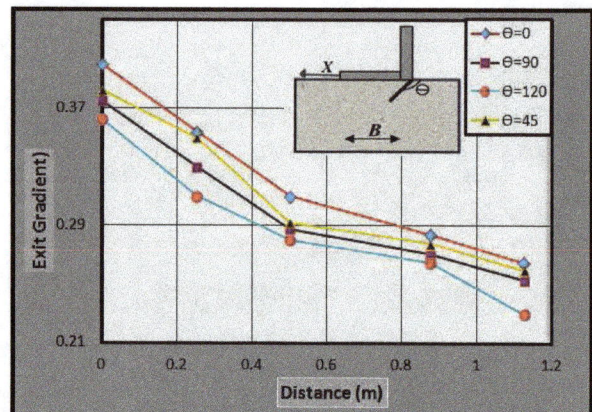

Figure (5). Exit gradient a range of (Θ) values for cutoff in U/S part of structure.

Table (3). The factor of safety against piping when cutoff at U/S.

U/S Cut-off Inclination	Max (exist gradient)	Fs
0°	0.4	2.47
45°	0.382	2.586
90°	0.375	2.63
120°	0.362	2.729

In figure(6) when the cutoff is put it in downstream side of hydraulic structure the exit gradient decreases as (Θ) increases toward the D/S side for Θ (90° and 120°) the maximum redaction in exist gradient according to the general case Θ=0 was (12.5% and 25%), respectively . So that the factor of safety against piping phenomenon of this case can be calculated as shown in table (4).These results show that using cutoff in D/S side inclination toward D/S Θ=120 increasing the factor of safety against piping.

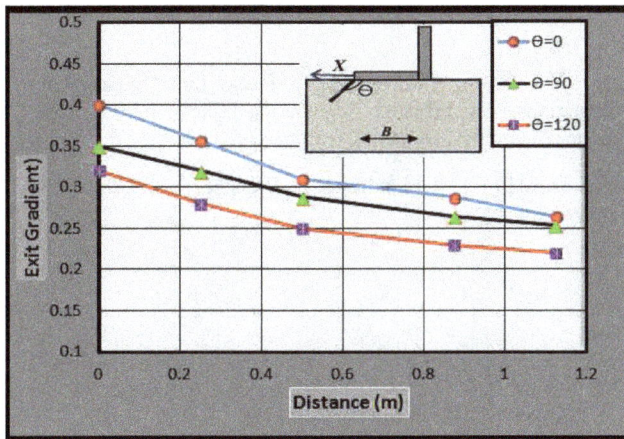

Figure (6). Exit gradient a range of (Θ) values for cutoff in U/S part of structure.

Table (4). The factor of safety against piping when cutoff at U/S.

U/S Cut-off Inclination	Max (exist gradient)	Fs
0°	0.4	2.47
90°	0.35	2.82
120°	0.31	3.18

When the Cutoff in up and down stream side of structure the exit gradient in the bed level of structure decreases compare with general case as shown in figure (7).

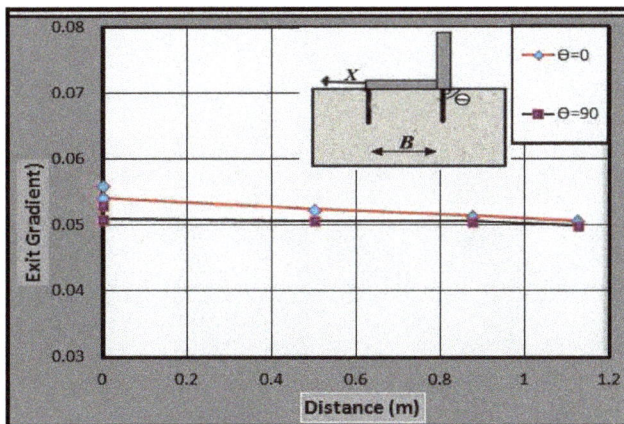

Figure (7). Exit gradient a range of (Θ) values for cutoff in U/S part of structure.

4.3. Effect of Cutoff Position and Inclination on the Seepage Quantity

When the Cutoff is in upstream side of hydraulic structure, as shown in figure(8), it's found that the seepage decreases while (Θ) decrease toward U/S , and the least quantity of seepage occurred when (Θ) value around (45).

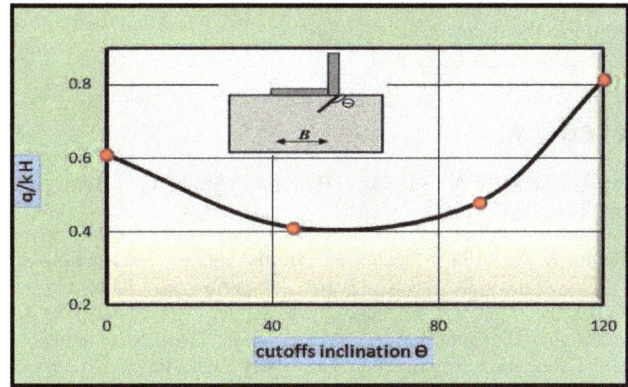

Figure (8). The Seepage quantity a range of different values (Θ) for cutoff in U/S part of structure.

In figure(9) When the Cutoff is in upstream side of hydraulic structure the seepage decreases while (Θ) increases, and the least quantity of seepage occurred when (Θ) value around (120), then the seepage increases rapidly for (Θ ≤ 90).

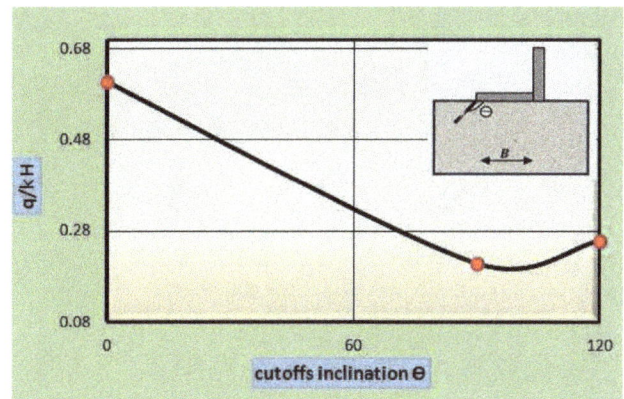

Figure (9). The Seepage quantity a range of different values (Θ) for cutoff in D/S part of structure.

5. Conclusions

The following main conclusions can be drawn from the results presented in this research:

The shapes of the flow net depend upon many factors such as diameter of the particle of the soil and location of cutoff.

The minimum value of uplift pressure is obtained when the cutoff used in upstream part of hydraulic structure with angle of inclination (Θ= 45°). That given redaction in value of uplift pressure according to the general case Θ=0 was 53%.

The minimum value of exist gradient is obtained when the cutoff used in downstream part of hydraulic structure with angle of inclination (Θ= 120°) That given redaction in value of exist gradient according to the general case Θ=0 was 9.3%.

And increasing the factor of safety against piping to 2.06.

Placing the cutoff at the dam heel (cutoff U/S side) to reduce the piping phenomenon in downstream side is not recommended under any angle of inclination, because such placement has too small effect.

The best angle of inclination is 45° towards upstream where it reduced the uplift pressure to (43% according to general case $\Theta=0$) as compared with other type of inclination and reducing the quantity of seepage.

References

[1]   AL-Ganaini, M.A., (1984), "Hydraulic Structure", Beirut, pp47-60, (In Arabic).

[2]   Selim, M.A., (1947) "Dams on Porous Media", Transaction ASCE, Vol. 1 Roy, S.K. (2010). "Experimental Study On Different Types Of Seepage Flow Under The Sheet Pile Through Indigenous Model." M.Sc. Thesis Insoil Mechanics and Foundation Engineering, University Of Jadavpur.12, pp 488-526.

[3]   Arslan, C. A. And Mohammad, S. A. (2011)."Experimental and Theoretical Study for Pizometric Head Distribution under Hydraulic Structures." Department of Civil engineering; College of engineering, University of kirkuk Volume 6, No.

[4]   EL-Fitiany, M. A. Abourohim, R. I. And El-Dakak, A. Y. (2003). "Three dimensional ground water seepage around a simple hydraulic structure." Alexandria Engineering Journal, Volume 42, Issue 5, September.

[5]   Roy, S.K. (2010). "Experimental Study On Different Types Of Seepage Flow Under The Sheet Pile Through Indigenous Model.", M.Sc. Thesis Insoil Mechanics and Foundation Engineering, University Of Jadavpur

[6]   Aziz, L. J. (2008). "Lateral Resistance of Single Pile Embedded in Sand with Cavities.", D.Ph Thesis, University of Technology, Iraq.

[7]   Khasaf, S. I.(1998). "Numerical Analysis of Seepage Problems with Flow Control Devices Underneath Hydraulic Structures." Ph.D. Thesis in Water Resources Engineering, University of Technology.

[8]   U.S. Bureau of Reclamation.(1977). " Design of Small Dams." A Water Resources Technical Publication, Washington, DC.

[9]   HM 169 (2011). "Drainage and Seepage Tank."Experiment Instructions Equipment for Engineering Education G.U.N.T. Gerätebau GmbH, Barsbüttel, Germany.

[10] Harr, M.E.(1962). "Groundwater and seepage." McGraw-Hill Book Company.

4

An Overhead Costs Assessment for Construction Projects at Gaza Strip

Nabil I. El-Sawalhi[1], Ahmed El-Riyati[2]

[1]Civil Engineering Department, The Islamic University, Gaza, Palestinian Territories
[2]United Nations Development Program, Gaza, Palestinian Territories

Email address:
nsawalhi@iugaza.edu.ps (N. I. El-Sawalhi), ariyati@undp.org (A. El-Riyati)

Abstract: An overhead cost of projects has significant role and influence on the performance of the construction companies. In addition, maintaining a good performance and quality of work away from possible loss. The purpose of the research is to examine the overhead costs management practice in the construction industry in the Gaza Strip. The awareness of construction companies and techniques followed by the contractors in estimating the overhead costs, affecting factors and mitigation measures was examined. A structured questionnaire supported by personal interviews was used to collect information. Sixty-three contracting companies working in the construction industry in the Gaza Strip were surveyed. The analysis shows a good awareness of the concept of overheads costs and its components. Gaza Strip Contractors follow precise practices to estimate the overheads costs. Contractors are taking several precautionary measures to minimize any risks associated with overheads costs. Factors affecting the estimation of overhead costs are including the management capacity of the company and its policy, led by the company's experience and ability to implement the project within the time frame. There are other factors related to the working environment of the project, especially the closure of commercial crossings of the Gaza.

Keywords: Project Overhead, Home Office Overhead, Construction

1. Background

The percentage of overhead in cost estimation is considered principal parameter in estimating the financial value of bid offer [1]. Many contractors take the risk and not consider the actual cost of overhead, especially the home office overhead in order to win the tender. Hence, neglecting overhead cost has forced some contractors out of business [2]. In construction industry, any increased overhead costs will include both site overhead and home-office overhead [3]. In Gaza Strip, the disturbance in commercial market resulted from closing the trade terminal for long periods, prevention of selected raw material and delaying materials entrance has increased the risk of contractors. These actions have created many problems and delay which is out of control of the two parties of contracting. Hence, the overhead of construction companies and the clients has increased became tangible. In order to verify overhead cost in the tenders, few contracts have separate items for overhead and profit. Most contracts tell contractor to spread or allocate their overhead and profit costs across all pay items in the schedule of values [4].

Under these circumstances, it is not surprising that the number of disputes and claims within the construction industry continues to increase because of increasing the overhead. Thus, in many times suspension of projects and delays in execution are considered from the main causes of increasing the overhead costs. In many countries, cases of overhead disputes between the two parties are carried to the arbitration or court. Therefore, it is very essential to pay attention to have solid evidences and documentation instead of foggy presentation [5].

The construction industry in Gaza Strip became indeed to judicious management can face all obstacles and be able to minimize the disputes between the parties. Minimization of these disputes need tight and fair contract clauses in addition to presenting guidance for the risks should be in consideration within any contractual process regarding the overhead cost. On the other hand, the contractor should be aware about the important of accurate estimation of the overhead in their projects to avoid any damages which may occur. The main aim of this research is to assess the overhead cost in constructions projects and explore the factors effecting on projects overheads costs.

2. Construction Overhead

General expenses of the construction project are defined as the additional, indirect costs that are necessary for the facilitation of the construction project. However, this definition inadvertently causes confusion by linking indirect costs and general expenses [4, 6, 7]. Overhead cost can be defined as a cost that cannot be identified with or charged to a construction project or to a unit of construction production [8]. Therefore, overhead costs generally are divided into two categories: general overhead costs and job overhead costs [9].

General overhead costs are those costs that cannot be identified readily with a specific project. General overhead costs are items that represent the cost of doing business and often are considered as fixed expenses that must be paid by the contractor [2].

2.1. Field Overhead

Field overheads are defined as the general cost or direct cost of the project for providing general plant and site-based services like insurance, site accommodation, etc [10]. It mainly consists of the costs expended to manage and administer a specific project (e.g., the cost of providing a job site office) [11]. In other words, it is used to quantify overhead costs that are incurred in the field [12]. Table 1 shows items that might qualify as field overhead costs as specified by [12].

Table 1. *Field overhead items.*

Possible Field Overhead Items	
Office Trash Removal	Airfare - Home Office Personnel
Office/Field Water Ice	Builders Risk Insurance
Portable Toilets	Cell Phones
Postage & Shipping	Engineers' Office Rent
Safety Supplies	Field Office Expenses
Telephones	Insurances Required by contract
Utilities	Lodging - Home Office Personnel
Yard Rent	Miscellaneous Expenses
Yard Tools & Supplies	Office Security
Office Trailer Rental	

Field overhead costs include items that can be identified with a particular job, but are not materials, labor, or production equipment. Job overhead includes expenses that cannot be charged directly to a particular branch of work, but are required to construct the project [2]. Job overhead (field overhead) is similar to general overhead but it must be distributed over the associated project, since it cannot be allocated to specific work packages [13].

An increase in site overhead expenses is usually easier to quantify. It requires the contractor to disclose its buildup of site preliminaries, showing detailed costs for all items considered as general site items (site infrastructure, cranes, and other general site equipment) [3]. According to reference [14], Indirect costs allocable to contracts include the costs of indirect labor, contract supervision, tools and equipment, supplies, quality control and inspection, insurance, repairs and maintenance, depreciation and amortization, and, in some circumstances, support costs, such as central preparation and processing of payrolls.

2.2. Home Office Overhead (HOOH)

General overhead expenses include the general business expenses that are incurred by the home office in support of the company construction program [15]. In other words, home office overhead represents the costs of the activities of the Contractor's home, or corporate, office necessary to run the business and to support the projects in the field [16].

General overhead costs (main-office or home-office expenses) are intended to include all those expenses incurred by the home office that cannot be tied directly to a given project such as home-office building rental, clerical, or utilities [13]. Therefore, these costs are distributed over all company projects by some basis [6]. It should be made clear that general and administrative costs are also referred to an overhead and in fact are part of the contractor's total overhead [14].

HOOH is generally described as company costs incurred by the contractor for the benefit of all projects in progress. This is the actual cost, which is an essential part of the cost of doing business. These costs cannot be directly allocated to a project. Contractors are reasonably free to account for such costs in whatever manner they choose. They must, however, use the same system at all times and on all contracts [4, 14,16].

Home office overhead costs include but are not limited to [11]: Rent; Utilities; Furnishings; Office equipment; Executive staff; Support and clerical staff not assigned to the field; Estimators and schedulers not assigned to field staff; Mortgage costs; Real estate taxes; Automobile maintenance and travel costs for home office personnel; Non-project-related bond or insurance expenses; Depreciation of equipment and other assets; Advertising; Marketing; Office supplies (paper, staples, etc.); Interest; Legal services; Accounting and data processing; and Professional fees/registrations.

2.3. Technique of Overhead Estimation in Project

Reference [17] mentions that estimating the project indirect cost is a time-consuming and inexact task, and hence contractors often apply a percentage of direct costs as an estimate of indirect cost. The method of applying a fixed percentage to the total value for project overheads allowance is particularly common in case of small-scaled, repetitive works. However, this may result in under-estimation, as many preliminary items bear no linear relationship to the value of works [10]. Reference [9] criticized that some contractors will multiply the direct cost by a certain percentage to get the overhead cost. However, such a quick method may not be sufficiently accurate for most estimates.

More sophisticated estimation models using ANN [18, 19], fuzzy modeling [20], and simulation models [21] are proposed by many authors as being effective tools for cost estimation, but few are practiced in reality [22]. Whereas the estimation method for project overheads is more primitive due to lack of serious concerns [10].

In summary most of the estimators still relied on their experience to work out the estimates without much use of artificial intelligence. On the other hand, reference [10] mentioned that despite of advancement in IT in recent years, IT application in construction estimation by the contractors is still limited.

2.4. Factor Affecting Estimation of Overhead

The estimation of overhead percentage as presented by [23] was influenced by historical data of the projects; a forecast of future activity, the ratio between main contractors' and sub-contract work, competitive conditions, the size, nature and duration of the project and an evaluation of risk. Furthermore, reference [11] added that the profit potential of individual projects is driven by many factors, including the contract terms and the level of competition.

The project's duration, total contract value, projects type, special site preparation needs and project's location are identified as the top five factors that affect the value of the percentage of site overhead costs for building construction projects in Egypt[24].

3. Research Methodology

The researcher has reviewed the relevant literature of overhead costs in construction projects. A structured questionnaire addressed to owners and contractors was prepared. It aims to get their feedback about factors that affecting the overhead costs in construction projects and how to minimize these costs. It also explores how much they are aware of the parameters of overhead, and the risk related to the estimation of overhead in their projects. The questionnaire was distributed among the contractors to obtain their perspectives regarding the mentioned aspects of the overhead in construction projects in Gaza Strip.

Statistical analysis and tests has been conducted by using Statistical Program for Social Science (SPSS). In order to get an appropriate method of analysis, the level of measurement must be understood and simple. Ordinal scales were used. Based on Likart scale researcher has the following: 1= Not Important, 2= Low Importance, 3= Medium Importance, 4= Important and 5= Very Important.

To determine the Relative Importance Index (RII) of the factors, these scores were transformed to importance relative indices based on the formula:

$$\text{Relative Importance Index (RII)} = \frac{\sum w}{AN} = \frac{5n_5 + 4n_4 + 3n_3 + 2n_2 + 1n_1}{5N}$$

Where W is the weight given to each factor by the respondent, ranging from 1 to 5, (n1 = number of respondents for Very Important, n2 = number of respondents for Important, n3 = number of respondents for Medium Importance, n4 = number of respondents for Low Importance, n5 = number of respondents for No Importance). A is the highest weight (i.e. 5

in the study) and N is the total number of respondents. The RII equals ranges from 0 to 1. A one respondents t-test was used to test if the opinion of the respondents in the content of the sentences positive if RII $\geq$ 0.6 and the p-value less than 0.05, or neutral If the p-value greater than 0.05, or negative RII $\leq$ 0.6 and the p-value less than 0.05.

4. Results and Analysis

34.34% from the respondents companies are working in buildings, while 25.30% is roads, 22.89% is water and sewage, 13.25% is electro mechanical works, and 4.22% are working in other types. Furthermore, 72.1% of respondents are company owner, 26.2% is projects manager, and 1.6% from the respondents is office engineer. This result shows that the overhead management has pushed the seniors to participate and bring out their opinion in sensitive part of construction management. Hence, the participation of seniors will add more realistic and reliability to the coming results.

9.8% from the respondents have implemented projects with total value of one million dollars or less within the past five years, while 44.3 % from 1.1 to 3.0 million, 29.5% from 3.1 to 7.0 million, 13.1% from 7.1 to 15.0 million, and 3.3% more than 15.0 million. The above result shows that the majority of companies are located between one and seven million US Dollar. This results may refer to the small size of project that undertaken in Gaza Strip and the high competition between companies in Gaza.

The majority is located between three to seven projects, which are normal to the capacities of Gaza contracting companies that have low financial and administrative capabilities.

21.3% from the companies are working since four to nine years, while 44.3% are working since ten to fifteen years, and 34.4% from the companies are working for more than fifteen years. The results above show that the majority of companies were established after the establishment of Palestinian Authority in 1994, whereas intensive construction projects has implemented in the Palestinian territories.

4.1. Knowledge and Sources of Information About the Overhead Costs

The question was raised to get the feedback about how much are the contractors aware towards overheads concept and how much they have knowledge about the components of overheads. Results show that 16.39% replied with partially reasonable definition, and 31.15% replied by reasonable accurate definition. The results are close to the results of [1] in Saudi Arabia, whereas, 23% of the respondents has defend OH with reasonable accurate definition. Some of the reasonable definitions as specified by the respondents are " Everything related to the project costs other than the cost of raw materials, labor and workmanship";" All the expenses of the project staff and other needs during the period of the project which located outside the scope of construction, in addition to the requirements to run the company main office"; "The expenses necessary for the implementation and management of the project and often not directly mentioned as items in the bill of quantities, the costs of the supervision required for the project implementation, in addition to the administrative and technical cost for the company home office"; "The costs spent on project management and management of the company, and these costs is loaded on the projects items". The results show that most of contractors don't have specific definition or proper understanding of the OH term.

Table 2 show the responsible for estimating the overheads costs during the pricing of tenders. 65.6% from the respondents who usually estimate the overheads costs during the pricing of tenders is "company owner", while 27.9% is "specialized employee in pricing", and 6.6% shows that "senior employee" is usually estimate overheads costs during the pricing of tenders.

Table 2. Estimator for the overheads costs during the pricing of tenders.

Percent (%)	Frequency	Estimator
65.6	40	Company owner
27.9	17	Specialized employee in pricing
6.6	4	Senior employee
100	61	Total

These results indicate that overheads estimation is very important and critical to win tender. Seniors in a company is requested to evaluate the project nature and the work environment. Hence, they can judge the overheads accurately. It can be summarized that overheads cost is not like any work items which can be estimated depending on clear cost, but it depends on many factors that should be known well by the experience seniors in the company.

When asking about the sources used for estimating the overheads costs; Table 3 show that 6.6% has used the company records as a sources to estimate the overheads costs. While 31.1% has to use stakeholders such as insurance companies, banks, and services providers etc. 59.0% has used of both sources (company records & stakeholders) and 3.3% has used another sources.

Table 3. The sources used for estimating the overheads costs.

Percent (%)	Frequency	Sources
6.6	4	Records of the company
31.1	19	Use stakeholders
59.0	36	Both resources
3.3	2	Another resources
100	61	Total

The above results show that the majority of companies does not depend heavily on records only but also depends on the stakeholder information. Similar results were illustrated by [10], whereas 19% depends on company database and records, 31% depends on suppliers and stakeholders, and 50% referred to both.

Table 4 shows that 23.0% of companies depend on one year historical data to estimate prices, while 19.7% says they use two years. 18.0% says that they use three years records, 8.2% says they used four years or more, and 31.1% says that they do not depend on the historical data to estimate the

prices. Almost the same results founded by [6] where 30 % of respondents do not apply the historical data at all.

Table 4. Using the historical data records.

Percent (%)	Frequency	Number of years
23.0	14	One year
19.7	12	Two years
18.0	11	Three years
8.2	5	Four years or more
31.1	19	Not applicable
100	61	Total

Instability of political situation enforced the contractor to minimize the dependency on historical data. Whereas, about one third of the respondents do not depend the historical data and about one fourth of respondents just depend on one year historical data to estimate prices.

4.2. Methods Used to Determine the Field Overheads Costs

Table 5 shows that 70.5% says that the method followed by their company to determine the field overheads costs mainly depends on the detailed calculation which based on the contractual condition. While, 8.2% says that the method depends on adding a percentage to the total tender cost. 3.3% says that the method depends on a lump sum added to the value of the tender, 3.3% says that the method depends on totaling a percentage from the item direct cost, and 14.8% says that the method is vary from one project to another.

Table 5. The method used to determine the field overheads costs.

Percent (%)	Frequency	Method
70.5	43	Detailed calculation based on the contractual condition
8.2	5	Percentage of the total tender cost
3.3	2	A lump sum added to the value of the tender
3.3	2	Percentage of the item direct cost
14.8	9	Vary from one project to another
100.0	61	Total

The result shown above reveals that the majority of companies follow the detailed calculation based on the contractual condition; this result may be attributed to the diversity in contracts from one project to another. In Gaza Strip, you may find specific requirements vary from owner to another or from project to project. Hence, no specific amount or percentage could be applicable. Almost the same results cope with [1] whereas 71% of the survey respondents' estimate based on detailed calculation upon the contractual condition. Another survey in the USA by[25] revealed that 83% of responding contractors estimated field overheads in a detailed manner. In Hong Kong, 94% of the respondents estimate in details reference to the contract condition as stated by [10]. It seems that most of contractors spent their efforts and time in calculating the field overheads cost.

4.3. Method Used to Determine the HOOH Costs

Table 6 shows that 44.3% says that the method followed to determine the home office overheads costs depends mainly on detailed calculation which is based on the records and experience. While 23.0% says that the method depends on adding a percentage of the total tender cost. 3.3% says that the method depends on totaling a percentage of the item direct cost, and 29.5% says that the method is vary from one project to another.

Table 6. The method followed to determine the HOOH costs.

Percent (%)	Frequency	Method
44.3	27	Detailed calculation based on the records and experience
23.0	14	Percentage of the total tender cost
0.0	0	Percentage of the labors costs in the project
3.3	2	Percentage of the item direct cost
29.5	18	Vary from one project to another
100.0	61	Total

One of the main results has shown the importance of experience in deciding the value of HOOH in tender pricing. It is not static value or percentage but it is varying from project to project (29%), depending on the estimator experience. Furthermore, instability of political situation in Gaza Strip has created no clear trend. So, the apparent problem lies in the uncertainty in the volume of work that the company can successfully being awarded to decide the exact proportion.

4.4. The Distribution of Overheads Costs on Items

Table 7 shows that 73.8% says that overheads costs is equally distributed within each item proportionally to the total contract value. While, 8.2% says that overheads costs is distributed by loading on specific items. 1.6% says that overheads costs distributed in front loading method. 1.6% says that overheads costs distributed in back loading method, and 14.8% says that there is no specific method for the distribution of overheads costs.

Table 7. Distribution of overheads costs on the items.

Percent (%)	Frequency	The distribution of overheads costs
73.8	45	Equally distributed within each item Proportionally to the total contract
8.2	5	Loading on specific items
1.6	1	Front loading
1.6	1	Back loading
14.8	9	No specified method
100.0	61	Total

It is a good behavior that the majority of companies follow

the distribution of overhead cost on the items in proportion to the total contract. That behavior reflected positively on solving claims of suspension, delay by the owners, and termination. Hence, it helps in minimizing disputes with the owners in determining contractor financial damages.

4.5. Identifying the Factors That Affect the Overheads Costs

4.5.1. Factors Attributed to the Managerial Capacity and Policies

Table 8 shows the factors that affect on the overhead costs that attributed to the managerial capacity of the company and its policies in general. The factors are ranked according to RII:

1. The first factor was the expertise in the determination of the overheads costs percentage during the pricing of tenders with RII equal "0.93". It is concluded that the experience is very important in the process of estimating overhead costs because of its sensitivity and a decisive role in winning the tender.
2. The second factor was the company's ability to adhere to the implementation of projects according to the specification within the contractual period, whereas the RII equal "0.93". The company's ability to complete the project to the required specifications will help to reduce miss-technical works that may need remediation. Hence, delaying the project and increase overhead costs required to follow-up to mean while delay.
3. The third factor was the company's ability to cope with the problems during the implementation, whereas the RII equal" 0.90". In fact, the company's ability to address problems during the implementation period will therefore reduces the size of the required overhead expenses that may enter the contractor in losses.
4. The fourth factor was doing financial auditing for expenses and revenues in periodic and continuous manner, whereas the RII equal"0.87". Periodic financial auditing will but the contractor in view of his financial status. In addition, it will enhance the records which will be benefit in coming project.
5. The fifth factor was the company's ability to identify and expect risks, whereas the RII equal "0.86". The ability of contracting company in identifying the risk will guide the estimator in estimating of overheads accurately upon the degree of this risk.

Table 8. Factors related to the managerial capacity and policies.

Rank	p-value	T test	Relative index(%)	Std. Deviation	Mean	Influencing Factor
1	0.0	25.74	0.93	0.50	4.67	Expertise in the determination of the overheads costs percentage during the pricing of tenders
1	0.0	27.59	0.93	0.47	4.67	The company's ability to adhere to the implementation of projects according to the specification within the contractual period
3	0.0	19.58	0.90	0.59	4.49	Company's ability to cope with the problems during the implementation
4	0.0	12.58	0.87	0.83	4.34	Doing financial auditing for expenses and revenues in periodic and continuous manner
5	0.0	14.55	0.86	0.68	4.28	The company's ability to identify and expect risks
6	0.0	11.31	0.83	0.79	4.15	Company response in finding solutions for claims and disputes

4.5.2. Factors Related to the Work Environment and Project Circumstances

Table 9 shows the opinion of the respondents about the factors that affect on the overheads cost that related to the work environment and project circumstances. The first factor was "Closure and the inability to obtain materials", whereas the RII equal "0.94". It is normal to get closure on the rank number one, where the closure means delays in obtaining raw materials. Thus, increasing the duration of the project which would result in increased overhead costs, and putting the contractor in losses.

1. The second factor was "Financial liquidity of the company" with relative index "0.93". Financial liquidity will help the company to commit itself to the project duration. Hence, avoid delays. The desired goal come through the supplying of raw materials in a timely manner with more cheaper prices, bring the project within the required specifications, and provide an excellent crew to work, will thereby reduce the overhead costs to a minimum.

2. The third factor was "Mechanism of contractor financial dues (payments)" with relative index "0.91". Payments mechanism to the contractor will control the financial liquidity of the project is significant. Thus, how the contractor will manage the implementation of the project. As explained above, liquidity is clearly affecting the overhead costs. The contractor can determine the overheads expenses at the time of pricing depending on the payments policy, which will be followed by the client, for example, is there a an advance payment, the amount of the advance payment, and time for paying after submission the request.

3. The forth factor was "owner's commitment toward payments as scheduled" with relative index "0.90". The commitment toward payments schedule is differ from client to client. Sometimes you find delay in the payments disbursement to the contractor in contrast with periods specified by the contract, which negatively affects the project's progress. Hence, delay of the project schedule, and automatically becoming more and more overheads expenses than expected

4. The fifth factor was "The Company's experience in implementing similar projects" with relative index 0.88". The company's experience in implementing similar projects provide an opportunity for the contractor to determine accurately the amount of overheads expenses of the project planned to be priced .

Table 9. Factors related to the work environment and project circumstances.

Rank	p-value	T test	Relative index(%)	Std. Deviation	Mean	Influencing factor
1	0.00	25.82	0.94	0.52	4.72	Closure and the inability to obtain materials
2	0.00	22.14	0.93	0.57	4.64	Financial liquidity of the company
3	0.00	19.07	0.91	0.64	4.57	Mechanism of contractor financial dues (payments)
4	0.00	16.14	0.90	0.72	4.49	Owner's commitment toward payments as scheduled
5	0.00	17.14	0.88	0.64	4.41	The company's experience in implementing similar projects

Critical value of t at df "60" and significance level 0.05 equal 2

5. Conclusion

Majority of contractors in Gaza Strip are aware towards overheads concept and they have good knowledge about the components of overheads. Accordingly, companies' owners or senior mangers are usually estimate overheads costs during the pricing of tenders.

Around one third of contracting companies in Gaza Strip do not depend on historical data during pricing process.

The OH cost is calculated based on detailed calculation for all items required by contractual conditions. No specific amount or percentage could be applicable to be added. Furthermore, during the bidding stage the overheads costs is equally distributed within each item proportionally to the total contract value.

High competition in Gaza construction industry may enforce the contractors to reduce the HOOH percentage. Most of contractors believe that submission of overhead breakdown within their bids will give them opportunity to review the overheads accurately before submission.

The first five factors attributed to the managerial capacity and company's policies that affect the estimation of overheads were the experience in the determination of the overheads costs percentage during the pricing of tenders, the ability to adhere to the specification within the contractual period, the ability to cope with the problems during the implementation, doing financial auditing for expenses and revenues in periodic and continuous manner, and the ability to identify and expect risks.

The first five factors attributed to the work environment and project circumstances that affect the estimation of overheads through the pricing process were the repetitive closure of boarders and the inability to obtain materials through trading terminals, financial liquidity of the company, mechanism of contract payments, owner's commitment toward payments, the experience in implementing similar projects.

References

[1]   Assaf, S. A., Bubshait, A. A., Atiyah, S. and Al-Shehri, M. (2001), "The management of construction company overhead costs", International Journal of Project Management, Vol.19, pp. 295-303.

[2] Dagostino, F. R. (2002), "Estimating in building construction", 6th edition, Prentice-Hall, Englewood Cliffs, N.J.

[3] Abdul-Malak, M. A. U., El-Saadi, M. M. H. and Abou-Zeid, M. G. (2002), "Process model for administrating construction claims", Journal of Management in Engineering, ASCE, Vol.18 No.2, pp. 84-94.

[4] Zack, Jr. James G. (2002), "Calculation and recovery of home office overhead", Construction management Journal, International Cost Engineering Council, 3rd world Congress on cost Engineering, Project Management and Quantity Surveying, 6th Pacific Association of Quantity Surveyors congress, Melbourne, Australia, April 14-18.

[5] McKibbin, R. and Stokes, M. (2005), "Preparation and presentation of claims for delay", Precept Programme Management Limited, Stanford House, London.

[6] Holland, N. and Hobson, D. (1999), "Indirect cost categorization and allocation by construction contractors", Journal of Architectural Engineering, ASCE, Vol.5 No.2, pp. 49-56.

[7] Cilensek R. (1991), "Understanding contractor overhead", Cost Engineering, AACE, Vol.33 No. 12.

[8] Coombs, W. E., and Palmer, W. J. (1995), "Construction accounting and financial management", 5th edition, McGraw-Hill, New York.

[9] Peurifoy, R. L., and Oberlender, G. D. (2002), "Estimating construction costs", 5th edition, McGraw-Hill, New York.

[10] Chan, C. T. W. and Pasquire, C. (2002), " Estimation of project overheads: a contractor perspective", in Greenwood, D. (Ed.), 18th Annual ARCOM Conference, 2-4 September 2002, and University of Nothumbria, Vol. 1 Association of Researchers in Construction Management , pp. 53-62.

[11] Lowe, S., Bielek, R. and Burnham, R. (2003), "Compensation for contractors' home office overhead: A Synthesis of highway practice", American Association of State Highway and Transportation Officials, National Cooperative Highway Research Program, National Research Council (U.S.), Los Angeles, California

[12] Ruf, H. B. and Ruf S. (2007), "Documentation and presenting cost in underground construction claims", No-Dig Conference & Exhibition, San Diego, California April 16-19, 2007

[13] Neil, J. M. (1981), "Construction cost estimating for project control", Prentice-Hall, Englewood Cliffs, N.J.

[14] Shelton, F. and Brugh, M. (2002), "Indirect costs of contracts", Journal of Construction Accounting and Taxation, Vol. 12 No. 4, pp. 3-9.

[15] Clough, R., Sears, G. and Sears, K. (2000), "Construction project management", 4th edition, John Wiley & Sons, New York.

[16] Irwin II, W. (2005), "Current Issues to watch for in construction claims, part III: Overhead Claims",http://www.lorman.com/newsletters/article.php?article_id=81&newsletter_id=22&category_id=3&topic=CN, (accessed March 2012).

[17] Stewart, R. D., Wyskida, R. M., Johannes, J. D. (1995), "Construction cost estimating. In: cost estimator's reference manual", 2nd edition. John Wiley & Sons, Inc.

[18] Boussabaine, A.H. and Elhag, T. (2001), "Tender price estimation using artificial neural networks", Journal of Financial Management Property Construction, Vol. 63 No.3, pp.193-208.

[19] Hegazy T. and Ayed, A., (1998) "A Neural Network Model for Parametric Cost Estimation of Highway Projects," Jr. of Construction Engineering & Management, ASCE, 24(3), 210-218.

[20] Mason, A.K. and Kahn, D.J. (1997), "Estimating costs with fuzzy logic", Proceedings of the 1997 41st Annual Meeting of AACE International, July 13-16, Dallas, TX, 6 pages.

[21] Sha'ath, K. and Singh, G. (1993), "A Stochastic cost engineering system (SCENS) applied to estimating and tendering for bill of quantities contracts", In: Proceedings of 5th International Conference on Computing in Civil and Building Engineering, California, USA, 7-9 June, pp.70-77.

[22] Elhag, T. and Boussabaine, A.H. (1998), "Factors affecting cost and duration of construction projects", In: EPSRC Research Report, Phase (1)

[23] Eksteen, B. and Rosenberg, D. (2002), " The management of overhead costs in construction companies" in Greenwood, D. (Ed.), 18th Annual ARCOM Conference, 2-4 September 2002, and University of Nothumbria, Vol. 1 Association of Researchers in Construction Management, pp. 13-22.

[24] ElSawy, I., Hosny, H. and Abdel Razek, M. (2011), "A Neural Network Model for Construction Projects Site Overhead Cost Estimating in Egypt". IJCSI International Journal of Computer Science Issues, Vol. 8 No. 1, pp. 273-283.

[25] Hegazy, T. and Moselhi, O. (1995), "Elements of cost estimation: a Survey in Canada and the United States", Cost Engineering, Vol.37 No.5, pp.27–33.

5

Analysis of Seepage Under Hydraulic Structures Using Slide Program

Hala Kathem Taeh Alnealy, Najm Obaid Salim Alghazali

Civil Engineering Department, Babylon University, Babylon, Iraq

Email address:

drnajm59@gmail.com (N. O. S. Alghazali), Halhhalh300@yahoo.com (H. K. T. Alnealy)

Abstract: In this study, "SLIDE" program was used to analyze seepage flow under the hydraulic structure through single and multi- layers soils and its effect on structures with inclined cut-off at downstream, at upstream, and at both of them. The distribution curves of uplift pressure along the floor had been reached as well as the distribution of exit gradient at downstream. In the first experiment which included hydraulic structure based at single layers, results are compared with the general case (no cut-off) obtained when using single cutoff, minimum value of the uplift pressure and seepage quantity occurred when using cutoff at upstream side with Θ=45° which given decreasing 40.3%, 28.5% respectively and minimum value of exist gradient occur when using cutoff at downstream side with Θ 120° which given decreasing 8.03%. while using double cutoff minimum value of uplift pressure ,exist gradient ,seepage occur when using double inclined cutoff at upstream with Θ = 45° and downstream side with Θ 120° which given decreasing (42.1%, 8.03%, 30%) its contacted that using double cutoff given decreasing in each uplift pressure, exist gradient, seepage at the same time.

Keywords: Inclined Cutoff, Single Layer Soil, Multi Layers Soil, Finite Element Method

1. Introduction

The foundation soil of any structure should be given the greatest importance in analysis and design as compared with other parts of the structure, because failure in the foundation would destroy the whole structure. The differential head in water levels between the upstream and downstream affects the foundation and causes seepage flow (Selim 1947). The groundwater flow depends on the type of flow, the soil media and the boundary conditions. Seepage of water is one of the major problems which effects on hydraulic structures (Alsenousi and Mohamed 2008).The seepage problem can be analyzed and solved by using many methods solution such as the electrical analog models, empirical formulas, experimental works as well as numerical models (EL-Fitiany et al 2003). In this study a numerical model to determine flow net is adopted.

2. Aim of the Study

The main objectives of this work can be summarized bythe following points:

1. To steady the effect of position and inclination angle of cutoff wall on uplift pressure, exist gradient and seepage quantity in the foundation the hydraulic structure resting on single layer soil.

2. To steady the effect of position and inclination angle of cutoff wall on uplift pressure, exist gradient and seepage quantity in the foundation the hydraulic structure resting on multi layers soil.

3. General Case Study

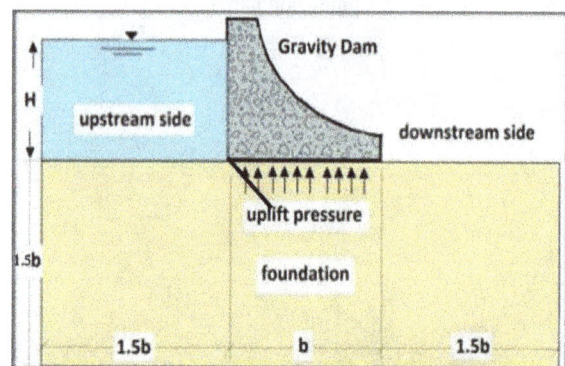

Fig. (1). *General case study and boundary conditions.*

Finite element method was used to analyze the general case study shown in Fig.(1) using (SLIDE.5.0) program.

4. Boundary Conditions

The boundary conditions should be specified beforestarting the solution. For the steady state of a confinedflow, the boundary conditions are defined as follows:

• Reservoir Boundaries

The height of the water above these boundaries has always a known value, so that the pressure on any point of these boundaries is also known; so, the piezometric head distribution along the reservoir boundaries is constant; that is:

$$H = H_O = \frac{p}{\gamma_w} + z$$

• Impervious Boundary

Impervious boundary has the perpendicular velocityfunction on the surface equal to zero ($\frac{\partial H}{\partial n} = 0$).

5. Finite Element Formulation of Seepage in Porous Medium

The finite element method is a very powerful numerical method. It requires the use of digital computer because of the large number of computations involved. In ground water flow problems, one could imagine that a region is subdivided into small elements, these elements may be two, or three-dimensional and joined to each other by nodes existing on the element boundaries. Such that for each element the flow is described in terms of the head in the nodal points, and that then a system of equations is obtained from the conditions that the flow must be continuous at each node (El-Katib, 2009The field variable model describing an approximate variation of piezometric head (He) within the element is:

$$H^e = \sum_{i=1}^{n} N_i H_i$$

Where:

Hi: Nodal value of head; H, of the element;

n: Number of nodes per element;

Ni: Shape function of the element

It is possible to write Equation (3-1) in matrix form as follows [Zienkiewicz,1966]

He =[Ni]{Hi} Where:

]Ni]: shape function matrix;

}Hi}: vector matrix of nodal values.

The approximate solution for head variation, H, over the whole domain is given as follows:

$$H = \sum_{e=1}^{n_e} [N_i]\{H_i\}$$

Where:

ne: is the total number of elements in the problem domain.

6. Computer Program

SLIDE v.5.0 it is analysis software built-in steady state groundwater analysis using finite element method .slide 5 is a steady state flow model and will compute the piezometric head value at each node of the finite element mesh. From these values, flow lines and equipotential lines are plotted showing the resulting seepage net.

7. Results and Discussion

The Hydraulic Structure Resting on Single Layer Soil

The problem of the effect of inclined cutoff and its location along the floor of hydraulic structure has been investigated on each of

7.1. Effect of Inclined Cutoff and Its Position on the Uplift Pressure

Figures (2) and (3) illustrates the relationship between the horizontal distance from the base of floor of hydraulic structure and pressure head, also this figures demonstrates of the influence of cutoff inclination and position on the uplift pressure generated along the base of hydraulic structure. When a cutoff subsist within upstream side with different angles inclination Θ(45° , 90°,120°) as shown in figure (4-32), the behavior of cutoff was noticed that the uplift pressure decreased as (Θ) decreased toward U/S side for Θ (45°, 90°, 120°) where the maximum redaction in uplift pressure as compared with the general case (no cutoff) was (40.3%, 31.2%, 24.6%) respectively.

So it is concluded that using upstream cutoff inclined towards the upstream side with Θ less than 90° is beneficial in decreasing the uplift head because of increasing the length of creep which in turn increase the head loss.

The effect of downstream cutoff inclination angle on uplift pressure head that demonstrates in figure (3) for Θ (90°, 120°) the maximum decreasing in uplift pressure as compared with the general case (no cutoff) was (3.75%, 10%) respectively. It was noticed that the reduce in the uplift pressure was small to their placing of the cutoff inclination, therefore it can be neglected.

When using double cutoff posited at heel with Θ=90⁰ and toe with Θ=90⁰as shown in figure(4), the uplift pressure reduced and the maximum decreasing in value of uplift pressure as compared with the general case (no cutoff) was 40.93%.

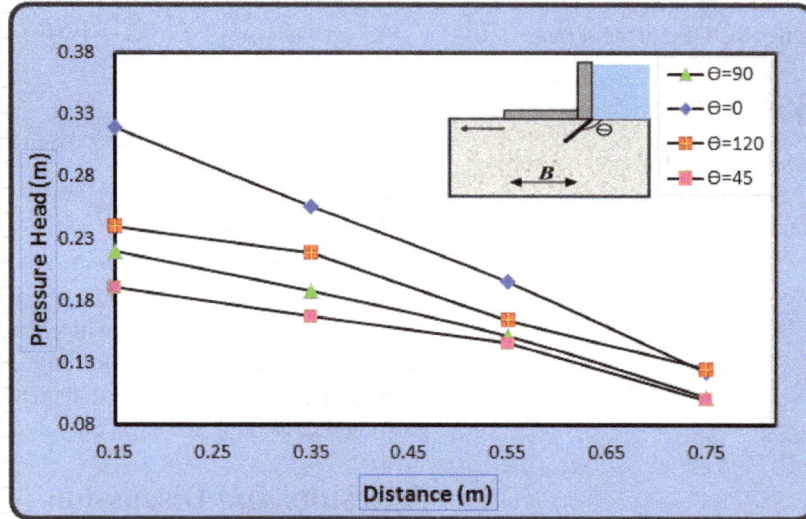

Fig. (2). *Variation of uplift head under hydraulic structureresting on single layer with (U/S) cut-off for different values of Θ.*

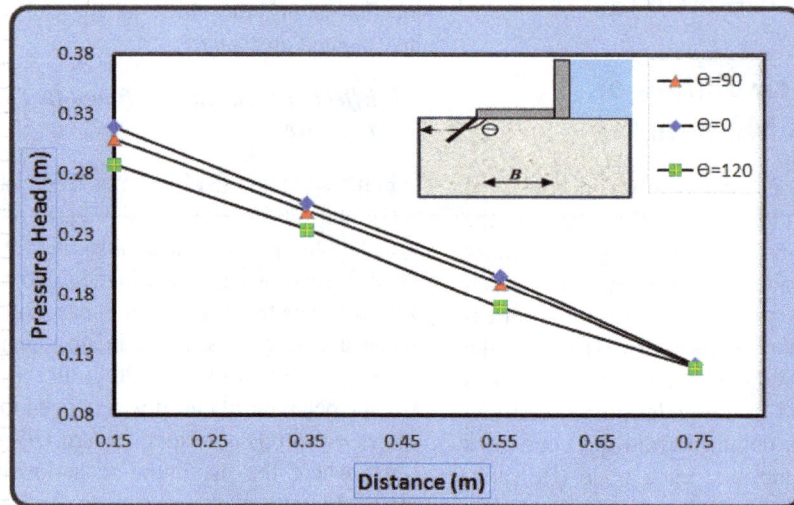

Fig. (3). *Variation of uplift head under hydraulic structure resting on single layer –with (D/S) cut-off for different values of Θ.*

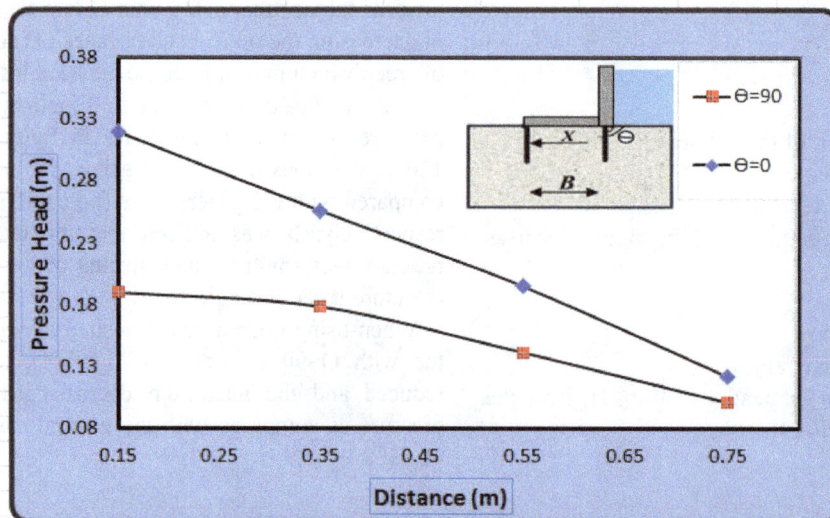

Fig. (4). *Variation of uplift head under hydraulic structure resting on single layer – with double cutoff U/S Θ=90 and D/S Θ=90.*

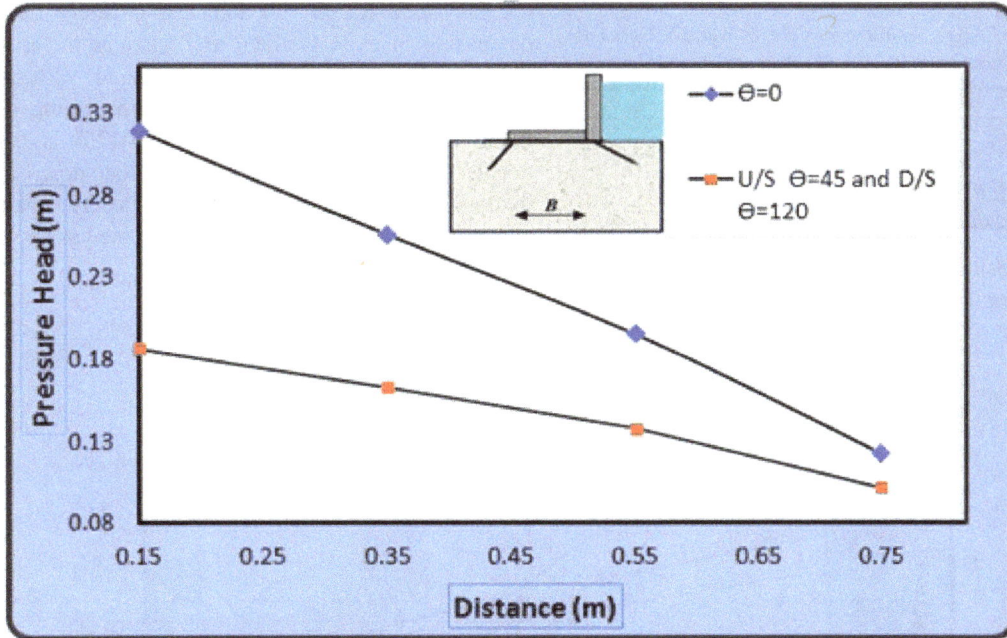

Fig. (5). *Variation of uplift head under hydraulic structureresting on single layer – with U/S Θ=45 and D/S Θ=120.*

The effect of using double inclined cutoff positioned at upstream side with Θ=45⁰ and downstream side with Θ=120⁰ is shown in figure (5). Which indicate that the value of uplift pressure decreases significantly to 42.01% as compared with the general case (no cutoff).

7.2. *Effect of Inclined Cutoff and Its Position on the Exit Gradient*

Figures (6) and (7) demonstrates the relationship between the horizontal distance from downstream side and exist gradient, also this figures show the influence of cutoff inclination and position on the exist gradient generated in the downstream side of hydraulic structure resting on sandy silty clay. In figure (6) illustrate the effect of cutoff inclination angle on exit gradient distribution along downstream side hydraulic structure with upstream inclined cutoff. It can be seen that values for exit gradient low decreased if the cut-off is inclined towards downstream side (Θ is more than 90°), were the redaction in values of exit gradient for Θ (45°,90°,120°), the maximum redaction in exist gradient compared with the general case (no cutoff), were (1.25% , 1.78% ,3.21%) respectively. It's shown that a slight decrease in the magnitude of the exist gradient when the cutoff posited in upstream side of the hydraulic structure. The factor of safety against piping calculated for this case in Table (1).

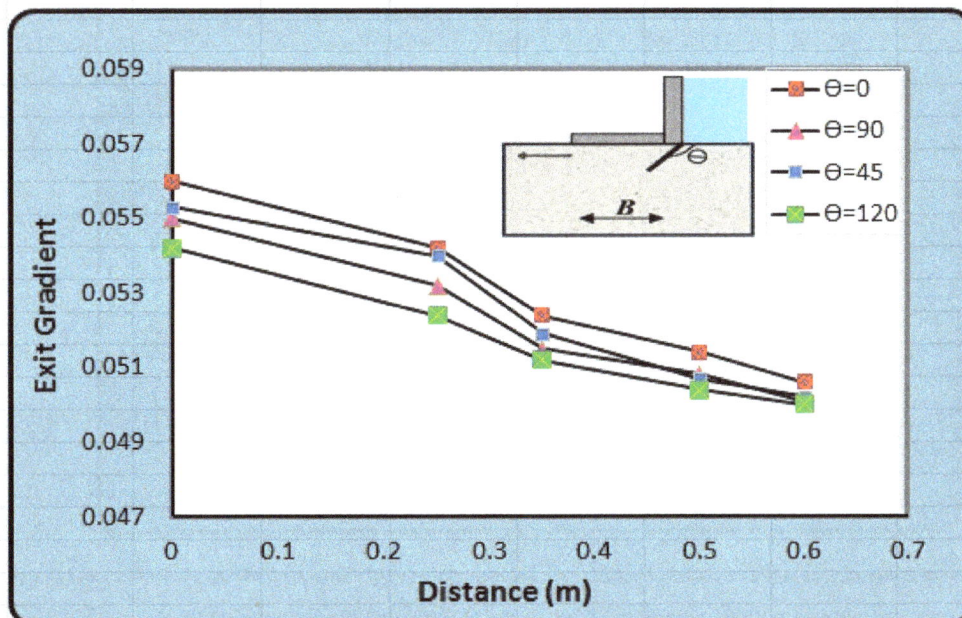

Fig. (6). *Variation of exit gradient for a hydraulic structure resting on single layer – with (U/S) cut-off for different values of Θ.*

Table (1). *The factor of safety against piping for a hydraulic structure resting on single layer – when cutoff at U/S with different values of Θ.*

U/S Cut-off Inclination	max (exist gradient)	Fs
0°	0.4	2.47
45°	0.382	2.586
90°	0.375	2.63
120°	0.362	2.729

As the cutoff posited in downstream side of hydraulic structure, the exit gradient decreased as (Θ) increases toward the D/S side for Θ (90° and 120°), where the maximum decreasing in exist gradient as illustrates in figure (7), was (3.57% 5.36%) respectively, as compared with the general case (no cutoff), So it is concluded that using downstream cutoff inclined towards toe with Θ=120⁰ is beneficial in decreasing the value of exit gradient along downstream of the hydraulic structure, the factor of safety against piping phenomenon of this case can be calculated as shown in Table (2).

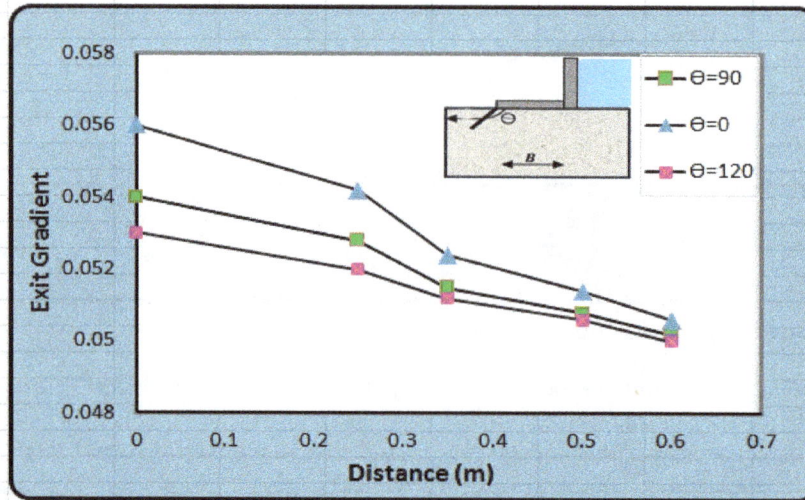

Fig. (7). *Variation of exit gradient for a hydraulic structure resting on single layer – with (D/S) cut-off for different values of Θ.*

Table (2). *The factor of safety against piping for a hydraulic structure resting on single layer – when cutoff at D/S with different values of Θ.*

U/S Cut-off Inclination	max (exist gradient)	Fs
0°	0.4	2.47
90°	0.35	2.82
120°	0.31	3.18

The percentage decline in the value of exist gradient when using double cutoff at heel with Θ=90 and toe of hydraulic structure with Θ=90 was 6.11%, as shown in figure (8) compared with the general case (no cutoff).

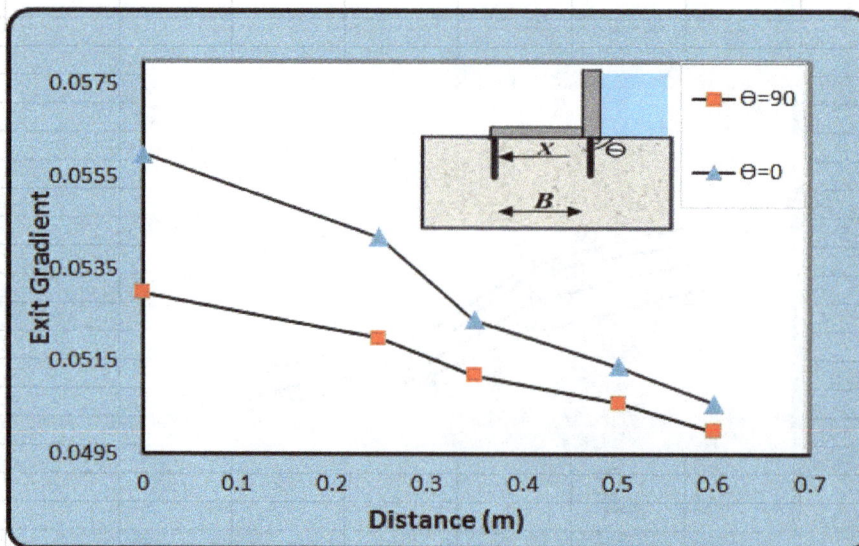

Fig. (8). *Variation of exit gradient along downstream hydraulic structure resting on single layer with inclined U/S Θ=90 and D/S Θ=90.*

High value in value factor of safety against piping occur when using double cutoff in the U/S with Θ=45⁰ and D/S with inclined $\Theta=120^0$ as shown in figure (9) were the decreasing in exist gradient was 8.03% as compare with general case(no cutoff). So it is concluded that using double inclined cutoff with $\Theta =45^0$ at heel and $120^{\,0}$ is beneficial to give perfect decreasing in exist gradient .

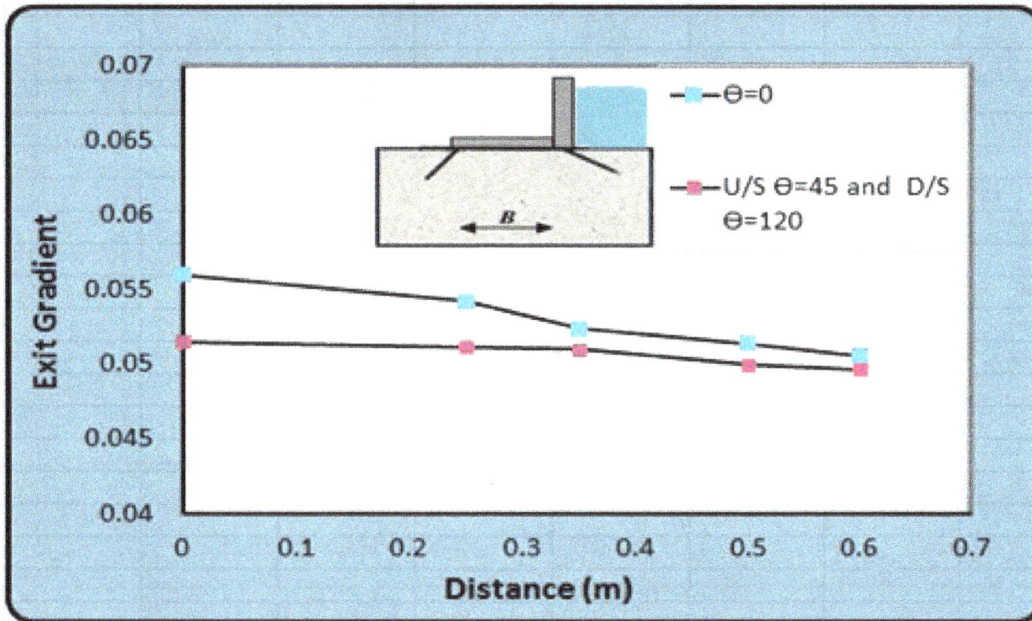

Fig. (9). Variation of exit gradient along downstream hydraulic structure resting on single layer –sandy silty clay with inclined U/S $\Theta=45$ and D/S $\Theta=120$.

7.3. Effect of Cutoff Position and Inclination on the Seepage Quantity

Figure (10) and (11) illustrate the influence of cutoff inclination on the seepage quantity, for the cutoff inclined in upstream side of hydraulic structure as shown in figure (10), noticed that the seepage under the hydraulic structure decreased as(Θ) decreased toward U/S side for $\Theta(45°$, $90°,120°)$ where the maximum decreasing in seepage compared with the general case (no cutoff) was (28.5% , 23% ,6.34%) respectively, so it is concluded that using upstream cutoff inclined towards the upstream side with Θ less than 90°($\Theta=45$) beneficial in decreasing the seepage quantity because of increasing the length of creep which in turn increase the head loss due to decrease the seepage quantity at the outlet .

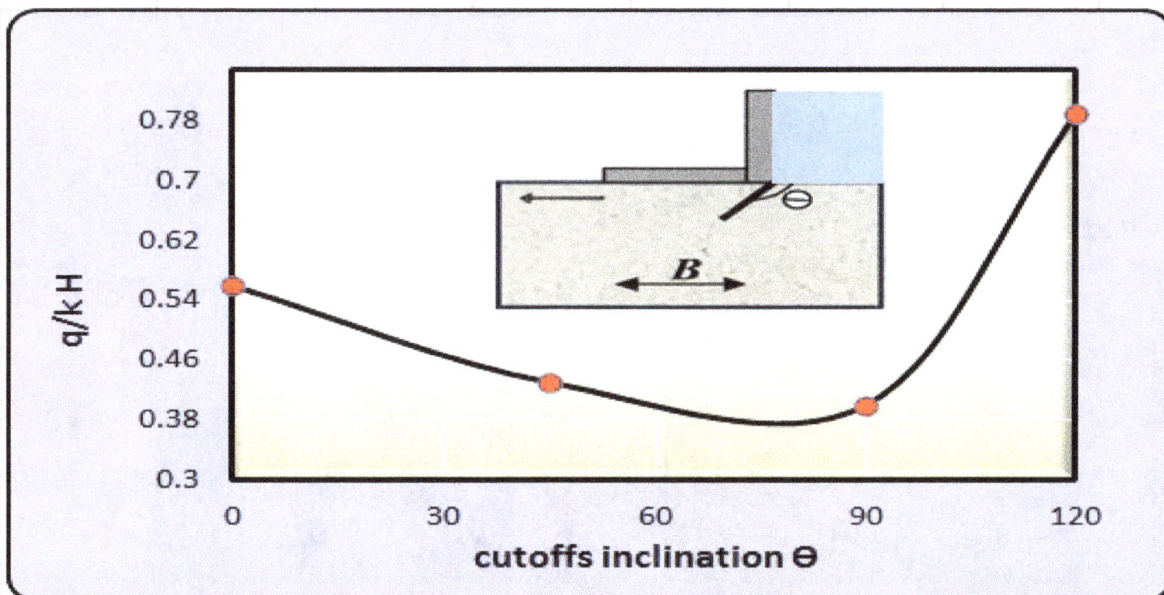

Fig. (10). Variation of Seepage quantity under hydraulic structure resting on single layer with (U/S) cut-off for different values of Θ.

In figure(11) when the cutoff posited at the toe, the value of seepage decreased as (Θ) increased toward D/S side for Θ

(90°, 120°) and the maximum decreasing in seepage compared with the general case (no cutoff) was (12.6% , 28.57%) respectively.

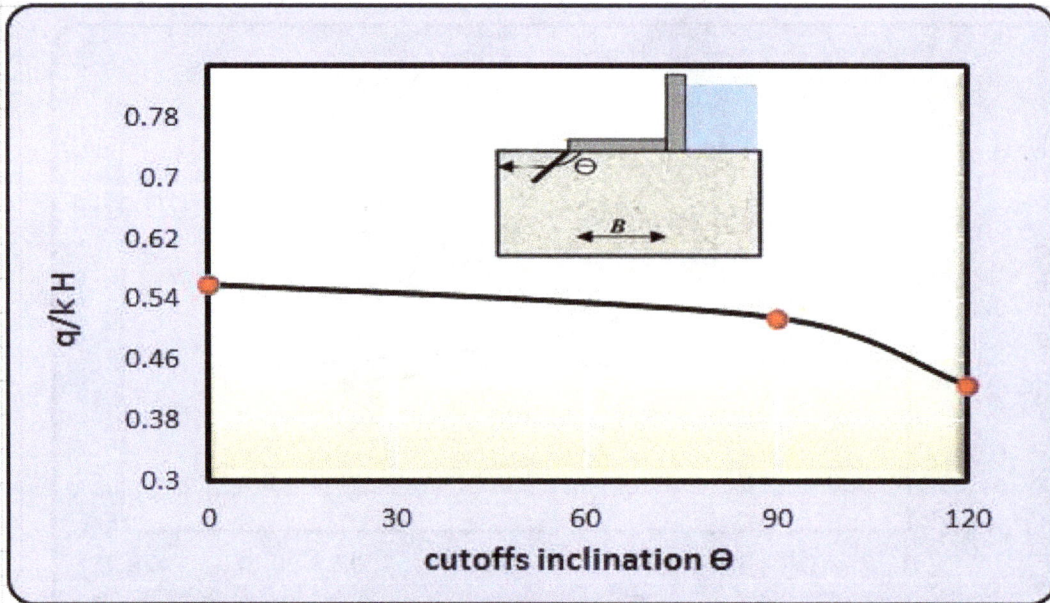

Fig.(11). Variation of Seepage quantity under hydraulic structure resting on single layer with (D/S) cut-off for different values of Θ.

When cutoff positioned in U/S with Θ=90 and D/S with Θ=90 the seepage quantity reduced to 30.156%, and when cutoff at U/S with Θ=45⁰ and D/S with Θ=120 ⁰ andseepage quantity reduced to 30.3% the compared with the general case (without cutoff).All results of above discussed cases were summarized in table (3).

Table (3). Effect of cut-off inclination on seepage controlfor hydraulic structure resting on single layer.

Maximum redaction according to the general case Θ=0(no cutoff)						
Cutoff at U/S			Cutoff at D/S		Double cutoff at U/S and D/S	
Θ=45⁰	Θ=90⁰	Θ=120⁰	Θ=90⁰	Θ=120⁰	U/S ,D/s Θ=90⁰	U/S Θ=45⁰,d/s Θ=120
Uplift pressure	40.31%	31.25%	24.68%	3.75%	10%	40.99%	42.01%
Exit gradient	1.25%	1.78%	3.21%	3.57%	5.35%	5.35%	8.03%
Seepage	28.5%	23%	6.34%	12.6%	28.75%	30.15%	30.3%

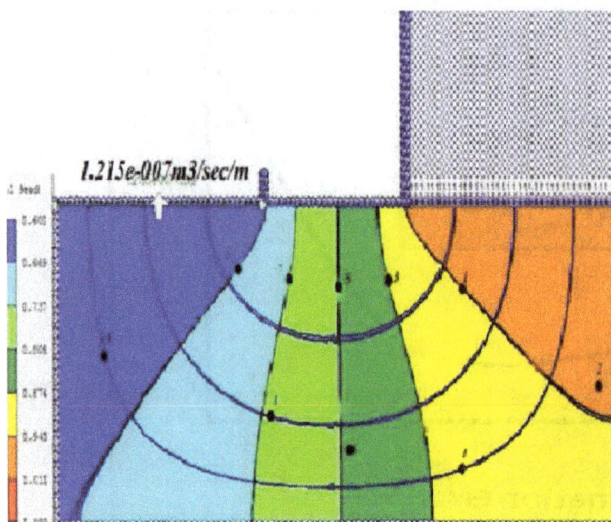

general case (no cutoff)

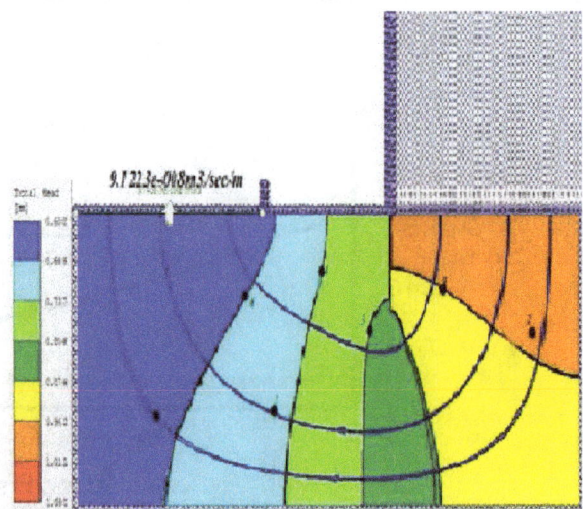

cutoff at the upstream side Θ=90

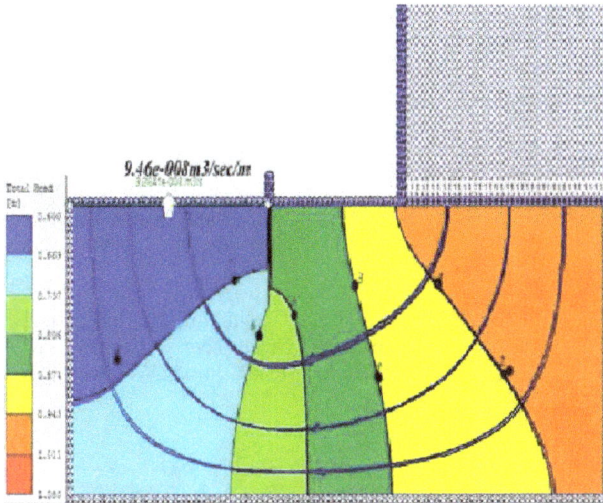

(cutoff at the downstream side Θ=90)

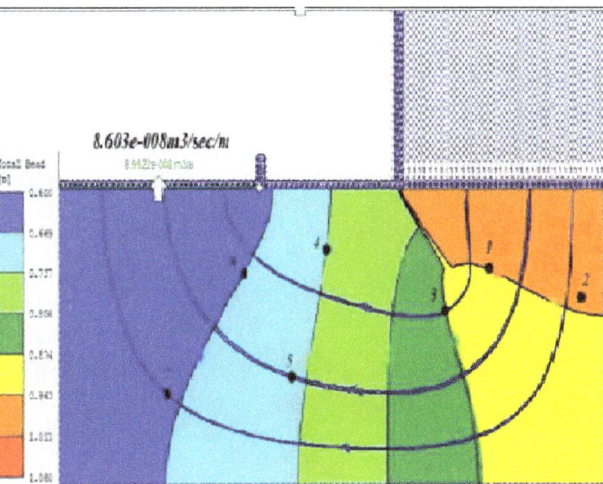

(cutoff at the upstream side Θ=45)

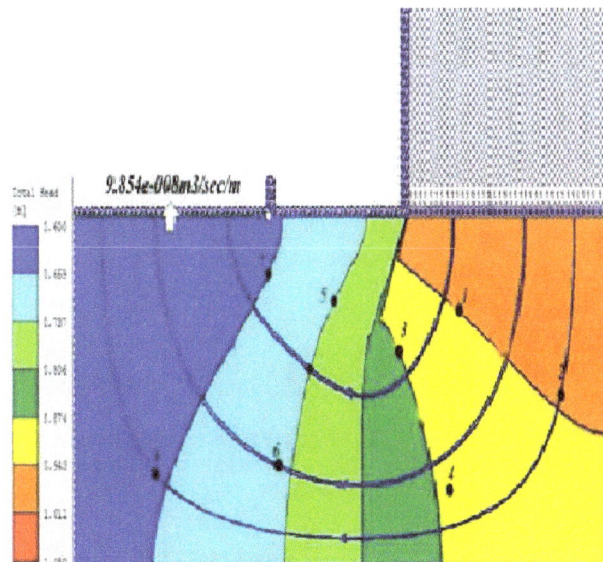

(cutoff at the upstream side Θ=120)

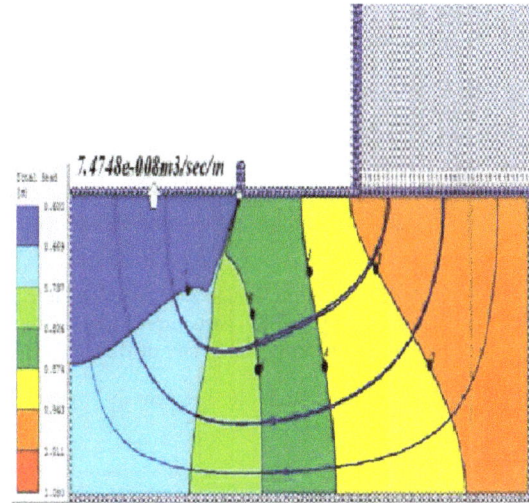

(cutoff at the downstream side Θ=120

Fig. (12). *The model images for second experiment included hydraulic structure resting on single layer .*

8. Conclusions and Recommendations

In the conducted research, finite element method used to analyze the seepage flow through soil foundation under hydraulic structures and control of seepage flow. Based on the theoretical application models, the following main conclusions can be explained

- Using an inclined cut-off towards the upstream side with Θ equal 45° beneficial in decreasing the uplift pressure to 40.3% , seepage quantity to 28.5%, as compared with the general case(no cutoff).
- Placing a cutoff at the hydraulic structure toe is not recommended under any angle of inclination to decreasing the value of uplift pressure.
- Using an inclined cutoff towards the downstream side with Θ = 120° beneficial in decreasing in value of exist gradient to 5.0% and increasing the safety factor against piping phenomenon to 3.18.
- The reduction in uplift pressure , exist gradient and seepage when using double cutoff at U/S with Θ=45 and D/S with Θ =120 is more than that of using double cutoff at the up and downstream at right angle that given decreasing in value of exist gradient to 8.03% , uplift pressure to 42.3% , seepage to 31.15% and increasing the safety factor against piping phenomenon to 2.43

References

[1] AL-Ganaini,M.A., (1984),"Hydraulic StructurBeirut,pp47-60, (In Arabic).

[2] Selim, M.A., (1947) "Dams on Porous Media", TransactionASCE, Vol. 1 Roy, S.K.(2010). "Experimental Study OnDifferent Types Of Seepage Flow Under The Sheet PileThrough Indigenous Model.", M.Sc. Thesis Insoil Mechanicsand Foundation Engineering, University Of Jadavpur.12, pp488-526.

[3] EL-Fitiany, M. A. Abourohim, R. I. and El-Dakak, A. Y.(2003). "Three dimensional ground water seepage around asimplehydraulic structure." Alexandria Engineering Journal,Volume 42 ,Issue 5, September.

[4] Arslan, C. A. and Mohammad, S. A. (2011)."Experimental andTheoretical Study for Pizometric Head Distribution underHydraulic Structures." Department of Civil engineering;College of engineering, University of kirkuk Volume 6,

[5] Musawi, W. H. 2002. Optimum Design of ControlDevices for Safe Seepage under Hydraulic Structures,M.Sc. Thesis, Department of Civil Engineering,University of Babylon.

[6] Zienkiewicz, O. C.(1982)."The Finite Element in Engineering Science" McGraw-Hill.

[7] Zienkiewicz, O.C. (1966): "Solution of anisotropic seepage by finite element", Journal of the Engineering Mechanic Division, Vol.(92), No.1, pp.111-120.ual, EM1110-2-1901, Chapter 4, Washington, D.C.,U.S.A.

[8] Selim, M.A. , (1947) "Dams on Porous Media", Transaction ASCE , Vol. 112, pp 488-526.

[9] Naamani, L., Turk, D. and Osman, H.(2002)."Finite Element analysis Versus Seepage Tank Model." Civil and Environmental Engineering.

[10] Ijam, A. Z.(2011)."Dams with an Inclined Cutoff." Civil and Environmental Engineering Department, Faculty of Engineering, Mu'tah University, Mu'tah, Jordan, Volume 16, p 1427-1440.

6

Experimental and Numerical Approaches to Overtopping Levee Breach Effects in a River and Floodplain

Md. Serazul Islam[1, *], Tetsuro Tsujimoto[2]

[1]School of Agriculture and Rural Development, Bangladesh Open University, Gazipur-1705, Bangladesh
[2]Department of Civil Engineering, Nagoya University, Nagoya, Japan

Email address:
seraz_bou@yahoo.com (Md. S. Islam)

Abstract: This paper described the results to identify, characterize, and simulate the levee breach effects in a river and floodplain by overtopping. One-side levee model is built in a laboratory experimental flume as well as numerical simulation using sand with proper compaction. An initial condition provided for the overflow breach is considered with partial crest opening. Small-scale laboratory experiments were performed to evaluate the effects of overtopping levee breaching and investigated simultaneous phenomena appears in a river, levee and floodplain, and validated the results with same scale numerical simulations; and the results of both approaches were in conformity. The failure behavior of an earthen levee focuses on the effects of material sizes, river bed slopes and bed variations relative to floodplain. According to the results, the higher bed level brings more rapid propagation of the levee breach and widening with more sediment deposition in the floodplain area as well as river bed degradation in the upstream of the levee breach point may cause further risk of the levee breach during the next flood. Using finer bed materials, river bed deformation and sediment deposition in the floodplain are clearly make differences with coarser materials, also it create the normal flow problem through the river in future.

Keywords: Overtopping, Levee Breach, Inundation with Sediment Deposition, Laboratory Experiment, Numerical Simulation

1. Introduction

Levees are constructed along water courses to provide protection against floods. The failure of such systems because of natural or manmade hazards can have monumental repercussions, sometimes with dramatic and unanticipated consequences on human life, property and the country's economy. Levee overtopping can be caused when flood waters simply exceed the lowest crest of the levee or if high winds begin to generate significant swells (a storm surge) in the ocean or river water to bring waves crashing over the levee. The top and middle overtopping of a levee are occurred due to wave action and the rising of the water level, respectively [1]. With an increase in urban development behind these levees, the risk to public health and safety from failure has increased. There are a number of mechanisms that would cause a levee to fail. Overtopping, surface erosion, internal erosion, and instabilities within embankment or foundation soils are some of them. While other mechanisms require more time to significantly damage a levee, overtopping and seepage would

erode the levee in a relatively shorter time, and the erosion would eventually lead to levee breach and failure. The failures of levees are mostly due to overtopping of the crest or piping [2]. According to Broich (1998), 43% of the failures are by overtopping and 40% due to piping [3]. Therefore, it is critical to investigate the breaching process and acquire the ability to assess how quickly a levee would fail due to overtopping.

In recent years, the frequency of abnormal floods in Bangladesh has increased substantially, causing serious damage to lives and property. Mostly, the levee breach disaster occurs in Bangladesh because of huge upstream catchments water and sediment load. Particularly, Bangladeshi river beds are aggrades very quickly due to continuous sedimentation, that changes in the river bed level can be observed during one's lifetime. An another problem is damming of the river, which reduces the power of water flow downstream from the dam, and the sediments carried by the river start to settle down faster on the riverbed; causing the river bed aggradations and in turn reducing the water carrying capacity of the river [4, 5], consequences as their banks an overflow, and the flow causes the levee breach. As for the example, due to the Farakka

Barrage on the Ganges has already caused tremendous damage to the agriculture, navigation, environment, and hydrodynamic equilibrium in Bangladesh [6-9].

Several studies have been conducted to understand the failure mechanism by overtopping. However, most of the research has been done on non-cohesive materials [10, 11]. Levee/dam break tests have been performed at different scales in the United States [12, 13], England [14] and Norway [15]. The scale of these tests varies from small laboratory flumes to large scale levee failures. Besides the lack of a complete dynamic similitude between models and prototype levees, experimental observations show similar characteristics and behavior during the breach formation process for similar soil composition [16].

The levee breach study is rare in the available literatures except for few experimental, numerical and field investigation have been conducted by Fujita et al., 1987 [17, 18]; Islam et al., 1994 [19]; Aureli and Mignosa, 2003 [20]; Tsujimoto et al., 2006 [21]; and Shimada et al., 2009, 2010 [22, 23]. In those studies; they investigated levee breach expansion process as well as floodplain sedimentation process but did not consider on the river bed height relative to floodplain and the subsequent phenomena appearing in the river bed and in the floodplain. On the other hand, Islam and Tsujimoto (2012) conducted a numerical study; they investigated breach evolution process and the risk of flood disasters in the low floodplain [24, 25]. The levee breaching phenomena appears not only at levee but also from the river to floodplain, and thus physical experiments are difficult while a numerical approach has not been well developed. In this study, the attempt have taken to conduct small-scale laboratory experiments and same condition numerical analyses using coarse and fine sand with steep and mild river bed slope, respectively. There had some difficulties in measurements during work in the laboratory and thus the numerical simulation is necessary for the conformity of this study. Therefore, the investigation have carried out utilizing both approaches to understand the breaching phenomena on the levee, and to evaluate the risk in the floodplain with different height of river bed to floodplain, various river bed materials and slopes.

2. Solution Approach

2.1. Experimental Set-up and Measurements Procedure

This section describes the laboratory preparation for the runs with experimental conditions to be maintained for different river bed height, and working procedures to fulfill the aforementioned objectives. The experiments are performed in a 20 m long, 2.2 m wide and 1.0 m deep concrete flume is located in the Hydraulic Engineering Laboratory of Nagoya University. The working section is made of wood and sand, which are 6 m long and 2.2 m wide (including river channel, levee, floodplain, and drainage channel). The levee slope is 1:2 for both sides, and levee height is 0.15 m from the floodplain. Same sizes of bed material in the river, levee and floodplain of d_{50}=1.00 mm (Run 1 to Run 3) and 0.13 mm

(Run 4 to Run 6) are used. Relative height of river bed and floodplain is set as follows: Run 1 and Run 4 (low river bed) z_b=-5 cm, Run 2 and Run 5 (river bed and floodplain at the same level) z_b=0 cm, and Run 3 and Run 6 (high river bed) z_b=5 cm, respectively. Figure 1 (a-b) is a schematic representation of the experiment setup, including the top view and the side view, respectively. In experiments, the inflow discharges (a) is supplied initially into an upstream inlet tank of the river channel from an underground water reservoir by a circulating pump. The fixed bed is made of wood (A, D) and the moving bed (B, C, E) is prepared by sand are used to construct the levee and floodplain. Initial breaching point is set at 2.5 m apart from upstream, and a notch (H) is prepared before starting the experiment. A downstream wall (e) of the floodplain is made of 2 cm height of the wooden board from the floodplain, and this wall is used to protect the movable floodplain, and as well as it maintained inundation depth into the floodplain. A 5 cm drainage channel (G) is provided at the downstream of the floodplain. The river inflow and outflow discharge are rectified (b) by using a steel wire, and the inflow water is passed through the river (F) over a rectangular weir (g). In order to keep the river water depth roughly to the uniform flow depth, a wooden weir (sill) (c) is installed at the downstream of the river channel. A wave metre (f) (CHT6-30 made by KENEK Co.) is put in front of a rectangular weir to collect the crest over flow water depth, and in the same way another wave metre (CHT6-40 made by KENEK Co.) is set near the downstream side triangular weir (d). During experiment, a video camera (GZ-HM350-B manufactured by JVC) is placed with moving carriage on top of the levee breach section to record the video footage of breach expansion and overflow by the breach. Levee breach expansion processes as well as topographic changes in the river, levee and floodplain are memorized by using a digital still camera (OptioS1manufactured by PENTAX). Two types of actuators (KMB-150A length 1.60 m and A30 length 1.0 m made by THK) along with laser sensor (IL-600 is made by KEYENCE) is placed lateral (Figure 2.a) and longitudinal (Figure 2.b) directions over the working area to survey floodplain topography and longitudinal length of the levee breach.

Before starting the experiment, the working section of the flume is prepared as shown in Figure 2.c, and then a notch (10×5 cm) is cut to provide the initial breach opening for the overflow experiment. Soil sample is collected from this notch section of the levee and analyzed the degree of compaction; we found it is reached nearly 100%. Then, the inlet and outlet tank is filled with water, and the wave meter reading is set at an initial condition (zero). Inflow discharge is allowed to enter gently in the river section and raised the river flow depth up to notch opening by putting a downstream sill properly. The early placed wave meter data are taken to estimate the inflow and outflow water discharge by using the equation for rectangular [27] and triangular weir [28], respectively. The electronic actuator with laser sensor is fixed with a moving carriage on the working area that is travels over the steel frame on both sides of the flume. During experiment, the longitudinal breach widening with time is measured. The river section and the floodplain are drained, and the bed is become dried; then the elevation of the bed is measured using computer-aided laser sensors for each run. The x-axis is the longitudinal direction with y=0 at the top of the levee crest, which is 2.20 m apart from the upstream end; and the final breach expansion is measured in the test area. The bed level changes in the river channel and in the levee, are measured along 32 longitudinal transects with 3 cm intervals, start at the center of the river channel (x=0) towards the floodplain. The floodplain topographic changes are measured along 64 laterals transects with 5 cm intervals are pointed from the left side of the floodplain with y=0 towards the right-side where the floodplain deposition is occurred and z start from the initial position of the floodplain.

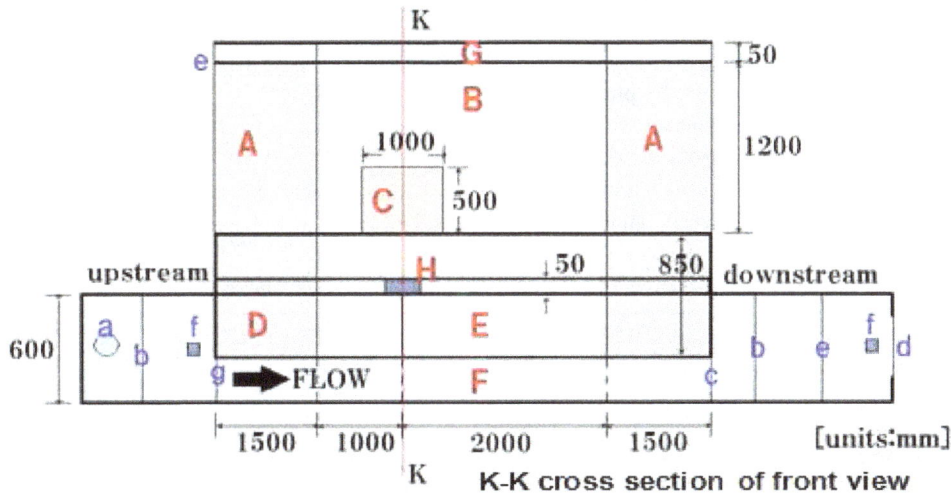

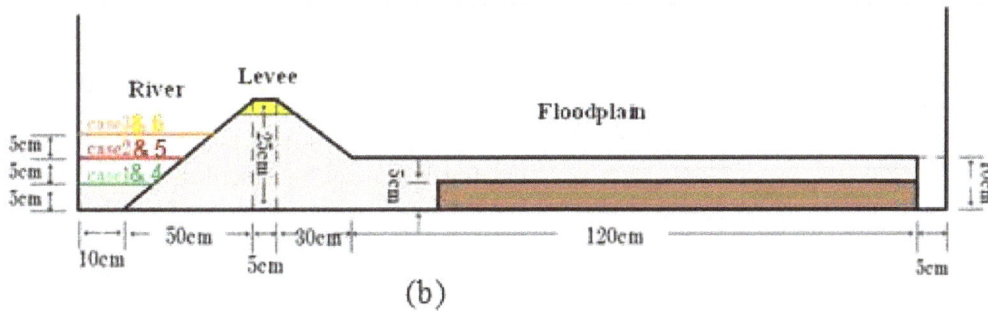

(a)

(b)

Figure 1. Experimental setup: (a) top view; (b) side view.

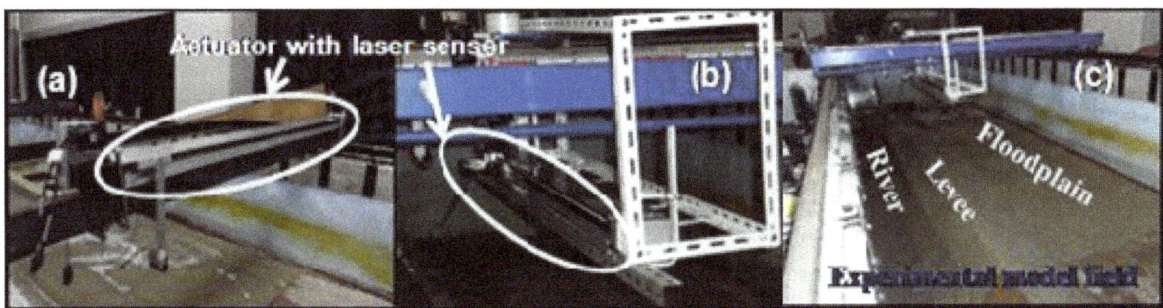

Figure 2. Electronic actuator with computer aided laser sensor (a) lateral, and (b) longitudinal direction; (c) experimental model field.

2.2. Numerical Set-Up and Measurements Procedure

The analyses have been made to observe the process appearing in the river, levee and floodplain in a same simulation scheme during the breach. Floodplain inundation with sediment and evolution process of the breach is studied with a numerical model. RIC-Nays (http://i-ric.org/nays/ja/sitmap.html), a two-dimensional (2D) model for the flood flow and morphology is utilized in this study. As for the simulation scheme, the river channel, levee, floodplain and the flow parameters are selected in the conformity with the typical field data. Schematic model area is spatially limited to a part of the actual fields. For all cases of simulation, computation reach is 6.00 m long and 2.20 m wide (river channel, levee and floodplain) with a bed slope of river channel is 1/500 (Runs 1 to 3) and 1/1000 (Runs 4 to 6) for the coarse and fine bed materials, respectively. Figure 3 depicts one of the model fields for simulation. Levee slope is considered as S_f=1:2 on both country side and river side. The levee height is taken as h_f=15 cm from the floodplain and 20 cm (Runs 1 and 4), 15 cm (Runs 2 and 5) and 10 cm (Runs 3 and 6) from the river bed as represent the low, same and high river bed, respectively. Idealized flow and sediment parameters are considered in the computation. Overflow starts from the hypothetical notch on top of the levee as a trigger of the breach, where an initial breach is 10 cm long (L_b) and 5 cm (h_c) deep from the top of the levee. Though the river discharge has a hydrograph in general, non-uniform discharge is correspond to the peak is assumed here by putting the downstream sill in the river. The solid boundary wall is imposed on the left-side of the floodplain to protect the direct flow through the floodplain. The inflow discharges (Q) and the corresponding river flow depth before the breach, and the

median sizes (d_m) of sediment are chosen, which are shown in Table 1.

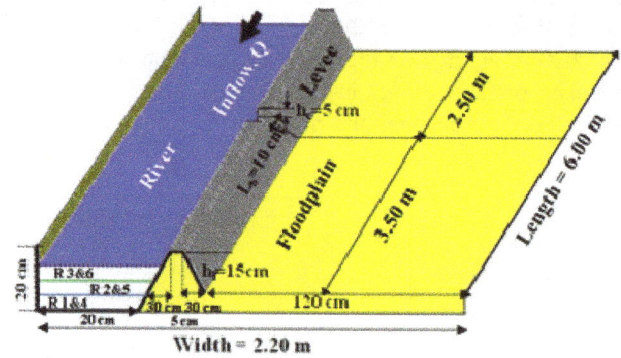

Figure 3. Simulated schematic model fields

Table 1. Condition for all experiments (Same discharge and bed material are used in numerical analysis)

Parameters	Coarser bed material with steep river bed slope			Finer bed material with mild river bed slope		
	R 1	R 2	R 3	R 4	R 5	R 6
Inflow Q (m³/hr)	32.22	31.36	31.28	34.16	31.86	17.75
River flow depth h_0 (m) (Exp.)	0.16	0.11	0.08	0.16	0.11	0.06
River flow depth h_0 (m) (Num.)	0.163	0.115	0.082	0.165	0.116	0.061
Mean velocity U (m/s)	0.16	0.19	0.23	0.18	0.22	0.22
Bed material size d_{50} (mm)	1.00			0.13		
Shields number τ_*	0.20	0.13	0.01	0.67	0.62	0.31
Froude number F_r	0.12	0.19	0.26	0.14	0.21	0.29
Sand Reynolds number R_{e*}	57	47	40	4.88	4.66	3.31

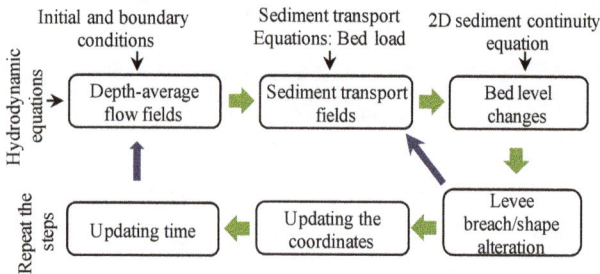

Figure 4. Outline of model computation steps

The flow model is based on the depth-averaged shallow-water equations. The equations expressed in a general coordinate system are solved on the boundary-fitted structured grids using the finite-difference method. Bed-load is calculated by Ashida and Michiue (1972) equations [29]; the effect of cross-gradient [30] and the influence of secondary flow [31] are taken into account. Finally, the bed deformation is determined using the 2D sediment continuity equation. Equations are solved for the unknown nodal values by an iterative process. The details of the model equations are discussed by Islam and Tsujimoto, 2012b. First, the flow field is computed utilizing initial and boundary conditions; then the sediment transport field is computed, to evaluate the rate of sedimentation, and followed by the bed topography changes. Figure 4 depicts the outline of the simulation steps for computation. The number of cell in the longitudinal and lateral direction is 120 and 44, respectively. In this study, the computation time step is used to 0.002 second, and the model

run is made in 10 minutes, when the temporal variations are considerably reduced. By numerical calculation, the breach propagation and the bed topography changes in the river, levee and floodplain can be described [see Figures. 5 (Sim.R1, Sim.R2, Sim.R3) and 7 (Sim.R4, Sim.R5, Sim.R6)], which is realized spatial characteristics of the levee breaching as well as disaster risk in the floodplain during the flood.

3. Results and Discussion

In this study, two sets of experiments and numerical analyses were conducted, and each had three runs. For the first set (Run 1 to 3), coarse bed materials with steep river bed slope were taken. The inflow discharges is provided nearly the same both in experiments and numerical throughout the all runs. The river flow capacity is reduced with the increased of the river bed height. Therefore, an initial overflow depth is lifted in case of the higher river bed level (3 cm for Run 3) than the lower ones (1 cm for Run 1). However, the second sets of experiments (Runs 4 to 6) have been carried out using fine bed material with mild river bed slope. The inflow discharges are reduced with the increased of the river bed height. The river inflow is higher (34.16 m³/hr) in Run 4 and lower (17.75 m³/hr) in the Run 6. Though, the small amounts of inflow discharges are provided in the Run 6, which is capable of an overflow levee breach. Considering the above criteria, this research has focused on the levee breaching phenomena and evaluates the disasters risk in the floodplain using both in experiments and numerical approaches.

3.1. Levee Breaching Process and Phenomena in River and Floodplain

3.1.1. Coarser Bed Materials and Steep River Bed Slope

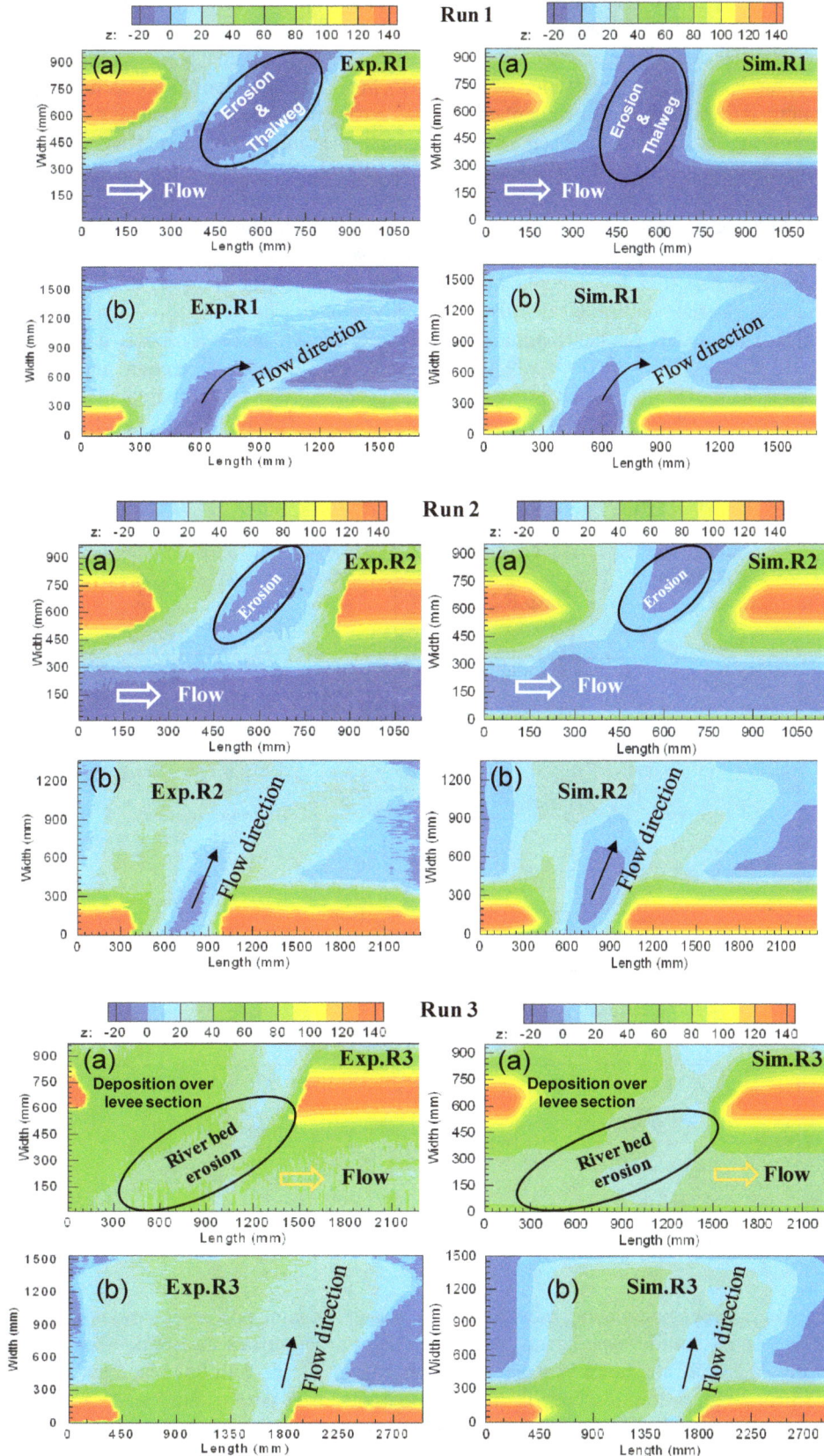

Figure 5. Experiments (Exp.R1, Exp.R2 and Exp.R3) and simulation (Sim.R1, Sim.R2 and Sim.R3) results of bed topographic changes (t=10 min): (a) River channel and levee section; (b) Floodplain.

The positions of the run are in a top, middle and below for the Run 1, Run 2 and Run 3, individually are depicts in Figure 5. The bed topographic pattern in river and levee section; and the floodplain are denoted by (a) and (b), respectively. After the beginning of overflow, an initial flow passes over the levee crest along with erosion on it near the floodplain, and afterwards, inundation water is spread over the floodplain with vertical erosion from the breach point. Then, the horizontal widening process starts by the collapse of the levee (Exp.R1). The more vertical erosion is observed on the levee section. Due to erosion in the levee as well as near the levee heel, a thalweg is formed along the flow direction from the river to the floodplain. Deposition pattern in the floodplain is smooth, because of coarse bed material, and it indicated that the flow is passes to the right-side direction in the floodplain (Exp.R1, Sim.R1).

In the Run 2, almost same nature of the erosion process appears initially; subsequently, the erosion process comes forward to the heel (inside edge of levee base at river side) of the levee section, and the levee material is washed out, then the horizontal widening process starts, but the rate is slower than the Run 1 (Exp.R2). A little erosion is observed on the levee section. The deposition pattern in the floodplain is exposed that the flow is moved all over the floodplain and had a little tendency to the right-side in the floodplain. The floodplain deposition thickness is observed high towards both sides of the flow direction (Exp.R2, Sim.R2).

Whereas in the Run 3, though the initial nature of the erosion is the same as Run 1 and Run 2, but the process is very quick, due to the large amount of inflow discharge, which provide high overflow depth and the level differences between the river beds to floodplain. The levee breach widening process starts in the horizontal direction with the higher rate than the other two runs (Exp.R3). The erosion is observed in the downstream side of the levee along with in the river bed. The early breach levee section is deposited by the eroded material from the levee section and the river bed. The sedimentation thickness in the floodplain is more than the Run 1 and Run 2. The higher bed level is more dangerous as because of the river bed deformation appears, and the bed material is eroded and deposited on the floodplain by the breach (Exp.R3, Sim.R3).

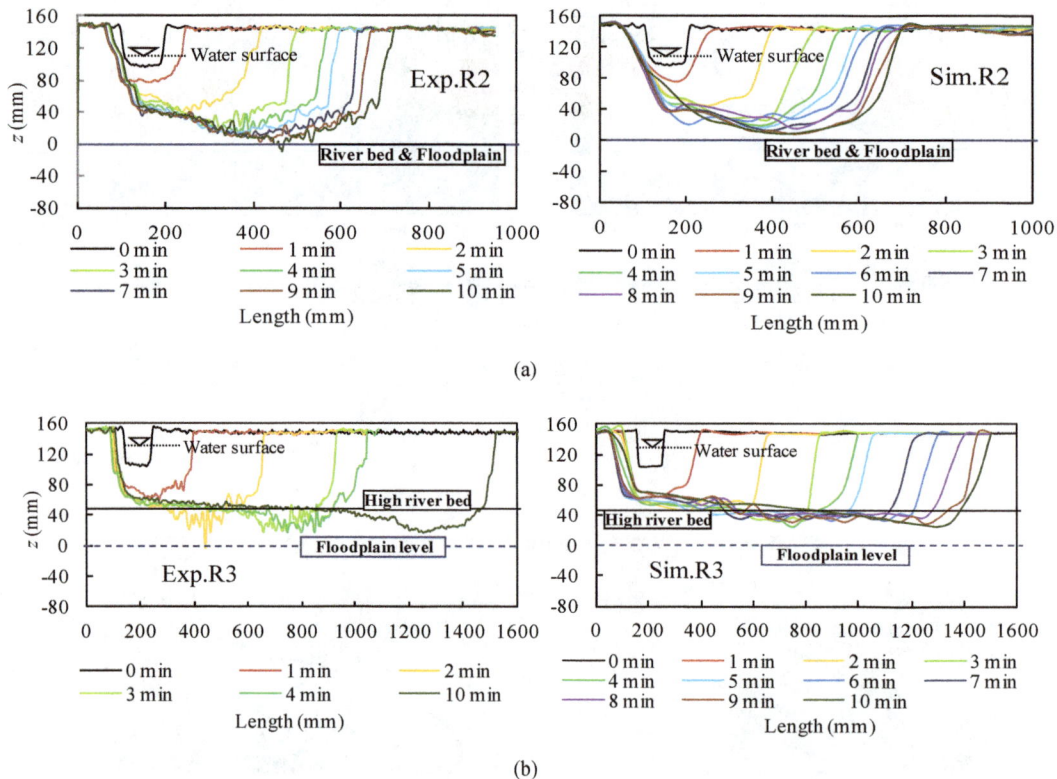

(a)

(b)

Figure 6. Longitudinal breach evolution processes of levee with time both in experiment and simulation: (a) Run 2 (Exp.R2, Sim.R2); and (b) Run 3 (Exp.R3, Sim.R3).

The longitudinal levee breach propagation along the river with time for the Run 2 and Run 3 both in experiments and simulation are shown in Figure 6 (a, b). In the early stage of overflow, the levee breach is progress towards both in the vertical, and in the horizontal direction along the downstream of the levee. Then, the sudden breach widening process is occurred in the longitudinal direction of the levee. After that, the breach widening process is slow, not only in the horizontal but also in the vertical direction (Exp.R2, Sim.R2). For the Run 3 (Exp.R3 and Sim.R3), the nature of the early erosion process is almost same as the Run 2. However, the horizontal breach widening is rapid, and the vertical erosion process is slow as compared to the Run 2. The total length of the breaches is double than the Run 2.

3.1.2. Finer Bed Materials and Steep River Bed Slope

In Figure 7; for the Run 4, the initial flow passes straight with the downstream of the floodplain, and then the erosion process starts in the floodplain near the levee toe (outside edge of levee base at the floodplain side). Afterwards, the erosion process comes forward to the centre of the levee with vertical erosion in the levee section, and the horizontal widening process starts by the collapse of the levee (Exp.R4). The large vertical erosion is observed in the levee section with little erosion in the river bed. The ripples and dunes of various dimensions are observed in the floodplain because of the fine bed material. The deposition pattern in the floodplain is indicated that the flow is passes to the right-side direction in the floodplain (Exp.R4 and Sim.R4).

However, in the Run 5, the different nature of the erosion process appears in the levee as compare to Run 4. The erosion process starts between the levee toe and the centre of the levee, and at the same time the levee section is eroded vertically. Suddenly, the erosion process dominates in the levee section with huge erosion of the levee material. Finally, the horizontal breach widening process starts by loss of the levee section. During the breach widening, the erosion process comes forward to the heel (inside edge of levee base at river side) of the levee as well as in the river bed (Exp.R5). Due to the erosion from the river bed, a thalweg is formed inside the river near the levee along the overflow direction. The river bed material is eroded, and it is deposited on the floodplain by the breach. The deposition pattern in the floodplain is exposed that the flow is moved all over the floodplain and had a little tendency to the right-side in the floodplain (Exp.R5 and Sim.R5).

Whereas, in Run 6, the inflow discharges through the river is smaller than Run 4 and Run 5, but the nature of the initial erosion process is rapid, though the erosion process starts at the levee toe as same as the Run 4. Because of the level difference between the river bed and floodplain, overflow water is quickly passed to the floodplain by the breach with huge vertical erosion in the levee section. Finally, the levee widening process starts in the horizontal direction at the higher rate than the other two runs (Exp.R6). The less vertical erosion is observed in the downstream side of the levee along with erosion in the river bed. In this case also (as like Run 3), the early breach levee section is deposited by the eroded material from the levee section and the river bed. The sedimentation thickness in the floodplain is higher than the Run 4 and Run 5 (Exp.R6 and Sim.R6).

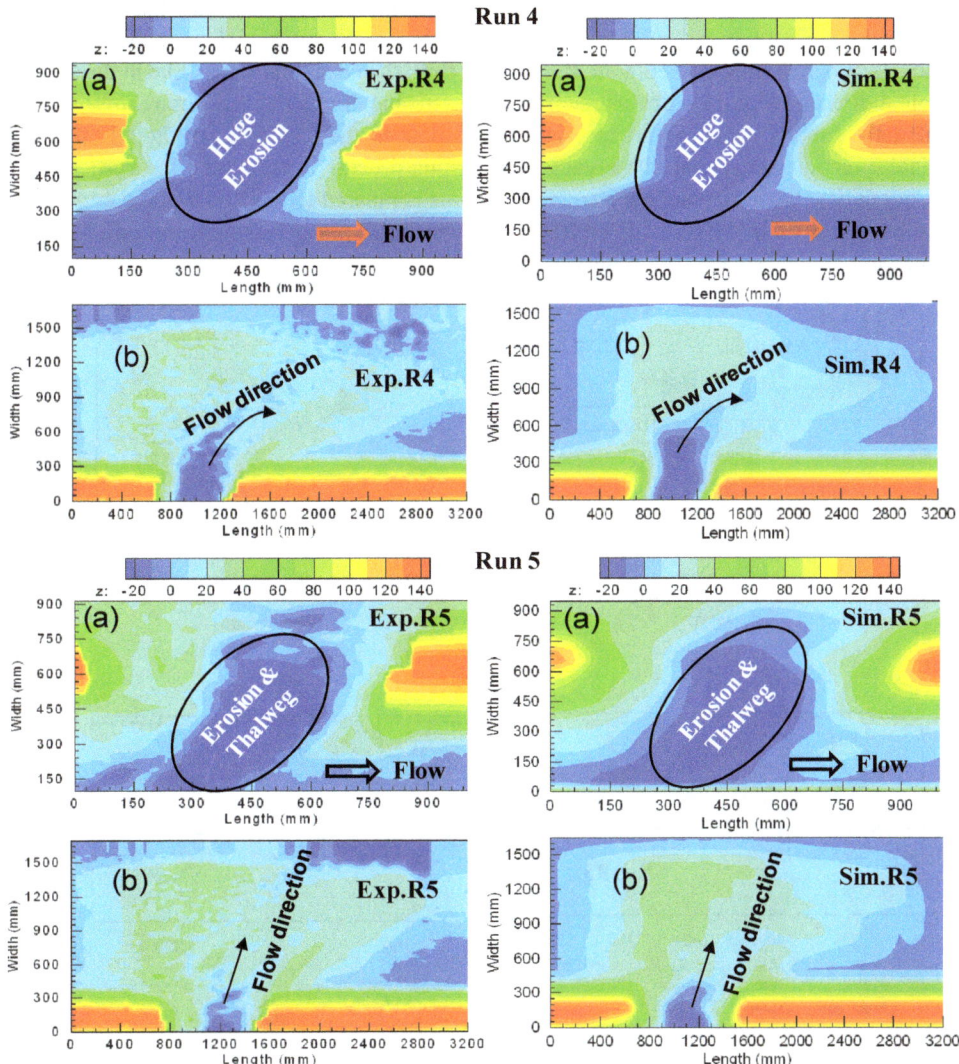

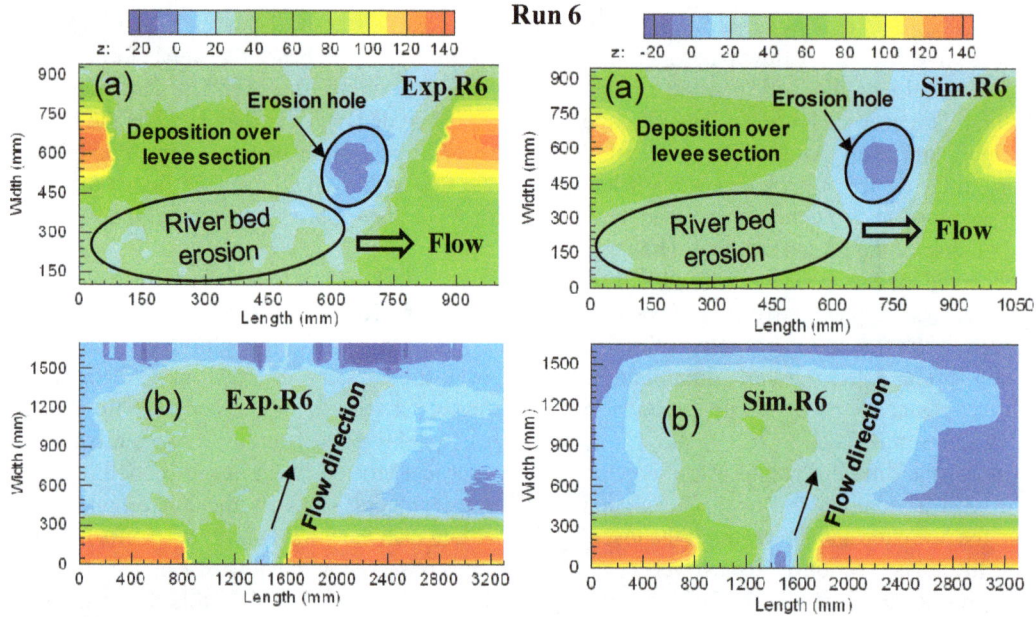

Figure 7. *Experiments (Exp.R4, Exp.R5 and Exp.R6) and simulation (Sim.R4, Sim.R5 and Sim.R6) results of bed topographic changes (t=10 min): (a) River channel and levee section; (b) Floodplain.*

The longitudinal levee breach propagation along the river with time for the Run 4, Run 5 and Run 6 both in experiments and simulation are shown in Figure 8 (a, b and c). At short duration, the levee breach is progress towards both in the vertical, and in the horizontal direction along the downstream of the levee. Subsequently, the breach widening process is occurred in the longitudinal direction. After that, the breach widening process is slow, not only in the horizontal but also in the vertical direction (Exp.R4, Sim.R4). In the Run 5 (Exp.R5 and Sim.R5), initially no horizontal erosion is observed

throughout the experiment, but the breach is progress towards both in the vertical and in the horizontal direction in simulation. Then, the breach widening process is same both in experiments and simulation as like in the Run 4. For the Run 6 (Exp.R6 and Sim.R6), the nature of the erosion is nearly equivalent as the Run 5. Even though, the horizontal breach widening process is rapid, and the vertical erosion process is slow as compared to the Run 4 and Run 5. The total length of the breaches is more than the Run 1 and near about same as the Run 2.

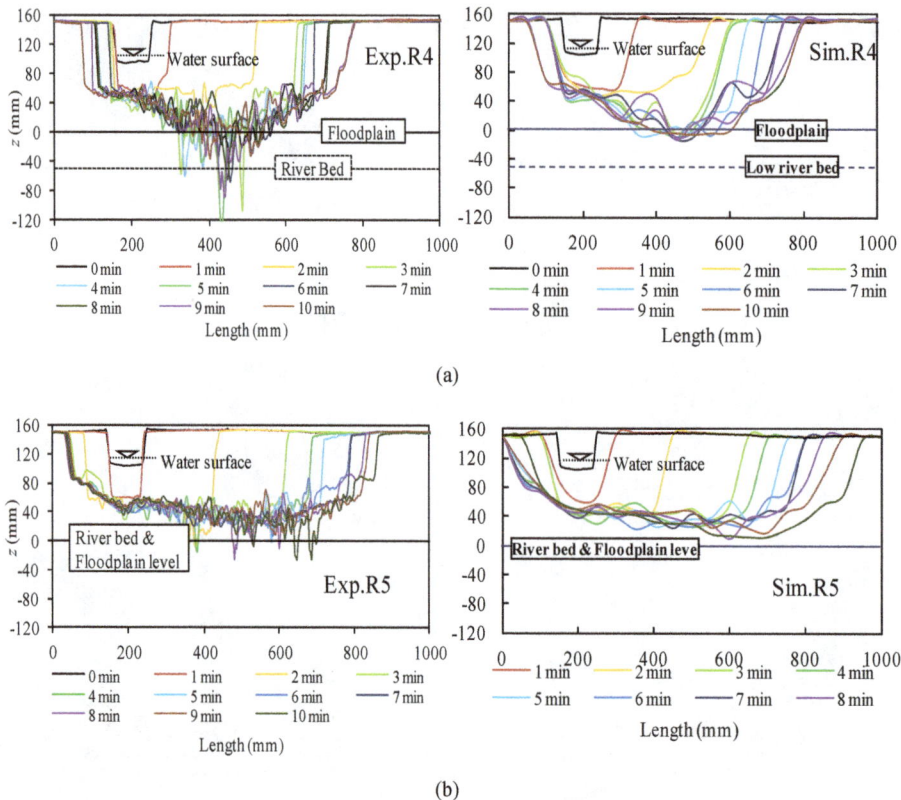

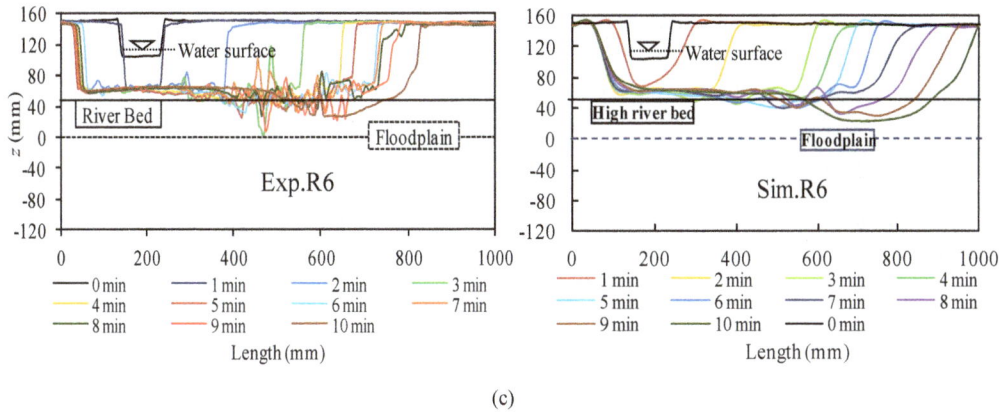

(c)

Figure 8. Longitudinal breach evolution processes of levee with time both in experiment and simulation: (a) Run 4 (Exp.R4 & Sim.R4); (b) Run 5 (Exp.R5 & Sim.R5); and (c) Run 6 (Exp.R6 & Sim.R6).

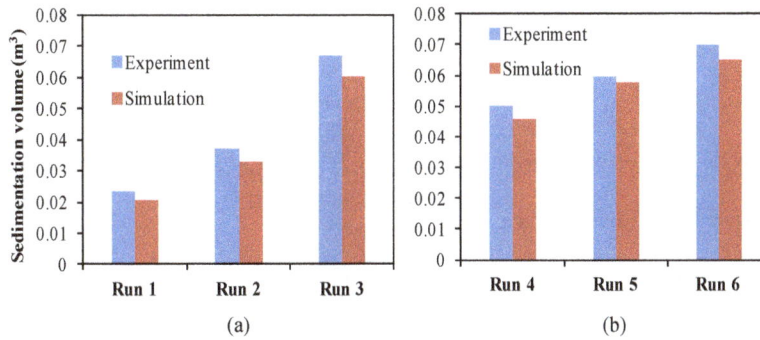

Figure 9. Comparisons of the volume of the floodplain sedimentation both in experiments and simulation Runs 1 to 6: (a) Coarse bed material and steep slope; (b) Fine bed material and mild slope.

Comparisons of the volume of the floodplain sedimentation at different river bed height are depicted in the Figure 9 (a-b) for both in the experiment and simulation. The sedimentation is less at the low river bed (Run 1, Run 4) as compare to the high river bed (Run 3, Run 6) level. The floodplain sedimentation is increased with increased to the river bed level, and the rate is more in the finer bed material due to the huge vertical erosion from the levee section and the river bed. It also shows that the higher river bed (Runs 3 and 6) with finer bed materials has the high risk of flood disasters in the floodplain considering with the sedimentation in the floodplain.

3.2. Differences in Levee Breach by River Bed Height Relative to Floodplain

The comparisons of the final length of the breach widening at different runs for both in the experiments and simulation are shown in the Figure 10 (a-b). The horizontal lengths of the widening are less in the Run 1 and Run 2 than in Run 3, but the vertical erosion is more in the Run 1 and Run 2. However, the larger widening is seen in the Run 6 as compare to the Run 4 and Run 5 but the vertical erosion is more in the Run 4 and Run 5. In case of the higher river bed with coarser material (Run 3), the horizontal widening is almost double than the lower and same river bed conditions. It happens due to the high river inflow with more overflow depth, and possesses the less bonding effect between the coarser particles. The horizontal widening is longer; it means the more amount of inundation flow passes to the floodplain along with sediment outflow by the breach. It can be concluded that, the higher river bed with coarser bed materials has the high risk of flood disasters in the floodplain.

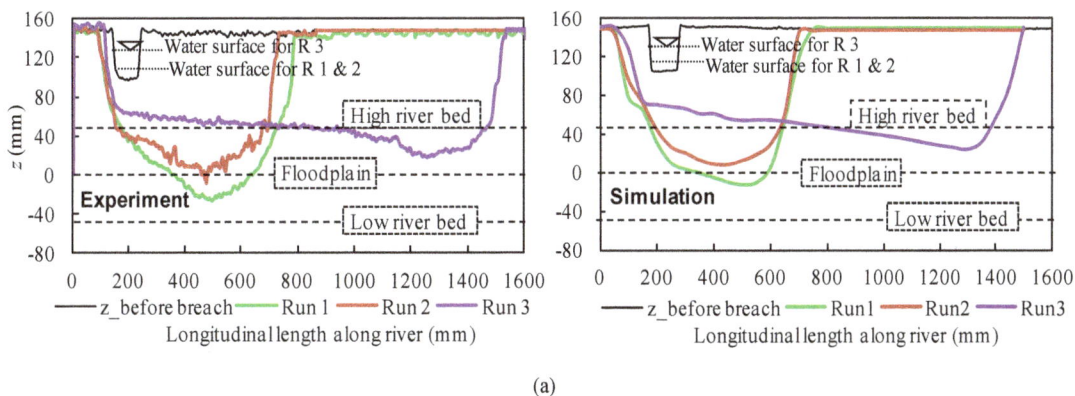

(a)

(b)

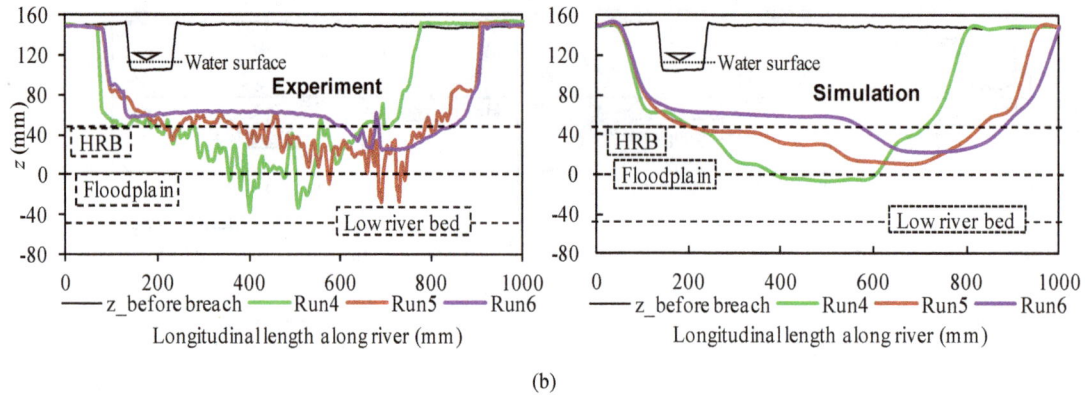

Figure 10. Comparisons of the final longitudinal lengths of breach along the river for experiments and simulation (t=10 minutes): (a) Runs 1 to 3; and (b) Runs 4 to 6.

3.3. River Bed Changes Accompanying Levee Breach

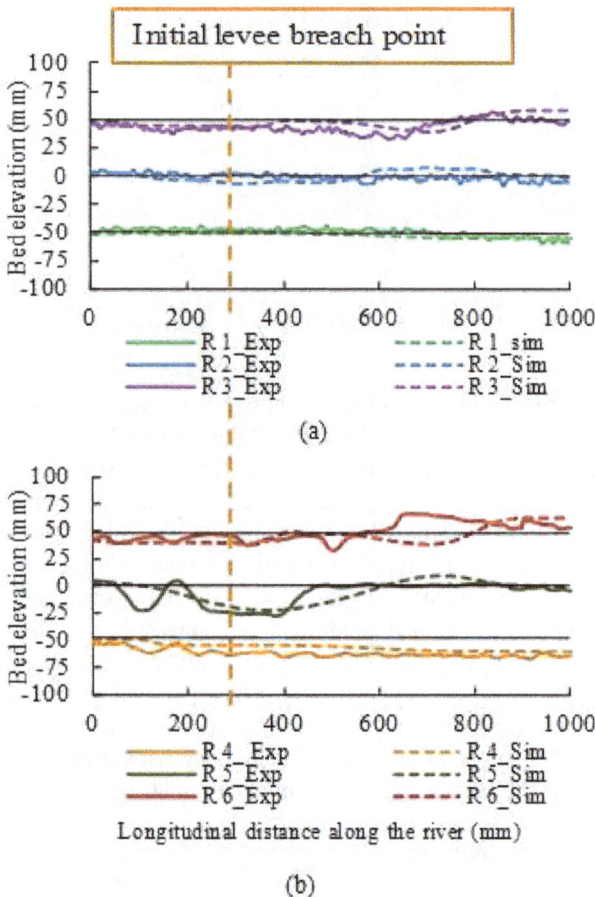

(a)

(b)

Figure 11. Comparisons of river bed variations along the river for experiments and simulation: (a) Runs 1 to 3; and (b) Runs 4 to 6.

The river bed deformation comparisons at different relative heights of river bed to floodplain are depicted in Figure 11 (a-b). Both in the experiments and simulation, for coarser bed material, the higher rates of changes are observed in the Run 3, as compared to the Run 1 and Run 2. Nevertheless, the overall deformation rate is more in the Run 5 and Run 6. Using finer bed material with the same and high river bed level are dangerous as because of more bed deformations are seen. The levee breaches with the high river bed has the problem, not

only in the rapid flow propagation with the larger amount of sediment outflow to the floodplain by the breach but also the river bed variation is remarkable, which brings further risk of the levee breach in the upstream reach across the river. The river bed material is eroded, and it is deposited on the floodplain by the breach as well as in the upstream of the levee breaching point.

4. Conclusions

This study have conducted using the different sets of experiments and same scenario numerical analyses, to understand the levee breach process and evaluates the risk in the floodplain with considering the effect of river bed height, bed material sizes and river bed slopes. The research result showed that the higher river bed not only influences the effect of levee breaching and floodplain deposition, but also it has unlike characteristics in the river bed variation using different bed materials. The overtopping levee breach study in this research was not cover with any vegetation. The conclusion can be drawn as follows:

1. Though there have some discrepancies between the experiments and same condition numerical analyses, both results showed reasonably good agreement.

2. In coarser bed material, the erosion process starts mainly on the levee crest, and the breach is progress by the washout of the levee material with flow; whereas to use finer bed material the different breach phenomena with huge vertical erosion in the levee along with more river bed deformation appears.

3. In coarser bed material, the higher river bed is exposed to levee breach with higher overflow depth and thus the widening rate of the levee breach is more rapid and inundation with more sediment volume to the floodplain not only from the levee but also from the river bed as compared to the finer bed material as well as to the lower and the same river bed height.

4. Using finer material, both in the same and the high river bed level, the river bed deformation is remarkable and the bed material is deposited not only in the floodplain but also into the downstream of river, which has the problem for the normal flow through the river in the future.

5. Furthermore, the levee breach with higher river bed is risky both in coarser and finer bed material, because of the rapid breach widening with more inundation and sediment outflow to the floodplain by the breach as compared to the lower river bed due to a difference to the level between the river bed to floodplain.

Acknowledgments

The authors are grateful to MEXT, Japan for the financial support required in the study. They wish to thank Dr. T. TASHIRO, Mr. YOSHIIKE and other students of the Hydraulic Engineering Laboratory of Nagoya University for their assistance during experiment setup and data measurements.

References

[1] Simm, J. and Wallis, M. (2012) *"International guidance on levee"* [Online]. Available: http://www.hrwallingford.com/projects/test-project. Date accessed: June 7, 2013.

[2] Singh, V.P. (1996) *"Dam-Breach Modeling Technology"*. Kluwer Academic Publishers, Dordrecht, The Netherlands, 242 pp.

[3] Broich, K. (1998) *"Mathematical Modelling of Dam Break Erosion Caused by Overtopping"*. Proc. of the CADAM Meeting, Munich, Germany.

[4] Khalequzzaman, M. (1994) *"Recent Floods in Bangladesh: Possible Causes and Solutions"*. Natural Hazards, 9: 65-80.

[5] Shalash, G. (1982) *"Sedimentation in the Aswan High Dam Reservoir"*. Hydrobiology, 92: 623-629.

[6] Shahjahan, M. (1983) *"Regional Co-operation in the Utilization of Water Resources of the Himalayan Rivers"*. In, Zaman, M. (ed.) River Basin Development: Dublin, Tycooly International Publishing Ltd., pp. 114-130.

[7] Siddiqui, M.F. (1983) *"Management of River System in the Ganges and Brahmaputra Basin for Development of Water Resources"*. In, Zaman, M. (ed.) River Basin Development: Dublin, Tycooly International Publishing Ltd., pp. 137-149.

[8] Broadus, J., Milliman, J. and Edwards, S. (1986) *"Rising Sea Level and Damming of Rivers; Possible Effects in Egypt and Bangladesh"*. In, Proc. of United Nations Environments Programme and the U.S. Environmental Protection Agency: Effects of Change in Stratospheric Ozone and Global Climate. New York, 4: 165-189.

[9] Khalequzzaman, M. (1989) *"Environmental Hazards in the Coastal Areas of Bangladesh: a Geologic Approach (summary)"*. In, S. Ferraras and G. Pararas-Carayannis (eds.), Natural and Man-Made Hazards, Proc. of the International Conference on Natural and Man-Made Coastal Hazards, August 14-21, Ensenada, Mexico, pp. 37-42.

[10] Visser, P.J., Zhu, Y. and Vrijling, J.K. (2006) *"Breaching of Dikes"*. Proc. of the 30th Conf. Coastal Eng., San Diego, USA, pp. 2893-2905.

[11] Chinnarasri, C., Tingsanchali,, T., Weesakul, S. and Wongwises, S. (2003) *"Flow Patterns and Damage of Dike Overtopping"*. Intl. J. of Sediment Research, 18 (4): 301-309.

[12] Hanson, G.J., Cook,K.R., and Britton, S.L. (2003). *"Evaluating Erosion Widening and Headcut Migration Rates for Embankment Overtopping Tests"*. ASAE International Meeting, Las Vegas, Nevada, USA.

[13] Sharif, Y. A. (2013) *"Experimental study on Piping failure of earthern levee and dams"*. [Unpublished PhD Dissertation], Accepted by the Collage of Engineering and Computing, University of South Carolona, USA. 73 pp.

[14] Hassan, M., Morris, M. and Hanson, G. J. (2004) *"Breach Formation: Laboratory and Numerical Modeling of Breach Formation"*. Proc. of the Annual Conference of the ASDSO, Phoenix, Arizona (in CD-ROM).

[15] Kjetil, A. V., Lovoll, M.A., Hoeg, K, Morris, J., Hassan, M. and Hanson G. (2004) *"Physical Modeling of Breach Formation; Large Scale Field Tests"*. Proc. of the Annual Conference of the ASDSO, Phoenix, Arizona, (CD-ROM)

[16] Hanson, G. J., Temple, D.M., Morris, M. and Hassan, M. and Cook, K. (2005) *"Simplified Breach Analysis Model for Homogeneous Embankments: Part II, Parameter Inputs and Variable Scale Model Comparisons"*. 25th USSD Annual Conference: Technologies to Enhance Dam Safety and the Environment, United States Society on Dams, U.S.A., pp. 163-174.

[17] Fujita, Y. and Tamura, T. (1987a) *"Enlargement of Breaches in Flood Levee on Alluvial Plains"*. J. of Natural Disaster Science, 9 (1): 37-60.

[18] Fujita, Y., Muramoto, Y. and Tamura, T. (1987b) *"On the Inflow of River Water and Sediment due to Levee Breach"*. Annual disasters prevention research Institute, Kyoto University, 30 (2): 527-549 (in Japanese).

[19] Islam, M. Z., Okubo, K. and Muramoto, Y. (1994) *"Embankment Failure and Sedimentation over the Flood Plain in Bangladesh: Field Investigation and Basic Model Experiments"*. J. of Natural Disaster Science, 16 (1): 27-53.

[20] Aureli, F. and Mignosa, P. (2001) *"Comparison between Experimental and Numerical Results of 2D Flows due to Levee-Breaking"*. XXIX IAHR Congress Proceedings, Theme C, September 16-21, Beijing, China.

[21] Tsujimoto, T., Mizoguchi, A. and Maeda, A. (2006) *"Levee Breach Process of a River by Overflow Erosion"*. River flow 2006, Fluvial Hydraulics, Proceedings of IAHR Symposium, Lisbon, Taylor & Francis, pp. 1547-1555.

[22] Shimada, T., Watanabe, Y., Yokoyama, H. and Tsuji, T. (2009) *"An Experiment on Overflow-Induced Cross-Levee Breach at the Chiyoda Experimental Channel"*. River, Coastal and Estuarine Morphodynamics, 1: 475-481 (in Japanese).

[23] Shimada, T., Hirai, Y. and Tsuji, T. (2010) *"Levee Breach Experiment by Overflow at the Chiyoda Experimental Channel"*. 9th Intl. Conference on Hydro-science and Engineering, IAHR August 2- 5.

[24] Islam, M.S. and Tsujimoto, T. (2012a) *"Comparisons of Levee Breach and Successive Disasters in Floodplain between Bangladesh and Japan"*. Procedia Engineering, Elsevier publication, 28: 860-865.

[25] Islam M. S. and Tsujimoto, T. (2012b) *"Numerical Approach to Levee Breach as a Key of Flood Disasters in Low Land"*. Int. J. of Civil Engineering, 4 (1): 23-39, India.

[26] Iwagaki, Y. (1956) *"Hydro-dynamical Study on Critical Tractive Force"*. Transactions of the Japan Society of Civil Engineers, 41:1-21(in Japanese).

[27] Itaya, S. and Tejima, T. (1951) *"Weir Flow Formula of a Rectangle with a basis of Reebok`s Formula"*. Proc. of the Society of Mechanical Engineers, 17 (56): 5-7 (in Japanese).

[28] Kurokawa, H. and Fuchizawa, T. (1942) *"Formula of Triangular Weir Flow"*. Proc. of the society of Mechanical Engineers, 7 (27), 5 (in Japanese).

[29] Ashida, K and Michiue, M. (1972) *"Study on hydraulic resistance and sediment transport rate in alluvial stream"*. Transactions, JSCE, 206: 55-69 (in Japanese).

[30] Hasegawa, K. and Yamaoka, S. (1980) *"The Effect of plane and bed forms of channels upon the meander development"*. J. of Hydraulic, Coastal and Environmental Engineering, JSCE, 29: 143-152 (in Japanese).

[31] Engelund, F. (1974) *"Flow and Bed Topography in Channel Bend"*. J. of Hydraulic Division, ASCE, 100(HY11): 1631-1648.

7

A Macroscopic Fundamental Diagram for Spatial Analysis of Traffic Flow: A Case Study of Nyeri Town, Kenya

Lekariap Edwin Mararo[1, *], Abiero Gariy[1], Mwatelah Josphat[2]

[1]Department of Civil, Construction and Environmental Engineering, Jomo Kenyatta University of Agriculture and Technology, Nairobi, Kenya

[2]Department of Geomatic Engineering and Geospatial Information Systems, Jomo Kenyatta University of Agriculture and Technology, Nairobi, Kenya

Email address:

englekariap@gmail.com (L. E. Mararo)

Abstract: Traffic flow analysis is an essential component of a town's traffic and transport systems since these flows could, and often do lead to the occurrence of congestion on our roads.Traffic congestion is a growing problem in Nyeri, Kenya, resulting from rapidly increasing population and the crowding of motorized traffic onto a limited street network. This research performed spatial analysis of traffic flows on some road links in Nyeri town. On those selected road links it also established fundamental traffic flow models and derived the flow characteristics associated with traffic operations in Nyeri town, determined the characteristics of a Macroscopic Fundamental Diagram (MFD) for Nyeri town and assessed to determine whether it is a property of the network infrastructure and control or of the travel demand. In this research, MetroCount Vehicle Classifier was used to collect traffic intensity and velocity data at different locations of the network over certain durations and at the same period between January 2015 and February 2015. The analysis of the data was performed by the MetroCount Traffic Executive MCReport. The MFD resulting from the study served as a road network performance indicator, which tells the performance levels of the town in general, in terms of traffic flow. The research was used to determine the capacity of the road network and the level of congestion in different links thereby determining the adequacy of the network. The speeds in Nyeri town are moderate, however, low and high speeds occur occasionally and the volumes of traffic in Nyeri town are not high hence congestions are rarely experienced. The results of this research study are anticipated to better traffic management, and also improve mobility and accessibility in Nyeri town.

Keywords: Microscopic Fundamental Diagram, Network Infrastructure, Nyeri Town, Spatial Analysis, Traffic Flow

1. Introduction

Traffic flow analysis is an essential component of a town's traffic and transport systems since these flows could, and often do, lead to the occurrence of congestion on our roads [1]. Global cities and towns face rising traffic congestion problems. This situation is getting worse and is becoming a major concern of the general public. Traffic congestion is a condition of traffic delay, because the number of vehicles using a road exceeds the operational capacity of the network to handle it [2]. Congestion has several causes such as: the volume of traffic being close to the maximum capacity of the road link and as a result of too many vehicles crowding available road space. Congestion has a number of negative effects: productive hours are lost and this has adverse effects

on the economy; it also contributes to air pollution (which has a debilitating effect on quality of life) and global warming. In view of these effects of congestion, there is the need to manage traffic congestion and help reduce its effects.

Congestion occurs on individual links within a network thereby making it a localised problem. The cause has to do with spatial - temporal distribution of demand and supply which therefore makes it possible to experience its effect when considering the performance of the entire network, making it a macroscopic issue. This then points out at the fact that network operators should be able to relate the effects of these localised congestion situations on road links to the entire network, which calls for appropriate indicators to be used to measure network performance [2]. Analysis of traffic flow and modeling of vehicular congestion has mainly relied on fundamental laws, inspired from physics using analogies with fluid mechanics, many particles systems and the like.

Nyeri town was initially the administrative headquarters of the country's former Central Province. Following the dissolution of the former provinces by Kenya's new constitution in 2010, Nyeri is now the largest town in the newly created Nyeri County, with a population of about 119, 273 (National Bureau of Statistics, 2009). Modern shopping centres and department stores that were found in much larger cities and towns have been opened in Nyeri as a result of the booming economic activities. Nyeri is served by a reasonably well - maintained tarmac road network connecting it to Nairobi, Nakuru, Nanyuki, Othaya and other surrounding towns. Most transportation of cargo to and from Nyeri is by road. The main mode of public passenger transport to, from, and within Nyeri is by way of fourteen - seater minibus taxis (matatus), though un - metered saloon car taxis are also widely used. The above - stated increase in population, business and economic activities also has effects on traffic flow in the town.

2. Statement of the Problem

In Nyeri town, despite the intensive road network expansion and the limited number of vehicle ownership compared to the other sub Saharan countries, traffic congestion has now become the threat in the town economic growth by restraining the commuters' mobility especially at peak hours. In addition to waiting time for the limited public transportation, both vehicle owners and public transport users are forced to delay within the congested traffic lane. Hence, late arrival to work places and appointments for social or business activities have become common. Despite the problem being recognized by all road users and transport professionals, there is no significant attempt for quantitative research done on the extent of the traffic flow and congestion in Nyeri. Traffic congestion has an economic cost on the productivity of the town's communities and economy. Primarily, traffic congestion is an outcome of insufficient traffic management in the town, insufficient capacity of the roads to cope up with the existing traffic volume and inadequate public transport, fixed working time, and poor land - use or transport - land - use planning integration. In addition, long travel time or delay to reach destination, affect business users' time productivity and increasing fuel consumption and wastage. These are the main impacts of vehicles congestion which are prevalent. Therefore, this research has been initiated to assess traffic congestion and the impact of the issue on travel time and fuel consumption.

3. Study Objective

To determine the characteristics of a Macroscopic Fundamental Diagram (MFD) for Nyeri town

4. Theoretical Framework

4.1. Traffic Flow Theory

Transportation, which is seen as a system that considers

the complex relationships between its core elements such as networks, nodes and demand, plays an essential role in our daily lives [3]. This relation gives rise to the flow of traffic on road networks. The mathematical representation of the interactions between vehicles, their operators and the infrastructure can be explained by traffic flow theories which seek to understand and develop optimal road networks that will allow the movement of traffic efficiently and help reduce congestion [4]. Road space is limited and traffic engineers have to maximise the capacity of the road as much as possible. The measurement of the capacity and what influences it, lie at the core of traffic flow theory [5]

It is asserted that the sole aim of transportation is to meet demand levels for mobility [3]. In this regard, travel demand modelling is viewed as an important factor in transport planning processes in that demand levels could directly be as a result of the outcome of varied economic activities, without which they would not occur; so there is the need to be able to predict demand levels so as to plan towards their effects.

Although traffic flow plays these requisite roles, their negative effects on the environment is one that should be well considered. An indirect effect of congestion caused by transportation is pollution (air and water); and this affects health standards of people. Also to talk about is its safety issues because growing traffic is linked to growing number of fatalities and accidents. With respect to its environmental effects, it is emphasized that" decisions relating to transport need to be evaluated taking into account the corresponding environmental costs" [3]. Other cost considerations which have effects on economies can also be seen from the delays spent in congestion and although transportation may have these negative effects, it plays very essential roles by supporting transport demands that are generated by the diversity of activities that are brought forth by the urban society.

4.2. The Fundamental Diagram Model

How effective a roadway system is can be evaluated based on a number of elements which include the number of vehicles that can travel on the road, the speeds at which these vehicles can travel, the density of vehicles along the roadway, the distances between these vehicles and the freedom to manoeuvre among lanes. These qualitative and quantitative measures affect each other in one way or the other and the derivation of the macroscopic parameters which relate to form the MFD have been shown below.

When vehicles move in a traffic stream, a relationship exists between spacing (s) and the density (k) of the stream of vehicles on a given length of a roadway. This is given by:

$$s = \frac{1}{k} \tag{1}$$

Also, the headway (h) between these vehicles in a stream is the inverse of traffic flow (q), thus:

$$h = \frac{1}{q} \tag{2}$$

And the headway *(h)* between two vehicles travelling at a spacing *(s)* with a speed *(u)* is given by:

$$h = \frac{s}{u} \qquad (3)$$

Substituting equations 4.21and 4.22 into 4.23 gives the relation between the macroscopic variables flow, speed and density as:

$$q = uk \qquad (4)$$

This equation represents the behaviour of one parameter with respect to the other and is the basis for the fundamental diagram since it involves a relation between traffic flow, traffic speed and traffic density. The sections below show how each of these parameters relate to the other and what kind of information can be obtained from them. As part of this research, the application of data from metro count devices was analysed to give rise to the MFD. Based on the interpretation from the diagram, information obtained from it can be used by planners and traffic engineers to further plan traffic circulation within a study area such as Nyeri town.

5. Empirical Review

5.1. Network Reliability

How reliable the road networks are, is becoming an increasingly important attribute of road networks and also a concern for planners and engineers in network design [6, 7]. This is because a network that is unreliable has effects on the lives of commuters and the economy of the nations, giving concern to studying the reliability of networks during area - wide studies. Reliability is defined from systems engineering point of view, as the degree of stability of the quality of service that a system normally offers; and in a transport system such as a road network, travel demand flows and the physical network may contribute to how reliable the network is. A similar idea is asserted that" reliability, by its nature, implies something about the certainty or stability of travel time of any particular trip under repetition".

In road network reliability [7, 8] share similar views that the level of stability of the transport network system can be related to its ability to respond and meet the expected demand levels under different circumstances (such as variability in flows and physical network capacities). A typical example where a network is reliable is where the network is able to cope with variations in demand over different days of a week by maintaining a constant average travel time between different origin - destination pairs [9].

The focus of this research is to determine factors that will ensure that the road network in Nyeri improves accessibility and mobility levels, thus reducing travel times and improving on its reliability, and that is what transport planners and engineers seek to do. Before improvements are made to the current network structure in terms of planning and engineering, network indicators will have to be used to assess present performance of the network.

5.2. Macroscopic Fundamental Diagram (MFD)

Macroscopic models common to traffic - related simulations have a long history, with the study of fundamental diagrams [10, 11]. The measurements and relationships that exist between macroscopic variables (speed, density and flow) have been broadly studied [12] for traffic streams theoretically and practically [13, 14]. Further works dealt with the development of macroscopic models for arterials [15, 16] and were later extended to general networks [18].

Further experimental findings show that MFDs can be used to control demand and improve mobility and accessibility within a city [19]. This improvement is seen to enhance the performance of the road network and show how the system responds to increasing travel demand levels. Once the MFD for a city is established and mechanisms are put in place to monitor the state of traffic flow, it is possible to know if the network is producing the desired accessibility levels at all times. Since the performance could be attributed to the way a city is planned, a typical MFD for a city/town may help explain if the city/town is well planned or not, or if some selected sections (of spatial interest) are performing better than others so as to know how to allocate resources for transport network improvements in an equitable manner.

Typically, an MFD can be put into four phases (A, B, C, D) as shown in Figure 1 and each of these phases shows the different state of the network with increasing traffic flow.

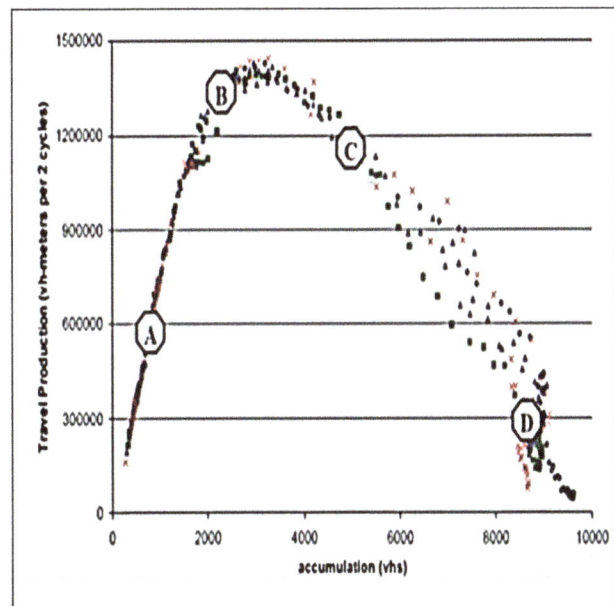

Figure 1. A Typical MFD for the City of San Francisco [19]

For the above diagram In Figure 2 - 5, phase A shows the condition when the system is under saturation and the average speed is about 25km/hr with an accumulation of about 952 vehicles. As demand increases the system moves to state B where the vehicle - miles travelled is near the maximum and the average speed is 17km/hr with an accumulation of 2143 vehicles. At this stage, engineers can tell at what accumulation the system experiences its maximum capacity, and this can be useful for planning purposes as well as knowing the land use activities that contribute to that accumulation. In phase C congestion is broad, long queues are observed and the average speed drops to 7km/hr (with an accumulation of 5337 vehicles). In state D the output is near to zero (at jam density), and the majority of vehicles are stopped (with accumulation of 8943 vehicles). Once again, knowledge about the jam density could assist in putting in management measures to manage the density within the town from getting to such high levels, thereby reducing congestion.

6. Conceptual Framework

The study was guided by a conceptual framework as shown in Figure 2 relating the dependent and independent variables.

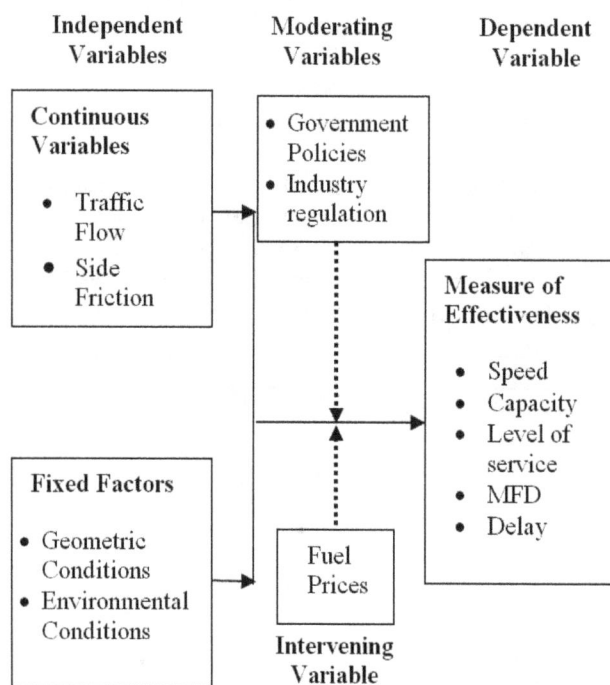

Figure 2. Conceptual Framework.

The independent variables constitute two sets of factors namely continuous and fixed factors. Continuous factors are traffic flow and side friction whereas fixed factors comprise of geometric conditions and environmental conditions. On the other hand, measure of effectiveness was presumed to be in terms of speed, capacity, level of service, MFD, and delay. The study hypothesized that both continuous and fixed factors do affect measures of effectiveness. More so, the study factored in moderating variables which included government policies and transport industry regulations.

7. Research Methodology

This section outlines the types of data, how they were collected. It also presents the instruments adopted in data collection and how the collected data was analyzed. The approach to establishing an area - based indicator for Nyeri town was based primarily on the relationship between average traffic flow and speed for the entire network. The data for this relation was collected by Metro Count Vehicular Classifier over a period of time. Developing very effective strategies to enhance traffic flow conditions on road networks require knowledge on the state of road at any point in time and space. This data was prepared and analysed in different ways to obtain the desired variables/ parameters that needed to be compared; for explanations, conclusions and recommendations to be made. The relationship between average traffic flow and speed for the network was an area of focus since this establishes the MFD, which was used to comment on the performance of the city's road network. Figure 3 outlines the methodological framework for analyzing the collected data.

8. Findings and Discussions

8.1. Introduction

Nyeri town is linked to other surrounding towns by several arterial roads and numerous collector roads that connect to the arterial roads. Data for this project was collected on the seven major arterial roads around Nyeri town. Table 1 shows the major links around Nyeri town

Table 1. Major Arterial Roads (Links) around Nyeri Town.

Link	Length (Km)	No. of travel lanes	Functionality class
Nyeri-King'ong'o	2	2	B5
Nyeri-Kiganjo	9.4	2	C75
Nyeri-Nyahururu	100.2	2	B5
Nyeri-Tetu	6	2	D434
Nyeri-Ruring'u	3.0	2	B5
Ruring'u-Marua	6.0	2	B5
Ruring'u-Othaya	30	2	C70

8.2. Traffic Flow Data

Figure 3 illustrates the average volume of traffic flow within Nyeri town.

Traffic data collected over a two week period for all the roads was averaged to obtain data for a virtual week and then a virtual day as shown in Figure 3. Analysis of the volume

data gave rise to the average volume graph of the town. The analysis of traffic flow for all the roads indicates that the morning peak occurs between 0800hrs and 0900hrs whilst the afternoon peak volume occurs between 1700hrs and 1800hrs. Afternoon peak periods recorded slightly higher volume than the morning peak periods. Nyeri - King'ong'o road recorded the highest traffic volume, that is, more than 1200v/hr.

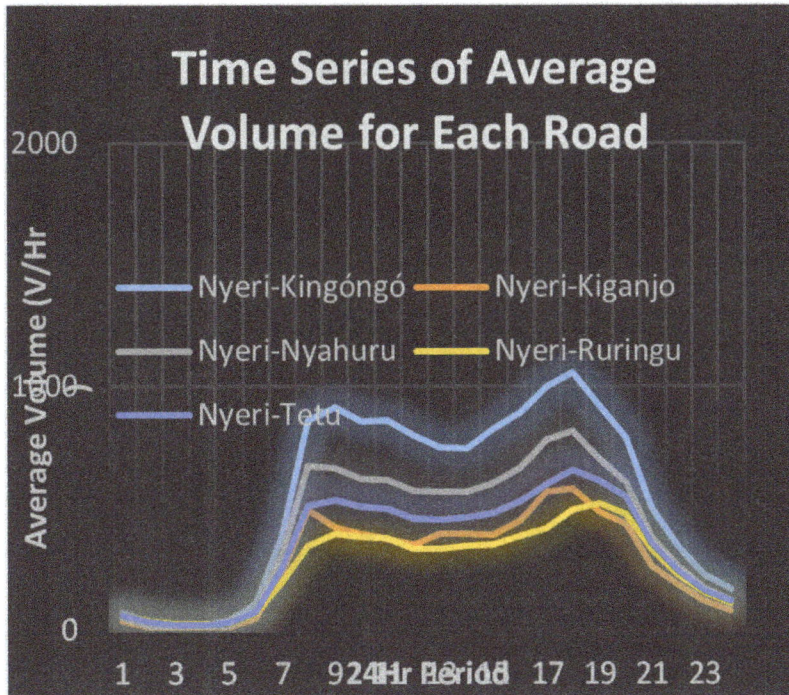

Figure 3. *Average Volume Graph.*

8.3. Speed Data

A time series analysis of speed was performed on all the links and the results of the analysis are as shown in Figure 4.

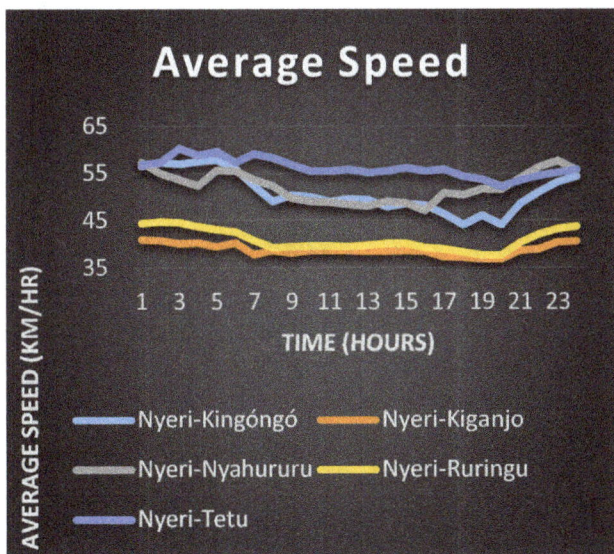

Figure 4. *Average Speed Graph.*

The average speed for the links was found to be 48km/hr. The highest speeds were recorded between 0300hrs and 0400hrs and 2200hrs and 2400hrs. These speed correspond with the times when traffic volume is the lowest hence there is little vehicle interactions. Speed in all the links decreases gradually from 0600hrs until 2000hrs when the lowest speed in the network is reached. The speed graph shows that morning peak periods are 0700hrs to 0800hrs and evening peaks occur between 1700hrs and 1800hrs just as the volume graph indicated. There was a clear repeatability of hourly variations on all roads. Vehicle volume was below 200v/hr before 0600hrs and then increased steadily up to 0800hrs. Volume varied slightly between 0900 hrs and 1900 hrs after which it decreased steadily.

8.4. Macroscopic Fundamental Diagrams

The traffic volume data collected and speed data were used to generate the Fundamental diagrams using Metro Count Executive Report software. The MFDs for each road link were first formulated then the general MFDs for the whole town. The diagrams aided in the qualitative and quantitative analysis of the entire road network.

8.4.1. Speed - Density Model

Vital road network characteristics like free flow speed, critical speed and jam density were obtained from the speed - density model. However, in this research study the speed - density model was used to obtain the free flow speed and critical speed. Table 2 shows the free flow speed and critical speed from the speed - density model.

Table 2. *Free Flow Speed and Critical Speed for all Road.*

Road	Free Flow Speed	Critical Speed
Nyeri-King'ong'o	75	55
Nyeri-Nyahururu	70	50
Nyeri-Kiganjo	55	40
Nyeri-Ruringu	60	40
Nyeri-Tetu	50	55
Ruringu-Othaya	75	40
Ruringu-Marua	55	55

Table 3. *Maximum Volume for Each Road Link.*

Road	Free Flow Speed (U0)	Critical Speed (Uc)	Max Vol. (qmax)
Nyeri-King'ong'o	75	55	2500
Nyeri-Nyahururu	70	50	600
Nyeri-Kiganjo	55	40	1400
Nyeri-Ruringu	60	40	1800
Nyeri-Tetu	50	55	900
Ruringu-Othaya	75	40	1500
Ruringu-Marua	55	55	1750

Table 4. *Jam Density for the Links.*

Road	Free Flow Speed (U0)	Critical Speed (Uc)	Max Vol (qmax)	Jam Density (Kmax)
Nyeri-King'ong'o	75	55	2000	40
Nyeri-Nyahururu	70	50	600	20
Nyeri-Kiganjo	55	40	1400	25
Nyeri-Ruringu	60	40	1800	25
Nyeri-Tetu	50	55	900	20
Ruringu-Othaya	75	40	1500	30
Ruringu-Marua	65	55	1750	25

8.4.2. Speed - Volume Model

The speed - volume model in this study was used mainly to obtain the critical volume (maximum volume) and to re - examine the critical speed and free flow speed values. The data obtained from the models are presented in Table 3.

8.4.3. Density - Volume Model

The density - speed volume for the roads (the jam density) was obtained from this model and the speed - density model. Table 4 shows the major characteristics of density - volume model.

9. Conclusions

It has been indicated that the MFD can be used to improve accessibility within towns. Information from the MFD can be channelled into enhancing accessibility levels to ensure effective traffic circulation within the town. This can generally be done with road pricing, rationing and/or perimeter control strategies based on neighbourhood accumulation and speeds, such as those proposed in previous studies [17, 19]. It was concluded that Nyeri town may not be congested enough to be given such measures but these can be thought of and applied where necessary in future instances.

Traffic volumes within the morning and evening peak periods were found to be of great interest to traffic engineers and planners in design and operational analysis since roads to carry traffic are designed adequately to meet such demand levels. These periods exhibit highest traffic volumes and lower speed levels throughout the 24 - hour period. Generally, these periods are estimated considering the work - trip patterns within a town. The analysis with the data suggested that the morning and evening peak periods for Nyeri town are 0800 hours - 0900 hours and 1700 hours - 1800 hours respectively. These periods were first established as a basis for all other analysis. Generally, the evening peaks

tend to show a slightly higher volume and a slightly reduced speed levels as compared to the morning peaks. It can further be investigated as to which land use activities within those corridors contribute to this variation. Such studies could enhance the effective distribution of resources in the town.

The study inferred that some road links indeed performed better than others and this is very significant to planners and engineers to further plan traffic flow within the town. All the considered links relatively performed worse between and during morning and evening peak periods. Performance levels were very good ($\approx$ 100%) from 2300 hours till 0530 hours the following day. The research also demonstrated the existence of a macroscopic fundamental diagram for Nyeri town. It was deduced that speed and traffic volumes are related. Nyeri town demonstrated a very high level of correlation between speed and volume and a model was established for the town to demonstrate the same. The model characterises the traffic flow in the town and may not be applicable to other cities as a result of possible different town and network structures. The strong relation also gave an indication that although other factors may contribute to speed variations on the network, traffic volumes accounted for the greatest percentage of variation of speed on the town's network.

Recommendations

It is recommended that further studies should be carried out into the other days where traffic variations for these days could be determined and further explored. For instance, traffic variations could be studied between weekdays and weekends so that traffic circulation within the town would be well understood and appropriate steps put in place to enhance traffic flow. It is further recommended that the designed capacities of the road segments be provided so that further analyses that pertain to comparing the performance levels of different links can be undertaken. In addition, it is suggested that the study could be replicated in other towns in Kenya in order to establish how a town's network structure (land use and transport planning) affects its MFD.

References

[1] J. Williams. Macroscopic flow models. Traffic Flow Theory. Washington, DC: US Federal Highway Administration, 1996, pp 6 - 1.

[2] G. Weisbrod, D. Vary, et al." Measuring economic costs of urban traffic congestion to business. Transportation Research Record," Journal of the Transportation Research Board 1839 (1): 98 - 106, 2003.

[3] J. Rodrigue, C. Comtois, et al." The geography of transport systems. London/New York. Salau, T. (1999). Spatial analysis of urban road system: A study of abeokuta. I. Transport Dev." Initiatives, vol. 1: 13 - 24, 2005.

[4] N. Gartner, C. Messer, et al." Traffic flow theory: A state of the art report." Transportation Research Board, 1992.

[5] C. O'Flaherty, and M. Bell," Transport planning and traffic engineering, John Wiley & Sons Inc, 1997.

[6] Y. Yin, and H. Ieda,"Optimal improvement scheme for network reliability. Transportation Research Record," Journal of the Transportation Research Board, vol. 1783 (1), 2002, 1 - 6.

[7] L. Dimitriou, T. Tsekeris, et al." Evolutionary game - theoretic model for performance reliability assessment of road networks" 2007.

[8] A. Sumalee, and D. Watling," Travel time reliability in a network with dependent link modes and partial driver response," Journal of the Eastern Asia Society for Transportation Studies, vol. 5, 2003, pp 1686 - 1701.

[9] A. Ang, and W. Tang," Probability Concepts in Engineering Planning and Design," Vol. II Decision, Risk, and Reliability, John Wiley & Sons, NY, 1990."

[10] B. Green shields," A study of traffic capacity. Proceedings of Highway Research Board" 1935.

[11] L. Muñoz," Macroscopic modeling and identification of freeway traffic flow," PhD Thesis, University Of California, 2004.

[12] V. Dixit, and E. Radwan,"Strategies to improve dissipation into destination network using macroscopic network flow models." Manuscript prepared for Symposium on the Fundamental Diagram: 75 years ("Green shields 75" Symposium)," 2007.

[13] L. Edie," Discussion of traffic stream measurements and definitions". Proceedings of the Second International Symposium on the Theory of Traffic Flow, 1963.

[14] D. Gazis,"Traffic Science," Wiley – Inter science, 1974.

[15] J. Wardrop,"Road Paper. Some theoretical aspects of road traffic research," Proceedings of the Institution of Civil Engineers, Thomas Telford - ICE Virtual Library, 1952.

[16] R. Smeed,"Traffic studies and urban congestion," Journal of Transport Economics and Policy, vol. 2, issue 1, 1968, pp 33 - 70.

[17] C. Daganzo,"Urban gridlock: Macroscopic modeling and mitigation approaches," Transportation Research Part B41, issue 1, 2007, pp 49 - 62.

[18] N. Geroliminis," A Macroscopic Fundamental Diagram of Urban Traffic: Recent Findings," Prepared for the Symposium on the Fundamental Diagram: 75 years, Woods Hole, Massachusetts, 8 - 10 July 2008.

[19] N. Geroliminis, and C. Daganzo,"Existence of urban - scale macroscopic fundamental diagrams: Some experimental findings," Transportation Research Part B42, issue 9, (2008). pp 759 - 770.

8

Analysis of Flanged Rectangular Waveguide Probe for Nondestructive Absorbing Materials Characterization Using FDTD Simulation

Abdulkadhim A. Hasan

Department of Electronics and Communications Engineering, Kufa University, Al-Najaf, Iraq

Email address:

Kadum56@ yahoo.com

Abstract: In this paper, the flanged open-ended rectangular waveguide probe technique is studied using Finite Difference Time-Domain simulation (FDTD). Both generally lossy and high loss electromagnetic materials are considered to investigate the influence of probe flange size, operating frequency and sample thickness on complex permittivity (ε_r) and permeability (μ_r) and thickness measurement. Variations in the probe flange size for different frequencies, material under test type and thickness are simulated. It is found that using of waveguide probe with finite flange affects probe input reflection coefficient substantially in some cases. To verify the obtained simulations results, a series of experiments are conducted for this purpose. Both ε_r and μ_r of material under test under different measurement conditions are extracted using FDTD modeling and compared with reference data. In order to evaluate the degree of accuracy of this technique, error analysis to various sources of errors and most importantly the effect of finite flange size are also demonstrated by using the measured data compared with the analytical model results. Simulations and measurements results have shown that the consideration of probe flange large enough for the practical purpose to be infinite is restricted by the constitutive parameters (ε_r and μ_r) and operating frequency as well as the thickness of the material under test. The FDTD simulations and experiments results are presented.

Keywords: FDTD, Numerical Analysis, Error Analysis, Complex Permittivity and Permeability, Rectangular Waveguide, Reflection Coefficient, EM Properties

1. Introduction

The technique using open-ended rectangular waveguide as a probe has its unique advantage for determination of electromagnetic properties of materials. It's diminishing for some restrictions to sample preparation and openness in structure makes the probe be readily used in nondestructive and *in-situ* testing. Moreover, waveguide is Powerful once the testing is going to be performed when high absorption material is to be characterized such as radar absorbing materials (RAM) because of high level radiating power from its opened aperture. This technique has attracted many researchers both in theoretical and in technical among others [1]-[4]. As nondestructive testing, most commonly, materials properties are derived from admittance measurement or its equivalent reflection coefficient by inverse problem, where a theoretical formulations for the probe aperture input admittance have been developed for this purpose using analytical procedure. Bakhtiari et al. [5]

examine a conductor-backed material for the purpose of determining the thickness of a lossy dielectric. He uses only the dominant mode to represent the aperture-field distribution. Maode et al. [6] examine a conductor-backed material with both magnetic and dielectric properties and obtains simultaneous extraction of permittivity and permeability. She employs an approximate variational method to determine the waveguide admittance and hence the reflection coefficient. Stewart et al. [7] perform parameter extraction of lossy conductor-backed materials using both single and dual-aperture probes. He employs a rigorous, full-wave integral-equation method of analysis and uses the two-thickness method to extract permittivity and permeability using a single-aperture probe.

In the aforementioned theoretical analysis of the probe, the formulations of the aperture input admittance are made based on the assumption that the waveguide opening has infinite conducting flange. Physically, an infinite flange can not be realized in practice. For low loss material characterization, the condition for infinite flange dimension is necessary in order to provide entire interaction between the fields and material under test. In fact , for high-loss materials, the electromagnetic (EM) fields are distributed around the aperture of the sensor over a limited distance consequently, for practical use, the flange with a definite dimension can be used. For this purpose, the flange dimension should be chosen at least as large as to the distance where the decay of field is near negligible to ensure measurement accuracy. The viability of any material characterization method is determined in part by a knowledge of how measurement uncertainty affects the extracted values of ε_r and μ_r. For the waveguide probe method, the uncertainty exists in geometrical factors is based on analytical procedure and errors introduced either by the measurement system or in the calculation of the theoretical reflection coefficient [8]. In this regard, the theoretical analysis of electromagnetic models is performed under assumption that the probe has an infinite flange while measurements are usually conducted using a finite-size flange. Consequently, the results of the model and those from measurements may not be sufficiently alike for accurate EM-properties evaluation. Hence, it is necessary in order to obtain a better approximate value in the measurement, the far field of the aperture field distribution should be analyze [6]; the other way is to calculate the fields of finite flanged aperture directly by using numerical methods. Potentially, the Finite Difference Time Domain technique has several advantages, which make it suitable for modeling complex geometries [9]-[12]. In previous investigations of rectangular waveguide probe, the size of the probe flange is not fully addressed. Many studies, for example, [6] and [13], either used the waveguide radiation pattern to determine the proper size of the flange or considered a very large flange. The work presented here investigates the effect of using an open-ended waveguide with finite-sized flange on the input reflection coefficient of rectangular waveguide terminated with the material under test using FDTD modeling under different test conditions. The variation in the flange size for different lossy materials with different thicknesses and frequencies will be studied. Furthermore, a detailed examinations of the effect of using a finite flange are presented and ultimately validated by measuring the complex permittivity and permeability of radar absorbing material samples. Also the error in evaluating both complex permittivity and permeability and thickness is analyzed. Simulation results are to be verified by comparing with published data and measurements to demonstrate the validity of the FDTD technique.

2. Method of Analysis

It is no lost generality to consider the case when the material under is backed by perfect conductor or it may be coated on metal. The geometry of the problem under consideration is shown in Fig.1(a and b). The measurement set-up consists of a flanged waveguide terminated with the material under test. The material under test is considered to be generally lossy or high loss electromagnetic materials with complex permittivity ε_r and complex permeability μ_r and thickness d . The analysis of this model is performed using two methods: the analytical method [5] previously derived based on the assumption that the probe flange is an infinite in extend as shown in Fig.1(a) and the FDTD method. In this paper, the FDTD method is used to model the problem for the case of finite size of probe flange as shown in Fig. 1(b). The object of this analysis is to study the influence of different test conditions such as probe flange size, material type, operating frequency and thickness of material under test on measurement accuracy of complex permittivity and permeability. In the analysis performed, the waveguide used is WR-90 type with square flange of different sizes. The dimensions (a and b) of the waveguide aperture are chosen such that only the dominant TE_{10} mode propagates within the band of interest (X-band of microwave frequency range).

2.1. Theoretical Modeling for Infinite Flange Case

The geometry of this case of the problem is shown in Fig. 1(a) with Cartesian coordinate system. The flanged open-ended rectangular waveguide probe is in close contact with the flat surface of the material under test. The energy radiated from the probe penetrates through the material under test and reflected back into the aperture. The reflection coefficient (Γ) carries the material properties and thicknesses information.

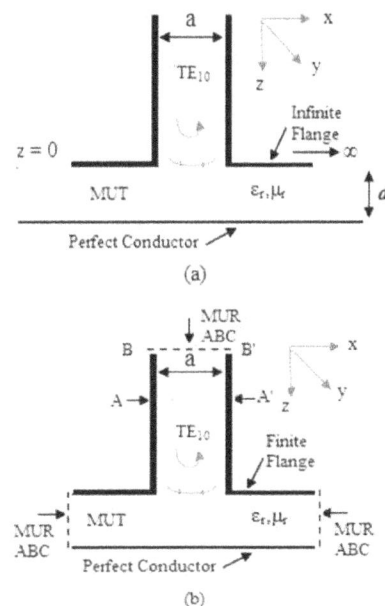

Figure 1. *Geometry of the problem for (a) Infinite Flange probe (b) Finite Flange probe.*

Considering only the dominant TE_{10} mode, the near-field interaction can be modeled by the following normalized admittance [5].

$$Y = \frac{j}{(2\pi)^2 \mu \sqrt{1 - \left(\frac{\lambda_o}{2a}\right)^2}} \int_{R=0}^{\infty} \int_{\theta=0}^{2\pi} \Im \left\{ (\varepsilon\mu - R^2 \cos^2\theta) \left(2C_\varphi + \frac{j\Im}{x_z}\right) \right\} R d\theta dR \tag{1}$$

$$\Im = 4\pi \sqrt{\frac{2A}{B}} \frac{\sin(\frac{x_y B}{2}) \cos(\frac{x_x A}{2})}{x_y [\pi^2 - (x_x A)^2]}$$

$$C_\phi = -\frac{\Im e^{jx_z D}}{2x \sin(x_z D)}$$

$$A = k_o a, \quad B = k_o b, \quad D = k_o d, \quad k_o = \omega\sqrt{\varepsilon_o \mu_o}$$

$$x_x = R\cos\theta, \quad x_y = R\sin\theta, \quad x_z = \sqrt{\varepsilon\mu - R^2}$$

$$\varepsilon = \varepsilon' - j\varepsilon'' = \varepsilon'(1 - j\tan\delta_\varepsilon)$$

$$\mu = \mu' - j\mu'' = \mu'(1 - j\tan\delta_\mu)$$

where ε, μ are the relative permittivity and permeability, a and b are the wide and narrow dimensions of the rectangular waveguide, and d is the thickness of the material under test. The reflection coefficient is a complicated function of the waveguide dimensions, test frequency (f), sample thickness and the parameters of material under test ε_r and μ_r such that $\Gamma(a, b, f, \varepsilon_r, \mu_r, d)$.

2.2. FDTD Modeling for Finite Flange Case

In this paper FDTD method is employed to model the geometry of the problem with finite flange probe as shown in Fig. 1(b). The main object is to calculate the probe aperture input reflection coefficient under different test conditions since it is quite difficult to use the analytical procedure to formulate aperture input admittance of rectangular waveguide with finite flange. This can be achieved by calculating different field's components within the probe and material under test and consequently, both complex permittivity and permeability of material under test can be extracted by inverse problem. FDTD method simulates structures in time-domain using a direct form of Maxwell curl equations. In this work, the classical FDTD approach, which is based on Yee's explicit formulation [14] is adopted. The model shown in Fig. 1(b) can be considered as combination of 3D cubes in Cartesian coordinates, which are called Yee's Cells in FDTD method . For an isotropic and homogenous medium with the medium permittivity ε and the medium permeability μ, we have

$$\nabla \times E = -\mu \frac{\partial H}{\partial t} \tag{1a}$$

$$\nabla \times H = -\varepsilon \frac{\partial E}{\partial t} \tag{1b}$$

rearranging these equations yield (2)

$$\frac{\partial H}{\partial t} = -\frac{1}{\mu} \nabla \times E \tag{2a}$$

$$\frac{\partial E}{\partial t} = -\frac{1}{\varepsilon} \nabla \times H \tag{2b}$$

The FDTD method uses a discretization in time and space (Δt, Δx, Δy, and Δz) to calculate a solution of Maxwell's curl equations. Evaluating the vector curl operator $(\nabla \times A)$ and employing central differencing in both time and space to approximate the partial derivatives, we obtain six update equations (one for each component of the electric and magnetic fields). The update equation for the E_x component is:

$$E_x^n(i,j,k) = \left[\frac{t}{t+\Delta t}\right] E_x^{n-1}(i,j,k) + \left[\frac{t}{t+\Delta t}\right]$$
$$\left[\frac{H_z^{m-1/2}(i,j,k) - H_z^{m-1/2}(i,j-1,k)}{\Delta y} - \frac{H_y^{m-1/2}(i,j,k) - H_y^{m-1/2}(i,j,k-1)}{\Delta z}\right] \tag{3}$$

The electromagnetic structure is modeled by approximating its geometry and composition with Yee cells of different material parameters; both permittivity and permeability of material under test and inside of the waveguide. In accordance with FDTD principles, to ensure adequacy and accuracy of computation, the model is meshed with cubic cells chosen so that the cell size is sufficiently smaller than λ in the material under test as well as it is in accordance with the Courant stability criterion [15]. In this work, FDTD available space was divided into many rectangular three dimension cells with the size Δx, Δy and Δz in x, y and z axis respectively. A fine spacing cell with dimensions of Δx =0.05mm, Δy =0.05mm, Δz =0.05 mm is used within the probe and the material under test to increase the accuracy of the calculation. Effective (average) values of constitutive parameters are used to calculate field components at different interfaces. The computational domain is truncated by first-order Mur absorbing boundary condition (ABC) [16] at plane B-B' of the waveguide and at the radial boundary of the problem space to limit the computational domain and increase the computational efficiency. An excitation of TE_{10} mode is then applied at the excitation plane A-A' to the computational model and the E-field and H-field computations are alternately matched through time from time zero to the desired stopping point. The field distribution in the sample and the probe are calculated using Yee cell procedure. The input admittance of the material under test is obtained by calculating field's components within material under test. According to the transmission line theory, the probe complex reflection coefficient, Γ_o, is calculated using (4) at sampling point located away from the aperture to avoid higher order modes that may exist at the aperture layered media interface boundary.

$$\Gamma_o = \frac{Y_a - Y_0}{Y_a + Y_0} \tag{4}$$

where Y_a and Y_0 are the aperture admittance and the equivalent characteristic admittance of the waveguide, respectively.

3. FDTD Modeling Validation

A 3D FDTD code is developed to calculate complex reflection coefficient of the probe terminated with the material under test as shown in Fig. 1(b). In order to validate the FDTD code; several simulations and tests were conducted for different flange sizes. Samples of different materials were used to calculate the input reflection coefficient of the probe over X-band (8.2 – 12.4 GHz) of microwave frequency range. Measurement corresponding to the selected cases of the flange sizes was performed using HP-8510B automatic network analyzer (ANA) with WR-90 waveguide probe over the given frequency range. Table I shows a comparison between the results of calculated reflection coefficient using FDTD modeling versus data obtained analytically previously published in [6]. The material under test used is radar absorbing material (RAM) MF 116 with complex permittivity of $\varepsilon_r = 16 - j0.96$ and complex permeability of $\mu_r = 1.5 - j1.02$ with thickness of 2.44 mm. The results obtained are calculated at frequency of 10 GHz where it may be seen that there is good agreement between them. Table II compares the measured and the calculated magnitudes of the reflection coefficients using FDTD method for selected dimensions of probe flange using the same material under test with thickness of 4.18 mm. The measured variation in the reflection coefficient closely follows the FDTD simulation results validating the computation tool.

4. Error Analysis

Radar absorbing materials are frequently encountered in many applications such as electromagnetic compatibility (EMC) and shielding. The characterization of these materials are usually performed nondestructively using flanged open-ended rectangular waveguide probe[6-8][17] since this technique is quite suitable for solid materials testing. In this method of measurement, there are many possible sources of error which can be broadly categorized into two groups: those associated with the measurement of the reflection coefficient (measurement error sources) and those associated with computing the theoretical reflection coefficient. The latter includes uncertainty in the modeling of the applicator. Moreover, the flange size is assumed to be an infinite in extend. Using of a waveguide with small flange size will significantly influence measured values and error factors are introduced/increased in the measurement of both complex permittivity and permeability which is one of stated goals of this paper.

Table I. Comparison Between the FDTD Results versus Reference Data [6] of the Reflection Coefficient.

Γ (Magnitude)		Γ (Phase in Degree)	
FDTD Model	*Reference [6]*	*FDTD Model*	*Reference [6]*
0.711	0.699	-68.10	-66.49

Table II. Comparison of the Measured and the Calculated Magnitude of the Reflection coefficient for Different Flange Sizes of the Probe.

Flange Size (mm)	40	50	60
Method			
FDTD Model	0.4358	0.4420	0.4542
Experiment	0.4198	0.4282	0.4481

4.1. Influence on Reflection Coefficient Measurement

To evaluate the effect of probe flange size on reflection coefficient measurement accuracy, a series of simulations are conducted for different flange dimensions. Samples of different materials are used to calculate the input reflection coefficient of the probe over the given frequency range. Different measurement conditions such as operating frequency, variation of constitutive parameters (ε_r, μ_r) values of material under test and the thickness of the material under test are considered for this purpose. Fig.2 shows variation of the calculated reflection coefficient of the probe terminated in lossy material with the operating frequency for different sizes of the flange. Two frequencies of 9 and 11 GHz are selected. It is clear from the figure that the variation in the magnitude of reflection coefficient is less at 11 GHz than at 9 GHz. This may be due to increasing of radiation losses at higher frequencies. The influence of constitutive parameters values of two different materials (lossy and high loss) on the measurement is also studied. In Fig. 3, the calculated percent of deviation in reflection coefficient (compared with that of probe with infinite flange size) due to variations in the flange size normalized to the wavelength in the material under test (λ_m) for this case are plotted. It is clear from the figure that the variation of reflection coefficient with the flange size decreases for high loss material. This is due to that for material with high loss, the fields decay faster than that of low loss materials. Fig. 4 shows the effect of material under test thickness on reflection coefficient measurement. As shown, for thick sample, the effect of flange size on reflection coefficient is less compared to thin sample. This is due to that for absorbing materials, the reflection decreases with increasing of sample thickness. For this reason the sensitivity of the thickness measurement of lossy materials is becoming poorer for thicker lossy material. From the FDTD analysis, according to the calculated probe reflection coefficient for different measurement conditions, it can be concluded that the choice of flange dimension of rectangular waveguide probe for generally lossy and high loss material characterization depends on sample thickness, ε_r and μ_r parameters of material under test and the operating frequency. In general, to ensure measurement accuracy, the flange dimension should be chosen at least as large as to the distance where the decay of the deviation of reflection coefficient is negligible and this issue will be further investigated in the following section.

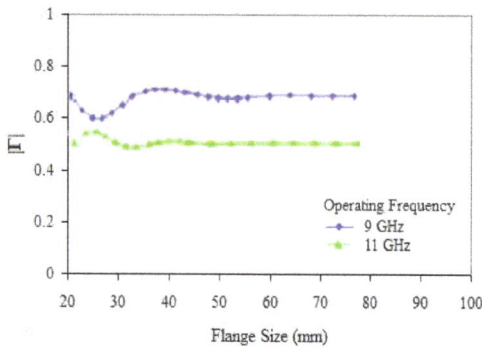

Figure 2. *Variation of magnitude of reflection coefficient versus flange size at different frequencies*

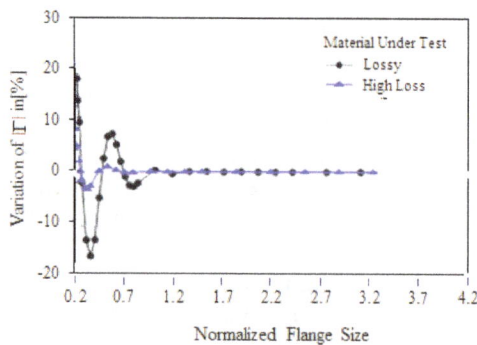

Figure 3. *Variation of magnitude of refection coefficient in [%] versus flange size normalized to the wave- length in the material under test for two different lossy materials*

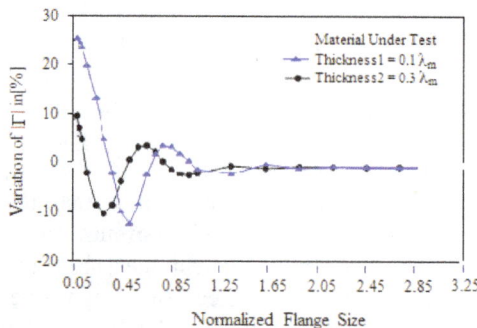

Figure 4. *Variation of magnitude of refection coefficient in [%] versus flange size normalized to the wave- length in the material under test for two different thicknesses of lossy materials*

For the cases examined of the lossy materials, the reflection coefficient had a damped oscillatory dependence on the probe flange size.

4.2. Influence on Constitutive Parameters Measurement

The viability of any material-characterization method is determined in part by a knowledge of how measurement uncertainty affects the extracted values of ε_r and μ_r. The sensitivities of ε_r and μ_r to the measured values of reflection coefficient must be understood in order to determine the conditions under which a given method may be used effectively. In this section, the analysis performed in the previous section based on FDTD modeling is to be verified experimentally in order to investigate the influence of

waveguide flange size on the measured complex permittivity and permeability. Since two complex quantities are to be determined (ε_r and μ_r) and the material under test is assumed to be conductor backed, an experimental procedure is required in which two-independent reflection coefficients are measured for a given probe flange size. Since frequency is an independent variable of the probe's reflection coefficient, which can be symbolized as $\Gamma(a, b, f, \varepsilon(f), \mu(f), d)$, swept-frequency technique is used to produce needed independent reflections [18]. Several experiments are conducted to investigate the influence of probe flange size and sample thickness on EM-properties measurement accuracy using waveguide probe with different flange dimensions over X-band of microwave frequency range (8.2-12.4) GHz. Table III shows variations of the measured permittivity and permeability (both real part and loss tangent) for three selected flange dimensions of 30 mm, 60 mm and for the case when the flange is an infinite in extend compared with reference data at frequency of 10 GHz. The material under test used is GEC-Marconi 9052 radar absorbing material with thickness of 2.08 mm. The frequency interval Δf used in the measurement is taken to be 0.3 GHz. For cases of 30 mm and 60 mm flange sizes, the results of both permittivity and permeability are obtained using FDTD modeling while for the case when the flange size is assumed to be an infinite in extend, the results are obtained analytically using (1). The reference data are given by the ECCOSORB MF Technical Bulletin and the Marconi Company [19]. It is clear from the table that the data obtained for the cases of 60 mm and infinite flange sizes and reference data are fairly consistent. As expected a large discrepancy is obtained between the measured results for the case of the flange with size of 30 mm (both in real parts and loss tangent) and reference data. This is due to that for a given lossy magnetic material, a 30-mm flange size is not enough to ensure adequate decay of the field resulting in radiated field from the material under test boundary. Furthermore, the edge-diffracted fields at the flange-edge boundary with the material under test produce a spurious reflection especially for the cases of measurement when the material under test is thin or low loss. As the flange size increases, the interaction between the EM fields and material under test increases by permitting a larger field into the material under test resulting improvement in measurement accuracy. Also, one can observe that a small deviation in the measured results of permittivity and permeability for the cases of using waveguide with 60 mm flange size and that obtained analytically under assumption that the flange size is infinite in extend as compared to the reference data. This is due to that for the second case the measurements are performed using standard X-band waveguide fitted with a square flange with side dimensions of 47.8 mm x 47,8 mm while in the theoretical formulation the flange is assumed to be infinite. For the first case, the probe flange size used in the measurement may be adequate to provide accuracy in the measurements in. Consequently, for practical use, the flange with a definite dimension must be used. This is also validated in [8].

Table III. Comparison Between the Measured Complex permiottivity and permeability of Lossy Material Under test For Different Probe Flange Size. Measurements are performed at f = 10 GHz

Flange Dimension (mm)	Measurement Results			
	ε_r	$\tan\delta_\varepsilon$	μ_r	$\tan\delta_\mu$
30 mm	18.52	0.041	1.95	1.198
60 mm	18.26	0.021	1.76	1.136
Analytical Method [5]	17.94	0.016	1.83	1.084
Reference Data	18.18	0.023	1.8	1.194

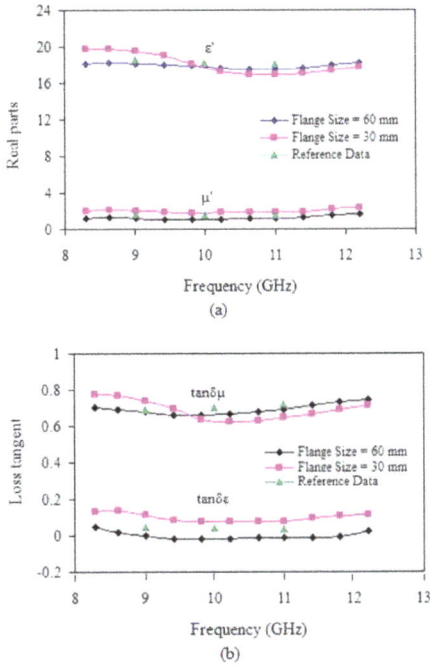

Figure 5. Variation of complex permittivity and permeability of radar absorbing material with frequency for different probe flange size

In order to get a better insight into the nature of the problem and estimate the main effective sources of error in the measurement, another set of measurements are performed over a given frequency range to investigate the influence of sample thickness on the extracted ε_r and μ_r. The same material under test is used for this purpose with thickness of 6.24 mm. The variations of the permittivity ε_r and permeability μ_r (both real parts and loss tangents) as a function of frequency are shown in Fig. 5 (a and b) respectively. Fig.5a shows variations of the real parts of permittivity (ε') and permeability (μ') while Fig. 5b shows variations of the loss tangents of permittivity ($\tan\delta_\varepsilon$) and permeability ($\tan\delta_\mu$). It can be seen from the figures that good agreement between the measured results and the reference data is achieved over the given range for the case when the flange size is 60 mm., but the thickness effect on sensitivity and accuracy is obvious especially for complex permittivity extraction . The results in Fig. 5b show a large discrepancy in the loss tangent (especially $\tan\delta_\varepsilon$) as compared to the reference data. Hence, the sample whose thickness is too thicker are dominant reasons. This is due to that if the thickness is too thick, the variation of reflection coefficient (both the amplitude and the phase) becomes duller, and finally tends to a point. This is because the sample is so thick that the decaying reflection wave from the short-circuit

plate can not influence the input wave on the flange plane. On the other hand, if the thickness is chosen to be too small, the spurious reflection coefficient would become too high especially for low loss material. It would lead to some difficulty in ensuring the testing accuracy. So it is important to choose a proper flange size and thickness for testing. Furthermore, the open-ended waveguide probe is at its best for measuring the high-loss materials such that it is difficult to get reasonable accuracy for $\tan_\delta$ which is less than 0.1 (here, the actual $\tan\delta_\varepsilon$ is about 0.06). The results obtained from this analysis show that generally lossy materials characterization is highly affected by probe flange dimensions, operating frequency and thickness of sample under test. In this work both complex permittivity and permeability of the material under test are extracted by inverse problem using iterative optimization technique from the calculated reflection coefficient (Γ_c) using FDTD modeling and measured reflection coefficient (Γ_m). Consequently, an iterative solution are sought to compute ε_r and μ_r for given $\Gamma_c(\varepsilon_r , \mu_r)$ and $\Gamma_m(\varepsilon_r , \mu_r)$ with an optimum objective function using (5)[6].

$$F = \sqrt{\sum_{k=1}^{4}[f_k(\varepsilon_r,\mu_r)]^2} \qquad (5)$$

From the results of the FDTD simulations and experiments, the desired solution of ε_r and μ_r can be obtained by minimizing the function F. At the beginning, initial values of ε_r and μ_r are assumed, then Γ_{C1} and Γ_{C2} are calculated from (4) using FDTD modeling and F can be obtained from (5). If F is greater than a desired tolerance ζ, new values will be updated automatically using iterative algorithms such as Newton-Raphson method or levenberg-Marquardt Method [20]. Then a new value F is evaluated from the new ε_r and μ_r This procedure is repeated until the value of F is less than ζ. These iterative algorithms guarantee a correct convergence in most cases but the converge process is very time-consuming especially for multiparameters measurement. A computer program is written to implement this optimization procedure. Figure 6 shows the algorithm for calculating both complex permittivity and permeability of material under test using Newton-Raphson method.

5. Evaluation of Error

The results of error analysis performed in the previous section showed that one major source of error in measurement of both complex permittivity and permeability using flanged rectangular waveguide probe can be due to the finite size of the flange and other test conditions. Since the choice of the material under test thickness depends on the parameters ε_r, μ_r and the operating frequency, therefore, it is necessary in order to obtain a adequate measurement accuracy, evaluating the percentage error due to the exclusion of finite size of the flange for different measurement conditions. Consequently, a general guidelines can be suggested for measurement set-up geometries. In this section, the evaluation of % error due to using finite flange size is investigated and discussed for the cases of measurement results obtained in the previous section.

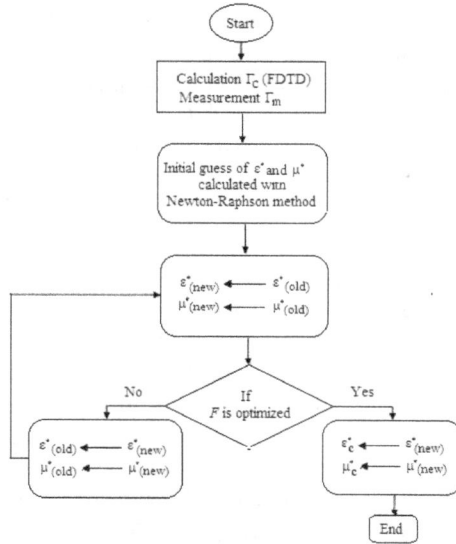

Figure 6. *Flowchart of calculation of complex permittivity and permeability*

The percent error, in real parts and loss tangents of both permittivity and permeability is defined using (6)

$$\% \, Error(x) = 100 \, \frac{x_{ref} - x_{fin}}{x_{ref}} \qquad (6)$$

where x is the real part or loss tangent of both complex permittivity and permeability, x_{ref} is the reference value of real part and loss tangent of both complex permittivity and permeability of the material under test and x_{fin} is the value of real part and loss tangent of both complex permittivity and permeability measured using finite flange probe. Computation of error is performed by comparing the results of complex permittivity and permeability obtained using FDTD method (for finite flange size) and the spectral domain analysis method using (1) for the case when the flange is considered to be an infinite in extend with the reference data for various material under test thicknesses of 2.08 mm and 6.24 mm respectively. Tables IV summarizes the calculated values of the % error at operating frequencies of 9 GHz for sample thickness of 2.08 mm while table V summarizes the calculated values of. the % error for sample thickness of 6.24 mm. It is clear from table IV and table V that the percentage error increases with decreasing probe flange size for both real and loss tangent of complex permittivity and permeability and this error decrease as flange size increases. The influence of sample thickness on measurement error is also obvious from the results of the two tables. For thin sample of 2.08 mm, the

percent error in the measurement increases especially in the real part of complex permittivity. On the other hand, one can observe a large discrepancy between the measured and the reference data of the loss tangents of both complex permittivity and permeability for sample thickness of 2.08 mm. For case of the sample with 6.24 mm thickness, the calculated % error decreases for measured real parts of both complex permittivity and permeability. As expected, the % error for the loss tangent of complex permittivity ($\tan\delta_\varepsilon$) increases for the reasons discussed in the previous section. Also the effect of probe dimensions and operating frequency on measurement accuracy is examined. Table VI shows the calculated values of the % error for the sample with thickness of 6.24 mm since this value of sample thickness represents the worst case in the measurement. The calculations are performed at operating frequency of 11GHz with different flange dimensions as illustrated in Fig. 5. and compared with the results of table V. It is clear from the results of the two tables that the % error for the case of 11 GHz is much lower for different flange dimensions than that obtained for the case of 9 GHz. This is reasonable since the radiation increase with increasing the frequency. the results of error analysis performed in this section show that there are three factors influence measurement accuracy when a flanged open-ended rectangular waveguide probe is used for nondestructive testing of generally lossy or high loss materials namely flange size, operating frequency and sample thickness. Among these three factors it is clear that flange size and operating frequency are key factors in measurement since the material under test is assumed to be coated or backed by metal with several millimeter thickness. In this regards, the imaginary part of complex permittivity is affected for thicker sample. From the results obtained, it is difficult to make a specific evaluation regarding using finite flange rectangular waveguide for nondestructive testing of generally loss and high loss materials since there are many factors should be taken into account. In general, from Figs. 2, 3 and 4, it can be concluded that for a thick lossy materials and low values of operating frequency , flange size larger than λ_m is sufficient for accurately measuring both complex permittivity and permeability and thickness since the error in the measurement caused by the non-infinite flange is very small. However, if the material under test is thin then a relatively large flange is required (larger than $2\lambda_m$) to have an accurate measurement of ε_r and μ_r. Hence, one can design the measurement setup with an adequate flange size to obtain the desired accuracy in the measurement.

Table IV. *Estimated values of complex permittivity and permeability of 2.08 mm thickness sample and errors as compared to the reference values of $\varepsilon_r = 18.64(1 - j0.045)$ and $\mu_r = 1.7(1 - j\,0.71)$ at $f = 9$ GHz.*

Flange Size (mm)	Complex permittivity $\varepsilon_r = \varepsilon'(1 - j \tan\delta_\varepsilon)$		Complex permeability $\mu_r = \mu'(1 - j \tan\delta_\mu)$	
	Real part error [%]	$\tan\delta_\varepsilon$ error [%]	Real part error [%]	$\tan\delta_\mu$ error [%]
30	-13.76	-52.32	-10.32	-7.32
60	-4.12	-39.62	3.63	-2.56
Infinite	-3.71	-45.74	-4.82	-4.15

Table V. Estimated values of complex permittivity and permeability of 6.24 mm thickness sample and errors as compared to the reference values of ε_r =18.64(1 –j0.045) and μ_r = 1.7(1-j 0.71) at f = 9 GHz

Flange Size (mm)	Complex permittivity $\varepsilon_r = \varepsilon'(1 - j \tan\delta_\varepsilon)$		Complex permeability $\mu_r = \mu'(1 - j \tan\delta_\mu)$	
	Real part error [%]	tan δ_ε error [%]	Real part error [%]	tan δ_μ error [%]
30	-6.54	-82.41	-5.84	-11.06
60	2.56	-64.24	4.10	-4.45
Infinite	-1.45	-71.12	-2.21	-6.98

Table VI. Estimated values of complex permittivity and permeability of 6.24 mm thickness sample and errors as compared to the reference values of ε_r =18.21(1 –j0.042) and μ_r = 1.62(1-j 0.72) at f = 11 GHz

Flange Size (mm)	Complex permittivity $\varepsilon_r = \varepsilon'(1 - j \tan\delta_\varepsilon)$		Complex permeability $\mu_r = \mu'(1 - j \tan\delta_\mu)$	
	Real part error [%]	tan δ_ε error [%]	Real part error [%]	tan δ_μ error [%]
30	3.54	-36.43	-5.54	3.88
60	1.54	-26.34	2.10	5.87
Infinite	2.32	-56.21	-1.26	1.46

6. Summery and Conclusions

In our study of using open-ended rectangular waveguide probe system for nondestructive testing, we analyzed the influences of several key parameters of measurement set-up on both complex permittivity and permeability and thickness measurement of generally lossy and high loss materials using FDTD simulation and measurement. The performed analysis has shown that the errors in measurement using this technique are highly affected by probe flange size and operating frequency especially for generally lossy materials. Also uncertainty of the material under test thickness does have some effects on the measured value of high loss materials complex permittivity and permeability, but this parameter is much easier to control, especially for solid samples. Simulation and measurement results with various flange configurations indicated that it is difficult to use the analytical models to include the effect of the finite flange size. In general, for the lossy materials, the material under test and the operating frequency as well as thickness restrict the consideration of flange large enough for the practical purpose to be infinite. By FDTD modeling, the required extend of the flange can be determined. The results presented in this paper may be useful in coating materials nondestructive testing and other applications such as thickness measurement of composite materials.

References

[1] E. Nyfors and P. Vainikainen, Industrial Microwave Sensors, Artech House, 1989.

[2] M. Hirano, M. Takahashi, and M. Abe, "A study on reflection coefficient from lossy dielectric by using flanged rectangular waveguide," IEICE Trans. Electron. vol. J82-C-I, no.5, pp.283–287, May 1999.

[3] Mazlumi, F., S. H. H. Sadeghi, and R. Moini, "Interaction of an open-ended rectangular waveguide probe with an arbitrary-shape surface crack in a lossy conductor," IEEE Trans. Microwave Theory Tech., Vol. 54, No. 10, pp. 3706,-3711, 2006.

[4] Hyde IV, M. W., J. W. Stewart, M. J. Havrilla, W. P. Baker, E. J. Rothwell, and D. P. Nyquist, "Nondestructive electromagnetic material characterization using a dual waveguide probe: A full wave solution," Radio Sci., Vol. 44, RS 3013, 2009.

[5] S. Bakhtiary, S. I. Ganchev and R. Zoughi, " Open-ended rectangular waveguide for nondestructive thickness measurement and variation detection of lossy dielectric slabs backed by a conducting plate," IEEE Trans. IM, vol. IM-42, pp. 19-42, 1993

[6] Maode, N., S. Yong, Y. Jinkui, F. Chrnpung, and X. Deming, "An improved open-ended waveguide measurement technique on parameters εr and μr of high- loss materials," IEEE Trans.Instrum. Meas., Vol. 47, No. 2, pp. 476-481, April 1999.

[7] Stewart, J. W. and M. J. Havrilla, "Electromagnetic characterization of a magnetic material using an open- ended waveguide probe and a rigorous full-wave multimode model," in Journal of Electromagnetic Waves and Applications, Vol. 20, No. 14, pp. 2037-2052, 2006.

[8] G. D. Dester" Error analysis of a two-layer method for the electromagnetic characterization of conductor- backed absorbing materials using an open-ended waveguide probe" Progress In Electromagnetics Research B, Vol. 26, pp. 1-21, 2010.

[9] Abdulkhadim A. Hasan, D. Xu, and Y. J.Zhang" Modeling and analysis of finite flange open-ended coaxial probe for planar and convex surface coating material testing by FDTD-method," Microwave and Optical Technology Letters, vol. 24 no. 2, pp. 117- 120, January 2000.

[10] K. Shibata O. Hashimoto, and R. K. Pokharel" Analysis of error due to exclusion of higher modes on complex permittivity measurement using waveguide with flange" IEICE Trans. Electron, Vol.E88–C, NO.1 January 2005 pp. 139-142

[11] Abdulkhadim A. Hasan, D. Xu, Z. Lin , and N. Maode" A Modified open- ended coaxial probe for concave surface coating materials testing," 2000 IEEE MTTs International Microwave Symposium Digest.

[12] F. Kung and H. T. Chuah' A Finite-Difference Time- Domain (FDTD) software for simulation of printed circuit board (PCB) assembly' Progress In Electromagnetics Research, PIER 50, pp. 299–335, 2005.

[13] C.-W. Chang, K.-U. Chen, and J. Qian, "Nondestructive determination of electromagnetic parameters of dielectric materials at X-band frequencies using a waveguide probe system," IEEE Trans. Instrum. Meas., vol. 46, no. 5, pp. 1084–1092, Oct. 1997.

[14] K. S. Yee,"Numerical solution of initial boundary-value problems involving Maxwell's equations in isotropic media", IEEE Trans Antenna Propag. AP, vol.14. pp. 302-307, 1966.

[15] A. Taflove and M. E. Brodwin, " Numerical solution of steady-state electromagnetic scattering problems using the time-dependent Maxwell's equations, "IEEE Trans, Microwave Theory Tech, vol, MTT-23, pp. 623-630, 1975.

[16] G. Mur,"Absorbing boundary condition for the finite-difference approximation of the time-domain electroma- gnetic field equations", IEEE Trans. Eectromag. Compat. EMC-22, pp. 377-382, 1981.

[17] M. T. Ghasr, Devin Simms, and R. Zoughi," Multimodal solution for a waveguide radiating into multilayered structures—dielectric property and thickness evaluation" IEEE Trans. Inst & Meas, vol. 58, NO. 5, pp. 1505-1513, May 2009.

[18] S. Wang, Abdulkadhim A. Hasan, N. Moade and X. Deming," A Swept-frequency technique with an open-ended waveguide sensor for nondestructive, simultaneous determination of thickness, permittivity and permeability of radar absorbing coatings" 1998 Asia Pacific Microwave Conference, pp 129-132.

[19] Emerson & Cuming, Microwave Products, Inc., ECCOSORB R°FGM Permittivity & Permeability Data," 2007

[20] W. H. Press, S. A. Teukolsky, W. T. Vetterling and B. P. Flannery, Numerical Receips in C+, second Edition, Cambridge University Press, 1992

9

Experimental Testing and Analytical Modeling of Strip Footing in Reinforced Sandy Soil with Multi-Geogrid Layers Under Different Loading Conditions

Aram Mohammed Raheem[1, *], Mohammed Abdulsalam Abdulkarem[2]

[1]Civil Engineering Department, University of Kirkuk, Kirkuk, Iraq
[2]Geotechnical Engineer, Ministry of Construction and Housing, Kirkuk, Iraq

Email address:
engaram@yahoo.com (A. M. Raheem), En.muhammed@yahoo.com (M. A. Abdulkarem)

Abstract: In this study, large-scale physical models with dimensions of (0.9m * 0.9m * 0.55m) have been designed and constructed to investigate the behavior of strip footing in reinforced sandy soil with multi-geogrid layers under inclined and eccentric loading conditions. The effect of several parameters such as geogrid layers (N), soil relative density (RD), depth of the topmost geogrid layer (U/B), load inclination angle (α) and load eccentricity ratio (e/B) on the bearing capacity ratio (BCR) of reinforced soil have been investigated through 120 experimental tests. As the number of the geogrid layers increased from 0 to 4, the BCR increased by 255% for 15° load inclination angle and by 470% for 0.05 load eccentricity ratio in 60% RD. When the RD of the soil increased from 60% to 80%, the average decreases in horizontal displacement and footing tilting angle were about 35% and 21% respectively. Hyperbolic analytical model was used to predict the relationships of most of the studied parameters. However, p-q analytical model was suggested to model the relationship between the BCR versus U/B. Both suggested models (hyperbolic and p-q) were in a very good agreement with the experimental results.

Keywords: Strip Footing, Experimental Study, Sandy Soil, Geogrid, Analytical Models, Different Loading Condition

1. Introduction

Generally, a strip footing is used to transfer loads from superstructures to the supporting soils. Traditionally, theses footings might be under the impact of moments and shears in addition to vertical loads from different sources such as winds, earthquakes, earth pressure and water [1-3]. Thus, an eccentric load or an eccentric-inclined load can replace such forces or moments where the bearing capacity of a foundation with such loading conditions can be counted as one of the most essentials in geotechnical area. Eccentric loading could show substantial differential settlement causing the footing to tilt. Based on the loading eccentricity to the footing width ratio, the amount of footing tilt and the pressure distribution under the footing can change. Meyerhof [4] pointed out that the average bearing capacity of footing decreases parabolically with an increase in eccentricity. To reduce footing tilt, Mahiyar and Patel [5] examined an angle shaped footing exposed to eccentric loading. Reinforced soil

has been an ordinary practice in geotechnical engineering applications such as road construction, railway embankments, and stabilization of slopes and enhancement of soft ground properties [6]. It has been extensively expected that inducting reinforcements to a shallow foundation will considerably increase the bearing capacities [7-9]. Different types of reinforcement layers have been used to reinforce the underneath soil such as galvanized steel strips, geotextiles, and geogrids [10]. Essentially, it was reported that geogrids generally offer a higher interfacial shearing resistance than geotextiles [11]. The response of footings loaded over a reinforced soil bed by metal strips has been investigated by Binquet and Lee [12] and Fragaszy and Lawton [13]. Binquet and Lee [12] pointed out that the bearing capacity of shallow foundations could increase by (2 to 4) times when the underneath soil reinforced by galvanized steel strips. Laboratory model tests on square footing to quantify the bearing capacity of foundations reinforced with geogrids and geotextiles have been conducted by Guido et al. [7]. Khing et al. [14] examined the bearing capacity of a strip foundation placed over reinforced sandy soil. Multiple layers of geogrids have been used through laboratory tests [15-18].

Several numerical attempts have been done to study the stability of reinforced soil mass as a homogenous anisotropic material were analyzed through rigid plastic FEM [19, 20]. Furthermore, a numerical study was used through FLAC software to study the effect of geosynthetic reinforcement arrangement on two square footings on sandy soil [21-23]. Rarely, analytical models were used to investigate the behavior of reinforced sandy soil with geogrid layers.

2. Objectives

The overall objective of this study was to model the behavior of strip footing rested on reinforced sandy soil with different layers of geogrid under the influence of inclined and eccentric loading conditions. The specific objectives were as follows:

1. Perform large-scale laboratory testing of strip footing on reinforced sandy soil with geogrid layers.
2. Study the effects of geogrid layers (N), soil relative density (RD), depth of the topmost geogrid layer (U/B), load inclination angle (α) and load eccentricity ratio (e/B) on the bearing capacity ratio (BCR) of reinforced soil.
3. Investigate the validation of analytical models to predict BCR, horizontal displacement and tilting angle of strip footing over reinforced sandy soil under different loading conditions.

3. Materials and Methods

3.1. Laboratory Model Tests

3.1.1. Model Test Tank

The soil layers were prepared in a steel box with 0.9m × 0.9m and 0.55m dimensions made with a plate thickness of 6 mm supported by four steel channels as shown in Fig. 1. The inner faces of the steel box were painted to minimize the slide friction between the soil and steel box that may develop during experimental testing. Several lines were marked to identify carefully the required thickness of the soil layers and the location of the geogrid.

Fig. 1. Laboratory testing box.

3.1.2. Footing

A strip steel channel of 80 mm in the plan with a thickness of 4 mm was used to represent the tested footing as shown in Fig. 2. The transferred load to the footing was measured with a proving ring of 5 kN capacity. Both horizontal and vertical displacements were measured using three dial gauges (0.01 mm/ division). The footing size was made based on the size of the steel model tank and the zone of the influence. Detailed testing instrumentations including dial gauges, proving ring and strip footing are shown in Fig. 3.

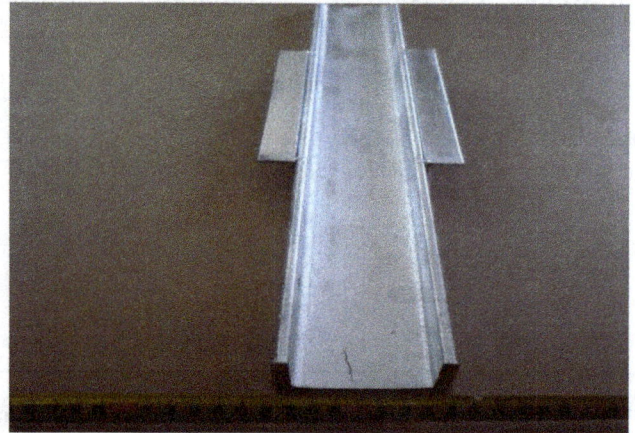

Fig. 2. Strip footing represented by steel channel.

Fig. 3. Detailed testing instrumentations.

3.2. Test Material

3.2.1. Sand Properties

A poorly graded sand passing sieve No.4 was used in this study. The sand was washed with running water to remove the dust as much as possible. Testing has been performed with dense and medium dense sand corresponding to approximately (16.9) kN/m^3 and (17.5) kN/m^3 consistent with relative densities of (60) % and (80) % respectively. The maximum and minimum dry unit weights of the sand were determined according to the ASTM (D4253-00) and ASTM (D4254-00), respectively.

The results have shown maximum and minimum dry unit

weight of sand as 18 kN/m^3 and 15.6 kN/m^3 respectively. The specific gravity of the sand was 2.59 and the test has been done based on the ASTM D-854. The grain size distribution analysis of the sand was performed according to the ASTM D-421 and it can be shown in Fig. 4. The sand was classified according to the unified soil classification system as poorly graded sand with coefficient of uniformity (C_u) = 3.0 and coefficient of curvature (C_C) = 1.0.

3.2.2. Geogrid

One type of commercially available geogrid type was used TriAx® TX140 Geogrid manufactured from a punched polypropylene sheet, which was oriented in three significantly equilateral directions so that the subsequent ribs shall have a high degree of molecular orientation. The properties influencing the performance of a mechanically stabilized layer are summarized in Table 1.

Table 1. Engineering properties of Tenax TT Samp geogrid.

Index Properties	Longitudinal	Diagonal	Transverse
Rib pitch, mm(in)	40(1.6)	40(1.6)	-
Mid-rib depth mm(in)	-	1.2(0.05)	1.2(0.05)
Mid-rib width mm(in)	-	1.1(0.04)	1.1(0.04)

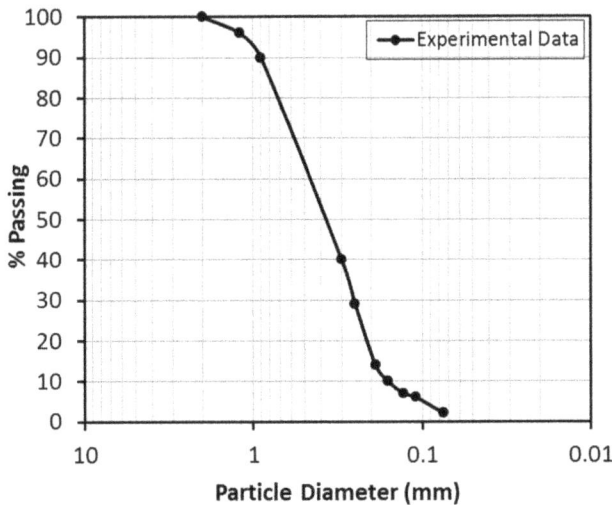

Fig. 4. Particle size distribution of tested sand.

4. Testing Program

Detailed testing program has been established to reach the aim of the study of the effect of load applied on strip footings on reinforced sand. The parameters were loads eccentricity (e/B), loads inclination (α), number of geogrids layers (N), depth of topmost layer (U/B) and relative density (RD) which varied from a test to another. For all the tests, the footing was rested on the surface of the sandy soil and the distance between consecutive layers kept constant with a value of 0.05 m. The maximum number of geogrid layers that used in this study was four. The embedment length for the geogrid layers was 0.8 m. A schematic diagram for the strip footing in the sandy soil is shown in Fig. 5.

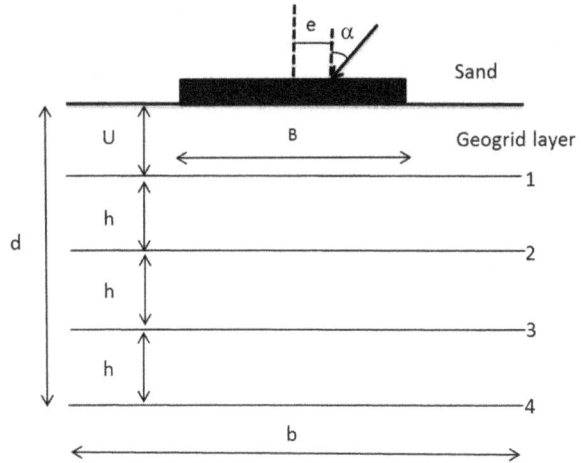

Fig. 5. Schematic diagram for tested strip footing on reinforced sandy soil.

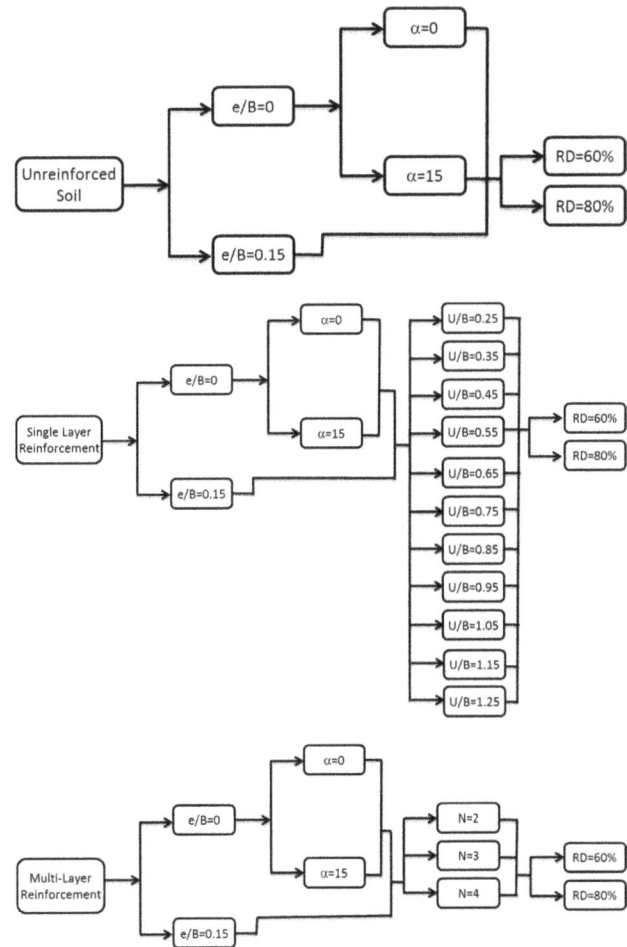

Fig. 6. Flow chart for the testing program.

Fig. 6 shows the flow chart of the testing program. The flow chart was divided into three parts, the first part included the experimental work of unreinforced soil where it can be used as a reference to compare the improvement of using a geogrid as a reinforcement. In addition, it was used to study the effect of changing the load inclination and eccentricity on the bearing capacity of unreinforced sand for two studied relative densities. The second part focused on a single layer of reinforcement where these tests were used to study and locate the optimum depth of the topmost layer of geogrid (U/B). The third part, which was the main emphasis of this study, showed the effect of the multi-reinforcement layer on the bearing capacity including the effect of the load inclination and eccentricity on the optimum number of the reinforcement layer. The term bearing capacity ratio (BCR) is used to express the combined effect of soil reinforcement with load inclined and eccentricity on bearing capacity and it can be shown as follows:

$$BCR = \frac{q_{ur}}{q_u} \qquad (1)$$

where q_{ur} is the ultimate bearing capacity of inclined and eccentrically load strip footing on reinforced sand while q_u is the ultimate baring capacity of strip footing on unreinforced sand.

Meyerhof [4] suggested an empirical relation to compute the ultimate bearing capacity of footings subjected to eccentric-inclined loads:

$$q = CN_c S_c D_c I_c + \gamma D N_q S_q D_q I_q + 0.5 \gamma B N_\gamma S_\gamma D_\gamma I_\gamma \qquad (2)$$

where C, γ are the soil cohesion and density respectively. N_c, N_q, and N_γ are bearing capacity factors. S_c, S_q, and S_γ are shape factors. D_c, D_q, and D_γ are depth factors. I_c, I_q, and I_γ are inclined load factors. B is the footing width. D is footing embedment depth.

5. Analytical Models and Prediction

Based on the experimental expectation results, the following analytical models can be used:

5.1. Hyperbolic Model

For reinforced soil, the bearing capacity ratio (BCR) versus number of geogrid layers (N) is expected to increase to a certain level where beyond that point no increase in BCR could be anticipated even if N increased. Such kind of relationships has been noticed in several engineering and environmental applications and modeled using hyperbolic model. In early stages of developing hyperbolic model, it was used to predict the amount of phenol leached from a solidified cement matrix [24]. Furthermore, hyperbolic model was used to represent the relationship between the changes in grouted sand properties with curing time [25]. Vipulanandan et al. [26] proposed hyperbolic relationship to characterize the variation of in-situ vertical stress and logarithmic undrained shear strength of the soft marine and deltaic clays. Hyperbolic relationship can be used to correlate the compressive strength variation with curing time for cemented sand [27]. Hyperbolic model were used for several other relationships such as fluid loss versus time in high-pressure high-temperature condition [28], shear strength versus solid content of ultra-soft soil [29], and changes in electrical resistivity versus salt content of bentonite drilling mud [30]. The hyperbolic model formulation is as follows:

$$BCR = \frac{N}{A + B*N} + 1.0 \qquad (3)$$

where A & B are model parameters while BCR and N are bearing capacity ratio and number of geogrid layers respectively.

5.2. P-q Model

It is expected that the bearing capacity ratio (BCR) will increase with increasing the depth of the topmost layer of geogrid (U/B) to the optimum value then it will start to decrease with increasing U/B value. Such nature of relationship can be modeled using p-q model. This model was first proposed by Mebarkia and Vipulanandan [31] to predict the stress-strain behavior of glass-fiber-reinforced polymer concrete. The original p-q model formulation is as follows:

$$\sigma = \left[\frac{\varepsilon/\varepsilon_c}{q + (1-p-q)\frac{\varepsilon}{\varepsilon_c} + p\left(\frac{\varepsilon}{\varepsilon_c}\right)^{\frac{(p+q)}{p}}} \right] \sigma_c \qquad (4)$$

where σ = compressive stress, σ_c, ε_c = compressive strength and its corresponding strain, p, q = model parameters.

This model has been adopted in this study to model the relationship between *BCR* and *U/B* as follows:

$$BCR = \left[\frac{(U/B)/(U/B)_c}{q + (1-p-q)\frac{(U/B)}{(U/B)_c} + p\left(\frac{(U/B)}{(U/B)_c}\right)^{\frac{(p+q)}{p}}} \right] BCR_C \qquad (5)$$

where *(BCR)_c*, *(U/B)_c* = maximum *(BCR)* and its corresponding *(U/B)*.

5.3. Comparison of Model Prediction

In order to determine the accuracy of the model predictions, both coefficient of determination (R^2) and the root mean square error (RMSE) in curve fitting as defined in Eqs. (6) and (7) were quantified.

$$R^2 = \left(\frac{\sum_i (x_i - \bar{x})(y_i - \bar{y})}{\sqrt{\sum_i (x_i - \bar{x})^2} \sqrt{\sum_i (y_i - \bar{y})^2}} \right)^2 \qquad (6)$$

$$RMSE = \sqrt{\frac{\sum_{i=1}^{n} (y_i - x_i)^2}{N}} \qquad (7)$$

where yi is the actual value; xi is the calculated value from the model; $\bar{y}$ is the mean of actual values; $\bar{x}$ is the mean of calculated values and N is the number of data points.

6. Results and Analysis

6.1. BCR Versus N

6.1.1. Load Inclination Angle Effect

In this section, the BCR versus N for load inclination angle (α) varied from 5° to 15° of two different RD (60% and 80%) can be identified clearly in Fig. 7 (a to d). In Fig. 7 (a) and 7 (b), the relationship between the BCR and N for zero load inclination for strip footing in 60% and 80% RD are shown respectively. As the RD of the soil increased from 60% to 80%, the average increase in BCR was about 17%. The hyperbolic model predicted the experimental data preciously with R^2 and RMSE of 0.99, 0.078, 0.98 and 0.131 for 60% and 80% RD respectively. At higher load inclination angle (α=15), the BCR decreased as the RD of the soil increased and the average decrease was about 43% as shown in Fig. 7 (c) and 7 (d). The R^2 and RMSE of the hyperbolic model were 0.94, 0.242, 0.99 and 0.09 for 60% and 80% RD respectively. The overall behavior implied that having higher number of geogrid layers increased the BCR of sandy soil while the load inclination effect was more at higher RD. As the geogrid layer increased from 0 to 4, the BCR increased by 210%, and 250% for zero load inclination and by 255%, and 100% for 15° load inclination for RD of 60% and 80% respectively. Hyperbolic model parameters that used to predict the relationship between BCR and N for different load inclination (Fig. 7) can be summarized in Table 2.

Table 2. *Model parameters of hyperbolic model (Fig. 7).*

α	RD(%)	A	B	R^2	RMSE
0	60	0.3	0.4	0.99	0.078
0	80	0.5	0.3	0.98	0.131
15	60	0.85	0.18	0.94	0.242
15	80	1	0.9	0.99	0.090

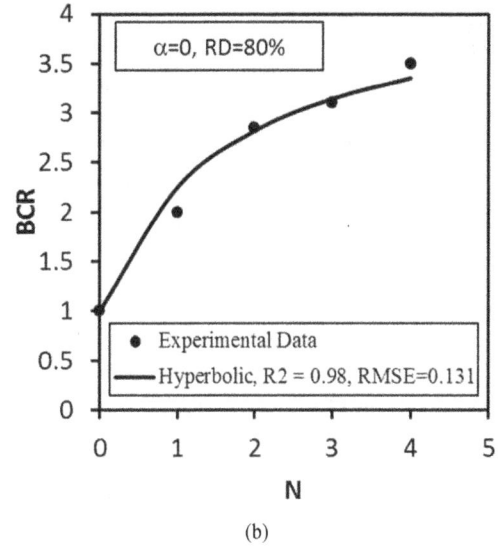

(a)

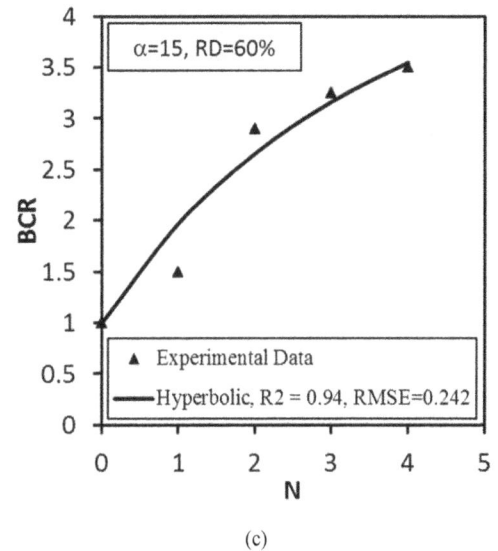

(b)

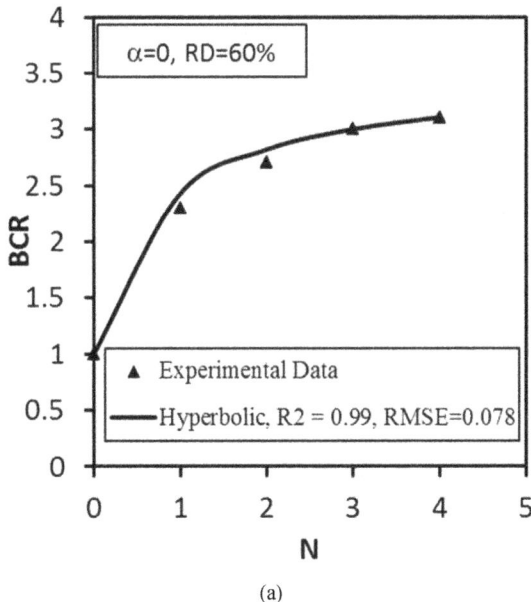

(c)

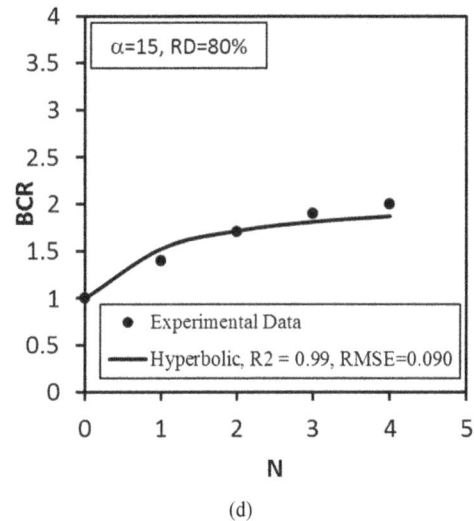

(d)

Fig. 7. *Modeling of BCR versus N relationship of strip footing rested on sandy soil.*

(a) α=0, RD=60%, (b) α=0, RD=80%, (c) α=15, RD=60%, and (d) α=15, RD=80%.

6.1.2. Load Eccentricity (e/B)

In this section, the BCR versus N for load eccentricity ratio (e/B) varied from 0 to 0.15 of two different RD (60% and 80%) can be displayed clearly in Fig. 8 (a to d). In Fig. 8 (a) and 8 (b), the relationship between the BCR and N for zero load eccentricity ratio of strip footing in 60% and 80% RD are shown respectively. As the RD of the soil increased from 60% to 80%, the average increase in BCR was about 23%. The hyperbolic model predicted the experimental data preciously with R^2 and RMSE of 0.97, 0.099, 0.98 and 0.135 for 60% and 80% RD respectively. At higher load eccentricity (e/B=0.15), the BCR decreased as the RD of the soil increased and the average decrease was about 45% as shown in Fig. 8 (c) and 8 (d). The R^2 and RMSE of the hyperbolic model were 0.96, 0.335, 0.98 and 0.250 for 60% and 80% RD respectively. As the geogrid layer increased from 0 to 4, the BCR increased by 180%, and 260% for zero load eccentricity and by 470%, and 185% for 0.05 load eccentricity for RD of 60% and 80% respectively. Hyperbolic model parameters that used to predict the relationship between BCR and N for different load eccentricity (Fig. 8) can be identified in Table 3.

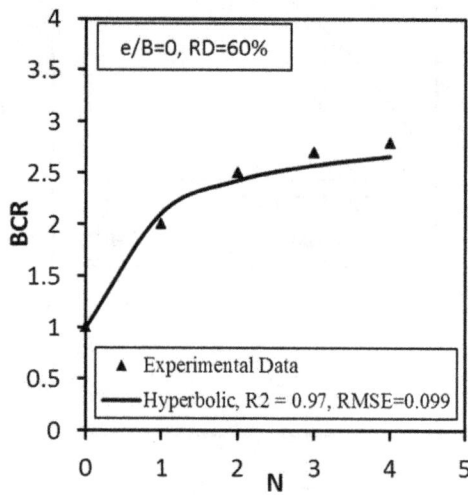

(a)

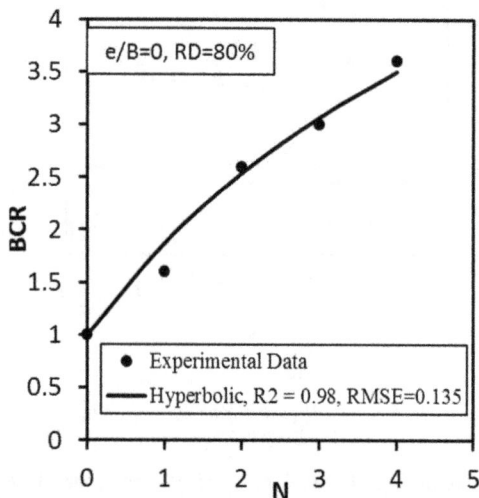

(b)

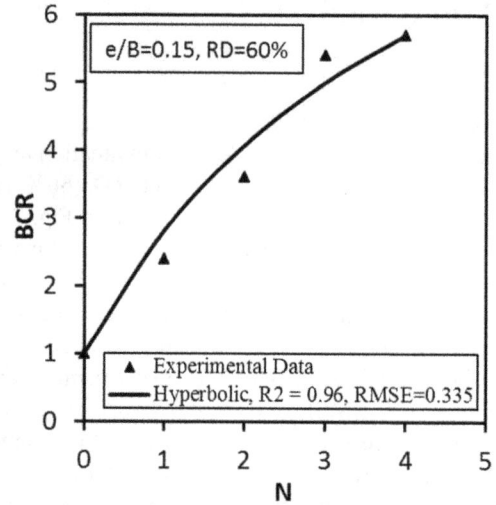

(c)

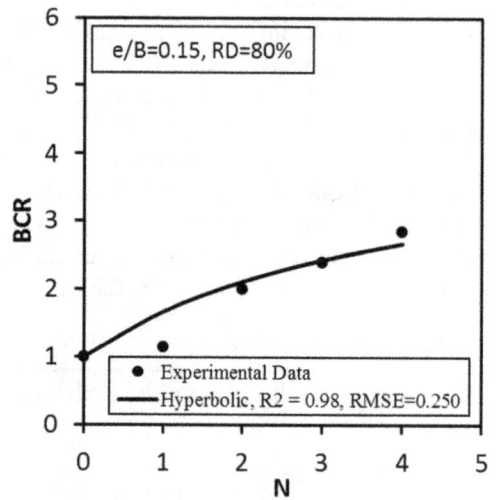

(d)

Fig. 8. *Modeling of BCR versus N relationship of strip footing rested on sandy soil.*

(a) e/B=0, RD=60%, (b) e/B=0, RD=80%, (c) e/B=0.15, RD=60%, and (d) e/B=0.15, RD=80%.

Table 3. *Model parameters of hyperbolic model (Fig. 8).*

e/B	RD(%)	A	B	R^2	RMSE
0	60	0.4	0.5	0.97	0.099
0	80	1	0.15	0.98	0.135
0.15	60	0.45	0.1	0.96	0.335
0.15	80	1.2	0.3	0.98	0.250

6.2. Horizontal Displacement Versus N

In this section, the horizontal displacement of the strip footing versus N for load inclination (α) varied from 0° to 15° of two different RD (60% and 80%) can be presented obviously in Fig. 9 (a to d). In Fig. 9 (a) and 9 (b), the relationship between the horizontal displacement and N for 5° load inclination of strip footing in 60% and 80% RD are shown respectively. As the RD of the soil increased from 60% to 80%, the average decrease in the horizontal displacement was about 35%. The hyperbolic model predicted the experimental data preciously with R^2 and RMSE of 0.99, 0.094 mm, 0.99 and 0.011 mm for 60% and 80% RD respectively. At higher load inclination ($\alpha=15°$), the horizontal displacement decreased as the RD of the soil increased and the average decrease was about 45% as shown in Fig. 9 (c) and 9 (d). The R^2 and RMSE of the hyperbolic model were 0.96, 0.335 mm, 0.98 and 0.250 mm for 60% and 80% RD respectively. The overall behavior suggested that the load eccentricity effect was further at higher RD. As the geogrid layer increased from 0 to 4, the horizontal displacement of the strip footing decreased by 62%, 14% for $5°$ load inclination and by 54%, 69% for $15°$ load inclination for RD of 60% and 80% respectively. Hyperbolic model parameters that used to predict the relationship between the horizontal displacement of the strip footing and N for different load inclination (Fig. 9) can be shown in Table 4.

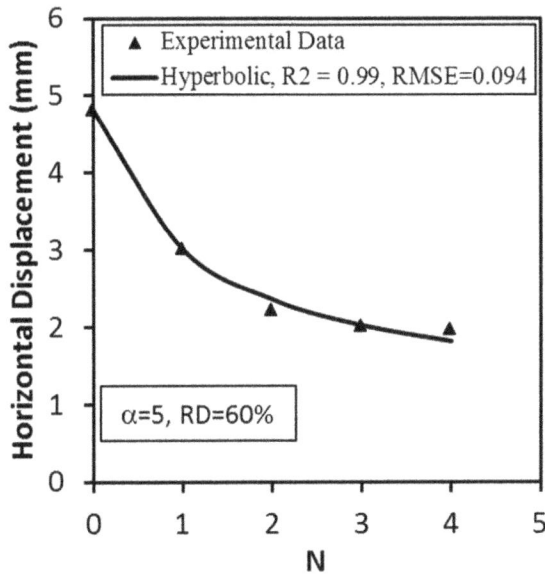

(c)

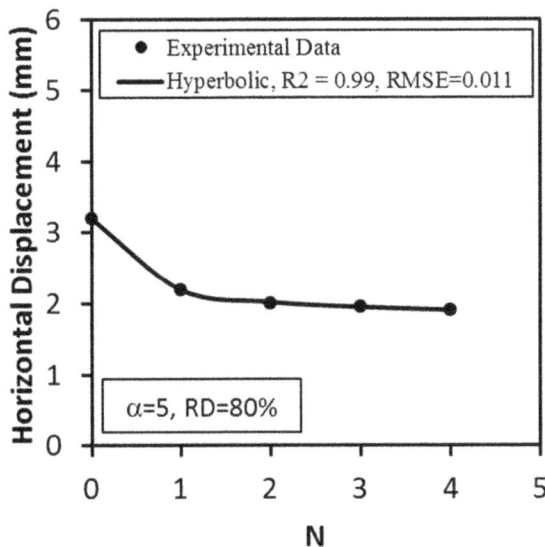

(a)

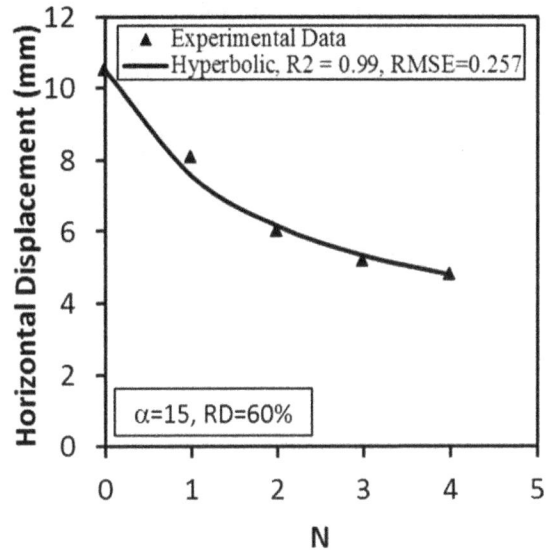

(d)

Fig. 9. *Modeling of horizontal displacement versus N relationship of strip footing rested on sandy soil (a) $\alpha=5$, RD=60%, (b) $\alpha=5$, RD=80%, (c) $\alpha=15$, RD=60%, and (d) $\alpha=15$, RD=80%.*

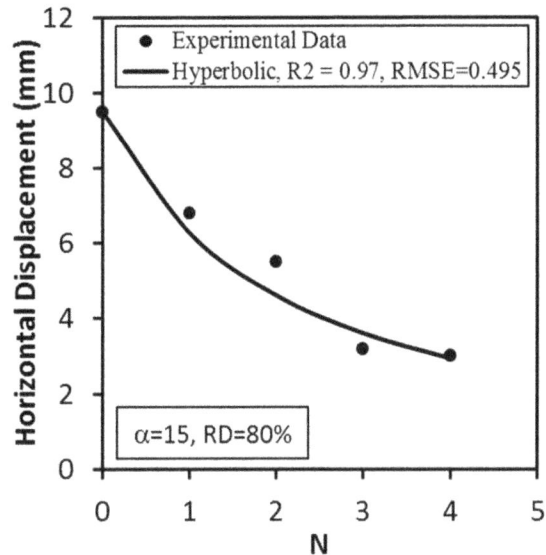

(b)

Table 4. *Model parameters of hyperbolic model (Fig. 9).*

α	RD(%)	A	B	R^2	RMSE (mm)
5	60	-0.3	-0.26	0.99	0.094
5	80	-0.3	-0.7	0.99	0.011
15	60	-0.22	-0.12	0.99	0.257
15	80	-0.21	-0.1	0.97	0.495

6.3. Tilting Angle Versus N

In this section, the strip footing tilting angle versus N for load eccentricity (e/B) varied from 0.05 to 0.15 of two different RD (60% and 80%) can be demonstrated obviously in Fig. 10 (a to d). In Fig. 10 (a) and 10 (b), the relationship between the strip footing tilting angle and N for 0.05 load eccentricity of strip footing in 60% and 80% RD are shown respectively. As the RD of the soil increased from 60% to

80%, the average decrease in the strip footing tilting angle was about 21%. The hyperbolic model predicted the experimental data preciously with R^2 and RMSE of 0.95, 0.081°, 0.98 and 0.116° for 60% and 80% RD respectively. At higher load eccentricity (e/B=0.15), the strip footing tilting decreased as the RD of the soil increased and the average decrease was about 17% as shown in Fig. 10 (c) and 10 (d). The R^2 and RMSE of the hyperbolic model were 0.98, 0.018°, 0.99 and 0.044° for 60% and 80% RD respectively. The overall behavior implied that having higher RD decreases the strip footing tilting angle regardless to the number of the geogrid layers. As the geogrid layer increased from 0 to 4, the strip footing titling angle increased by 90%, 275% for 0.05 load eccentricity and by 12%, 16% for 0.15 load eccentricity for RD of 60% and 80% respectively. Hyperbolic model parameters that used to predict the relationship between the footing tilting angle and N for different load eccentricity (Fig. 10) can be shown in Table 5.

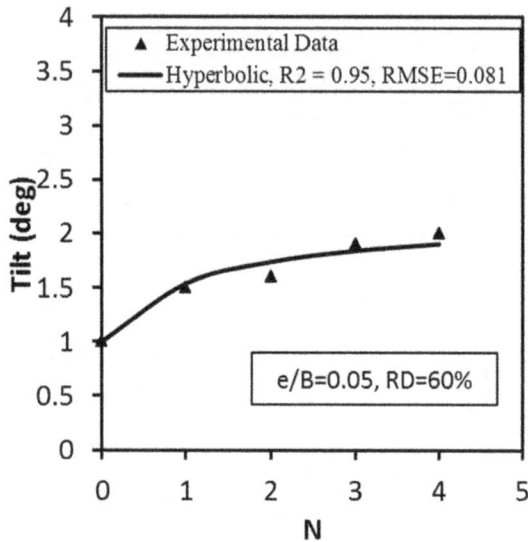

(a)

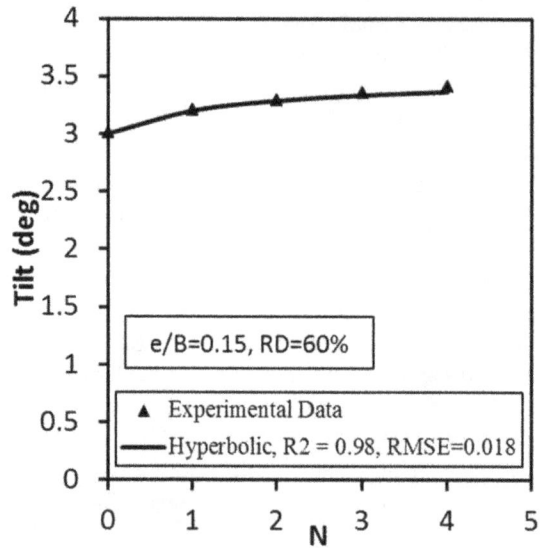

(b)

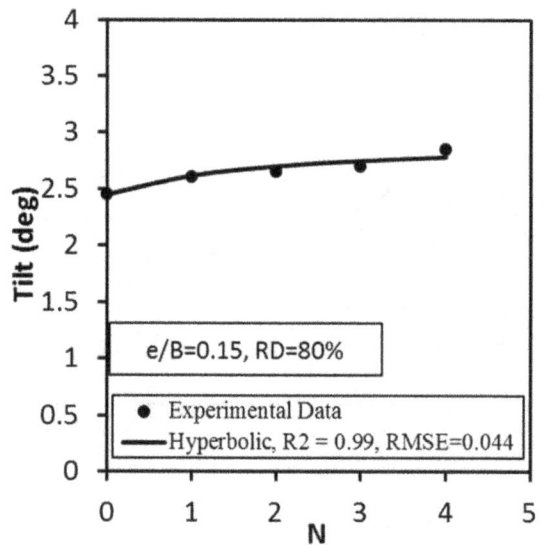

(c)

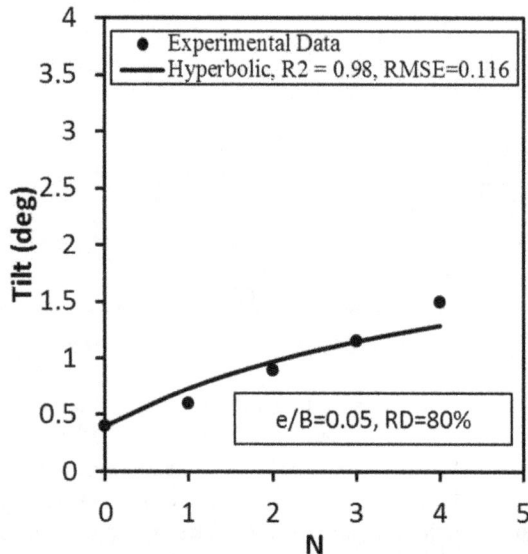

(d)

Fig. 10. *Modeling of footing tilting versus N relationship of strip footing rested on sandy soil (a) e/B=0.05, RD=60%, (b) e/B=0.05, RD=80%, (c) e/B=0.15, RD=60%, and (d) e/B=0.15, RD=80%.*

Table 5. *Model parameters of hyperbolic model (Fig. 10).*

e/B	RD(%)	A	B	R^2	RMSE (°)
0.05	60	1	0.86	0.95	0.081
0.05	80	2.5	0.5	0.98	0.116
0.15	60	3	2	0.98	0.018
0.15	80	4	2	0.99	0.044

6.4. BCR Versus (U/B)

In this section, the BCR versus U/B for load inclination angle (α) varied from 0° to 15° of two different RD (60% and 80%) can be presented clearly in Fig. 11 (a to d). In Fig. 11 (a) and 11 (b), the relationship between BCR and U/B for 0° load inclination angle of strip footing in 60% and 80% RD are shown respectively. As the RD of the soil increased from 60% to 80%, the maximum decrease in BCR was about 14%. The p-q model predicted the experimental data preciously with R^2 and RMSE of 0.97, 0.0238, 0.91 and 0.0219 for 60% and 80% RD respectively. At higher load inclination angle ($\alpha=15°$), the BCR decreased as the RD of the soil increased and the maximum decrease was about 22% as shown in Fig. 11 (c) and 11 (d). The R^2 and RMSE of the p-q model were 0.96, 0.0435, 0.96 and 0.0199 for 60% and 80% RD respectively. The overall behavior suggested that the load inclination angle effect was less at higher RD. P-q model parameters that used to predict the relationship between BCR and U/B for different load inclination (Fig. 11) can be shown in Table 6.

Table 6. Model parameters of p-q model (Fig. 11).

α	RD(%)	p	q	R^2	RMSE
0	60	1.1	1.15	0.97	0.0238
0	80	8	1.8	0.91	0.0219
15	60	0.9	1.2	0.96	0.0435
15	80	2	1.5	0.96	0.0199

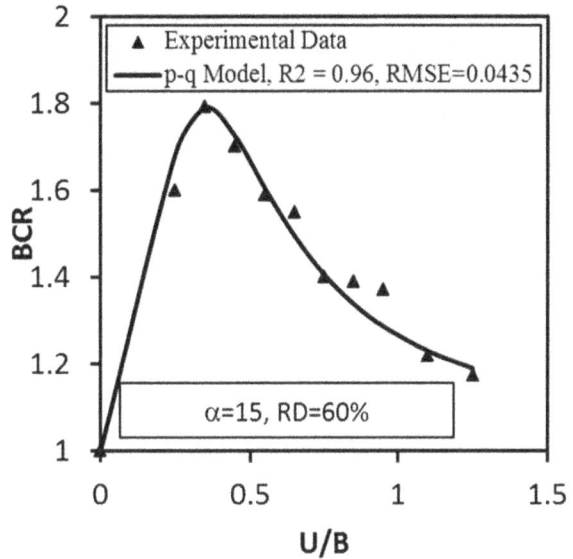

(a)

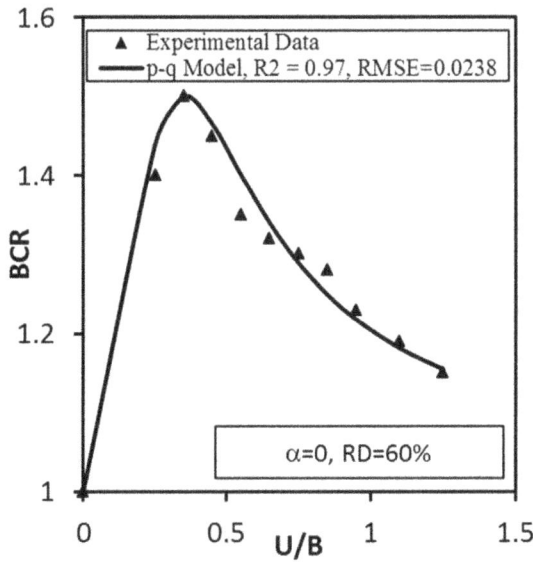

(b)

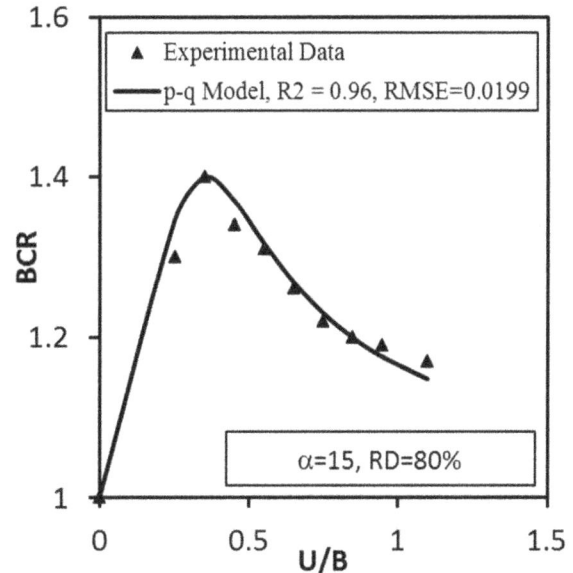

(c)

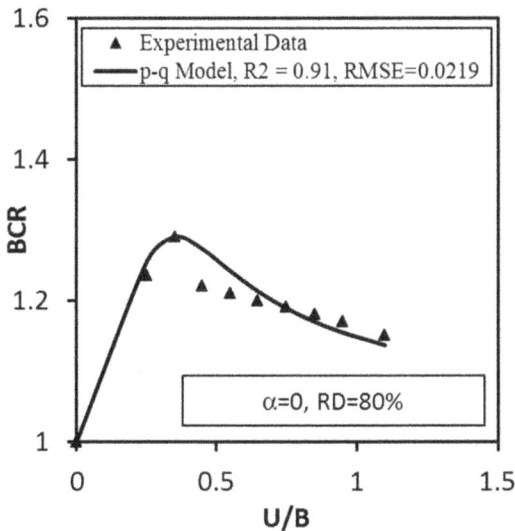

(d)

Fig. 11. Modeling of BCR versus U/B relationship of strip footing rested on sandy soil (a) α=0, RD=60%, (b) α=0, RD=80%, (c) α=15, RD=60%, and (d) α=15, RD=80%.

7. Conclusions

Based on the main results of the study, the following conclusions can be advanced:

1. Using geogrid for soil reinforcement has a significant impact in increasing the ultimate bearing capacity of the cohesionless soil.

2. Increasing the number of geogrid layers (N) increases the ultimate bearing capacity ratio notably and this increase reaches 255% for 15° load inclination angle and 470% for 0.05 load eccentricity ratio in 60% RD.

3. Having higher numbers of geogrid layers (N) decreased both the horizontal displacement and the footing tilt. When the RD of the soil increased from 60% to 80%,

the average decreases in the horizontal displacement and footing tilting were about 35% and 21% respectively.

4. The optimum value for U/B was about 0.5 and the BCR at this value (optimum) decreased as the RD increased. When the RD of the soil increased from 60% to 80%, the BCR at optimum U/B (0.5) decreased by 14% and 22% for 0° and 15° load inclination angle respectively.

5. The main factors affecting the ultimate bearing capacity of a strip footing under inclined and eccentric load on geogrid-reinforced sand can be addressed as follows:

a The load inclination angle (α)
  - Increasing (α) decreased the ultimate bearing capacity.
  - Increasing (α) increased the horizontal displacement of the footing.

b The load eccentricity ratio (e/B)
  - Increasing (e) decreased the ultimate bearing capacity.
  - Increasing (e) increased the tilting of the footing.

c *The Relative density (RD)*
  - Increasing (RD) increased the ultimate bearing capacity.
  - Increasing (RD) decreased the horizontal displacement of the footing.

Most of the studied relationships such as BCR versus N for different load inclination angle, BCR versus N for different load eccentricity ratio, footing horizontal displacement versus N and footing tilting versus N were modeled using hyperbolic model. However, p-q model was used to model the relationship between the BCR versus U/B. Both suggested models (hyperbolic and p-q) were in a very good agreement with the experimental results.

Acknowledgment

The civil engineering department at the University of Tikrit in Iraq supported the experimental part of this study. This support is gratefully acknowledged.

References

[1] El Sawwaf M. (2009). "Experimental and Numerical Study of Eccentrically Loaded Strip Footings Resting on Reinforced Sand," Journal of Geotechnical and Geoenvironmental Engineering ASCE, 135(10), 1509-1517, DOI: 10.1061/ASCEGT.1943-5606.0000093.

[2] LU Liang, Wang Zong-Jian and K. Arai (2014). "Numerical and Experimental Analyses for Bearing Capacity of Rigid Strip Footing Subjected to Eccentric Load," J. Cent. South Univ., 21, 3983–3992, DOI: 10.1007/s11771-014-2386-5.

[3] Dewaikar D. M., Guptha K. G. and Chore H. S. (2011). "Behavior of Eccentrically Loaded Model Square Footing on Reinforced Soil: An Experimental Investigation," Proceedings of Indian Geotechnical Conference, December 15-17, Kochi (Paper No. D-380).

[4] Meyerhof G. G. (1953). "The Bearing Capacity of Footings under Eccentric and Inclined Loads," Proc., 3rd Int. Conf. on Soil Mech. and Found. Engrg., 1, 440–445.

[5] Mahiyar H. and Patel A. N. (2000). "Analysis of Angle Shaped Footing under Eccentric Loading," J. Geotech. Geoenviron. Eng., 126(12), 1151–1156, DOI: 10.1061/(ASCE)1090-0241(2000)126:12(1151).

[6] Zhang M. X., Qiu C. C., Javadi A. A. and Zhang S. L. (2014)." Model Tests on Reinforced Sloped Embankment with Denti-Strip Inclusions under Monotonic Loading," KSCE Journal of Civil Engineering, 18(5), 1342-1350, DOI: 10.1007/s12205-014-0222-y.

[7] Guido V. A., Chang D. K., and Sweeney M. A. (1986). "Comparison of Geogrid and Geotextile Reinforced Earth Slabs," Can. Geotech. J., 23(4), 435–440., DOI: 10.1139/t86-073.

[8] Huang C. C. and Tatsuoka K. (1990). "Bearing capacity of reinforced horizontal sandy ground," Geotextile and Geomembranes, 9(1), 51-82.

[9] Yoo W., Kim B. and Cho W. (2015). "Model Test Study on the Behavior of Geotextile-Encased Sand Pile in Soft Clay Ground," KSCE Journal of Civil Engineering, 19(3), 592-601, DOI: 10.1007/s12205-012-0473-4.

[10] Chakraborty D. and Kumar J. (2014). "Bearing Capacity of Strip Foundations in Reinforced Soils," International Journal of Geomechanics, 14(1), February 1, 45-58. DOI: 10.1061/(ASCE)GM.1943- 5622.0000275.

[11] Yetimoglu T., Jonathan T. H. Wu and Ahmet S. (1994). "Bearing Capacity of Rectangular Footings on Geogrid-Reinforced Sand," Journal of Geotechnical Engineering, 120(12), 2083-2099, DOI: 10.1061/(ASCE)0733-9410(1994)120: 12(2083).

[12] Binquet J. and Lee K. L. (1975). "Bearing Capacity Tests on Reinforced Earth Slabs," J. Geotech. Engrg. Div., 101(12), 1241–1255.

[13] Fragaszy R. and Lawton E. (1984). "Bearing Capacity of Reinforced Sand Subgrades," J. Geotech. Engrg., 1500–1507, DOI: 10.1061/(ASCE)0733-9410(1984)110:10(1500).

[14] Khing K., Das B. M., Puri V. K., Cook E. E., and Yen S. C. (1993). "The Bearing Capacity of A Strip Foundation on Geogrid-Reinforced Sand," Geotext. Geomembr., 12(4), 351–361.

[15] Omar M. T., Das B. M., Puri V. K., and Yen S. C. (1993). "Ultimate Bearing Capacity of Shallow Foundations on Sand With Geogrid Reinforcement," Can. Geotech. J., 30(3), 545–549, DOI: 10.1139/t93-046.

[16] Shin E. C., Das B. M., Puri V. K., Yen S.-C. and Cook E. E. (1993). "Bearing Capacity of Strip Foundation on Geogrid-Reinforced Clay," J. ASTM Geotech Test., 16(4), 534–541, Paper ID GTJ10293J.

[17] Das B. M., Shin E. C. and Omar M. T. (1994). "The Bearing Capacity of Surface Strip Foundation on Geogrid-Reinforced Sand and Clay—A Comparative Study," Geotech. Geol. Eng., 12(1), 1–14.

[18] Das B. M. and Omar M. T. (1994). "The Effects of Foundation Width on Model Tests For The Bearing Capacity of Sand With Geogrid Reinforcement," Geotech. Geol. Eng., 12(2), 133–141.

[19] Asaoka A., Kodaka T. and Pokhaerl G. (1994). "Stability Analysis of Reinforced Soil Structures Using Rigid Plastic Finite Element Method," Soils Found., 34(1), 107–118.

[20] Ochiai H., Otani J., Hayashic S. and Hirai, T. (1996). "The Pull-Out Resistance of Geogrids in Reinforced Soil," Geotextiles and Geomembranes, 14(1), 19-42, DOI: S0266-1144(96)00027- I.

[21] Ghazavi M and Lavasan A. A. (2008). "Interference Effect of Shallow Foundations Constructed on Sand Reinforced With Geosynthetics," Geotext Geomembr, 26, 404–415, DOI: 10.1016/j.geotexmem.2008.02.003.

[22] Reza N. and Ebrahim M. (2014). "Bearing Capacity of Two Close Strip Footings on Soft Clay Reinforced With Geotextile," Arab J Geosci, 7, 623–639, DOI: 10.1007/s12517-012-0771-7.

[23] Won M. S., Ling H. I. and Kim Y. S. (2004). "A Study of the Deformation of Flexible Pipes Buried Under Model Reinforced Sand," KSCE Journal of Civil Engineering, 8(4), 377-385, DOI: 10.1007/BF02829161.

[24] Vipulanandan C. and Kirshnan S. (1993). "XRD Analysis and Leachability of Solidified Phenol–Cement Mixtures," Cem. Concr. Res., 23,792–802, DOI: 10.1016/0008-8846(93)90033-6.

[25] Ata A. and Vipulanandan C. (1998). "Cohesive and Adhesive Properties of Silicate Grout on Grouted – Sand Behavior," J. Geotech. Geoenviron. Eng., 124(1), 38–44, DOI: 10.1061/(ASCE)1090-0241(1998)124:1(38)).

[26] Vipulanandan C., AhossinY. J. and Bilgin O. (2007). "Geotechnical Properties of Marine and Deltaic Soft Clays," GSP173 Adv. Meas. Model. Soil Behav., 1–13, DOI: 10.1061/40917(236)5.

[27] Usluogullari O., VipulanandanC. (2011). "Stress–Strain Behavior and California Bearing Ratio of Artificially Cemented Sand." J. Test. Eval., 39(4), 1–9, Paper ID JTE103165.

[28] Vipulanandan C., Raheem A. M., Basirat B., Mohammed A. and Richardson D. (2014)." New Kinetic Model to Characterize the Filter Cake Formation and Fluid Loss in HPHT Process", OTC, 25100-MS, Houston, TX, 5-8 May, 1-17, DOI: 10.4043/25100-MS.

[29] Raheem A. M and Vipulanandan C. (2014). "Effect of Salt Contamination on the Bentonite Drilling Mud Shear Strength and Electrical Resistivity," THC Proceedings Conference & Exhibition, Houston, TX, USA.

[30] Vipulanandan C. and Raheem A. M. (2015). "Rapid Detection of Salt Contamination in Bentonite Drilling Mud in Deep Oil Well Applications," AADE National Technical Conference and Exhibition, San Antonio, Texas, April 8-9, 15-NTCE-30, pp. 1-7.

[31] Mebarkia S. and Vipulanandan C. (1992). "Compressive Behavior of Glass-Fiber-Reinforced Polymer Concrete," J Mater Civ Eng, 4(1), 91–105, DOI: 10.1061/(ASCE)0899-1561(1992)4:1(91).

10

Determination the Flow Net through Multi Layers Soil by Using the Hydraulic Modeling Method

Hala Kathem Taeh Alnealy

Civil Engineering Department, Babylon University, Iraq

Email address:

halhhalh300@yahoo.com

Abstract: In this research the experimental method by using Hydraulic modelingused todetermination the flow net in order to analyses seepage flow through multi- layer soil foundation underneath hydraulic structure.as well as steady theconsequence ofthe cut-off inclination angle on exit gradient, factor of safety, uplift pressure and quantity of seepage by using seepage tank were designed in the laboratory with proper dimensions with two cutoffs. The physical model (seepage tank) was designed in two downstream cutoff angles, which are (90, and 120°) and upstream cutoff angles (90, 45, 120°). After steady state flow the flow line is constructed by dye injection in the soil from the upstream side in front view of the seepage tank, and the equipotential line can be constructed by piezometer fixed to measure the total head. From the result It is concluded that using downstream cut-off inclined towards the downstream side with Θ equal 90° that given value of redaction (25%) is beneficial in increasing the safety factor against the piping phenomenon.using upstream cut-off inclined towards the downstream side with Θ equal 90° that given value of redaction (31%) is beneficial in decreasing uplift pressure and quantity of seepage.

Keywords: Flow Net, Inclined Cut-Off, Hydraulic Structures

1. Introduction

Hydraulic structures like barrages, dams, regulators, weirs, sheet pile wall.etc, area specific type of engineeringstructures designed and executed in such a way in order to operate them to control natural water or save industrial sources to guarantee optimumuse of water. These structures are frequently build on soil materials and the foundation thickness must be thick so as to be safe against uplift pressure (AL-Ganaini 1984*).*The analysis and design of foundation, as compared with other parts of the structure, should be given greatest importance because failure in the foundation would destroy the whole structure.the differential head in water levels between theupstream and downstream acts on the foundation and causes seepage flow (Selim 1947). Different methods solution can be used to analysis the seepage problem such as experimental works using physical model as well as numerical models (EL-Fitiany et al 2003). A flow net is in fact a solution of Laplace's equation in two dimensions. The model of seepage tank (sand tank)is very useful in studying the conditions of fluid flow under the hydraulic structure (Roy 2010). The hydraulic modeling method to determine the flow net in this study represented by

a physical model was built to study the phenomenon of seepage through soil using different types of soil under the hydraulic structure foundation (multi layers soil). And the factor of safeties will be found for each position.

2. Literature Review

Various theories and investigations were put forward to predict seepage phenomenon and determination flownet by experimental and numerical methods.Abourohim (1992) investigate the effect of seepage underneath the structures that generated uplift pressure acting on structures with: i) simple floor and ii) floor having an intermediate sheet pile by using a sand model. He concluded that such an effect becomes negligible when the canal width exceeds 2.6 times the length of the floor of the structure. Desai and Christian, (1977): studied the seepage through a two layered foundation of a dam, within each layer the soil was assumed homogenous,they used the finite element method. The computed values from the finite element method were compared with those from graphical solutions. Zheng-yi. and Jonathan, (2006): used the finite elementmethod to analyze seepage through a two-layer soil system. The program SEEP was employed to analyze flow characteristics of an impervious dam with sheet pile on a layered soil. The results were reduced to simple charts, The chart curves allow a designer to obtain solutions to the seepage problem. and can be extended to a soil system comprising more than two layers.Arslan and Mohammad, (2011)measured the pizometric head distribution under hydraulic structures and studied the effect of upstream, intermediate and downstream sheet piles inclination. using experimental method. The study consisted 12 separated case of these inclined sheet piles with changing the direction of this inclination. they foundthe optimum case of the uplift pressure reduction.

3. Objectives of the Study

The main objectives of this work can be summarized by the following points:

1. Equipotential lines and flow lines were located by conducting experiments on a known sample in a seepage tank and for multi-layers soil used in research work.
2. Study the effect of inclined cutoffs at different angles of inclination on exit gradient, uplift pressure underneath the hydraulic structure quantity of seepage and find the best inclination angle of cut-off for upstream and downstream side of hydraulic structure for all types of soils placed in different position under the foundation of hydraulic structure used in research work.

4. Experimental Work

The results obtained by the present hydraulic model using the seepage tank that designed and carried out at the hydraulic laboratory of the Engineering College at Babylon University. The major purpose of the physical model adopted in the present research is to study the flow net and calculate the values of uplift pressure underneath the hydraulic structure, distribution of exit gradient, quantity of seepage for different types of soil at different position under the hydraulic structure foundation.

5. Engineering Properties of Soil Used in Research

The experimental soil sample was taken from Hilla city region. The soil profile of any region contains many soil horizons, the difference between these horizons is marginal (no-homogenous soil). In the model tests, the profile is assumed to consist of one horizon (homogenous soil) (Aziz 2008). The tests used here to analyze the soil specimen in order to determine soil distribution and other engineering properties were conducted as per the Unified Soil Classification System, also the hydraulic conductivity values of the soil samples was measured inlaboratory method.Table 1 shows a summary of the physical properties that measured of the soils which consist of three layers arranged in descending order.used in the present study.figure (1) shown the arrangedfor type of soil used in present study.

Table 1. Physical properties of used soils.

No of layer.	Type off soil Arrange in descending order	Void ratio	Unit weightKN/m3	Gs	Value of Hydraulic Conductivity m/sec
1	Well-graded clayey Sandy silt	0.77	19.5	2.66	$8.32*10-5$
2	Poorly-graded sand	1.35	16.73	2.63	$9.453*10-4$
3	Sandy silt	0.58	20.2	2.68	$4.899*10-4$

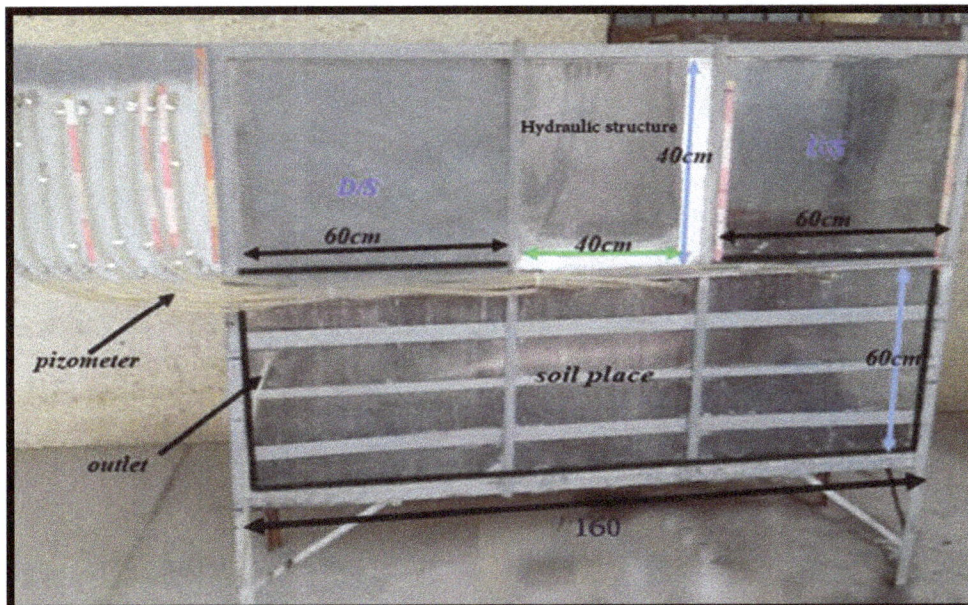

Figure (1). the seepage tank used in present study (front view).

Laboratory experiments have been conducted in a seepage tank that has been designed with hypnotically dimensions of 1.6 m long, 0.5m width and 1.1m height The bottom and sidewalls of this tank were made of Acrylic of (10mm) thickness. Figure (1)shows the seepage tank used in present study (front view). The bottom of the tank was filled with this material of soil to a depth of 60 cm. Acrylic walls were used to build the body of the superstructure which consists of two parts. The first part simulated as foundation of a structure (40 cm long × 50 cm wide).This base is connected with the upstream and downstream cutoffs by gluing rubber strips. 10 piezometers were placed at the right side of the tank at different location. All piezometers are fixing to the board.

6. The Experimental Procedures

1. Taking the datum to be at the bottom of the tank, Install the soil in the form of three layers each layer thickness of 0.2m and it is monitoring the process of using. the metal cylinder
2. Feeding the water to the seepage tank through the inlet hose until the water level in the upstream region reached the overflow hose level previously adjusted to meet the desired upstream water level.

3. After reachingsteady-state flow dye is injected from dye bottles which placed in the specific points , after a period of time flow lines were drawn Flux, which represents how the flow of water within the soil particles
4. After drawing flow lines, the vertical piezometers were installed transparent glass vertically into the soil to measure the total head in the points to draw the equipotential line.
5. Measure the discharge of drained water collected from the downstream funnel using the volumetric method by using jar.
6. Record the reading of the piezometric head of all installed piezometers under the base and downstream side.
7. Put the cutoffat upstream side with the angle of inclination for upstream(Θ=45, Θ=120, Θ=90) and repeat the step (7-8-9)to find the best angleto gave less value of uplift pressure and quantity of seepage.
8. Put the cutoffat downstream side with the angle of inclination (Θ=90, Θ=120) and repeat the step (7-8-9) to find the best angle to give max value factor of safety.
9. figure (2) shown the type of testing that made on multi - layer soil

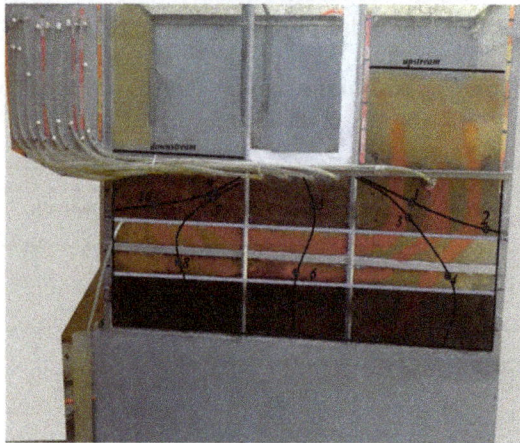

general case (without any cutoff)

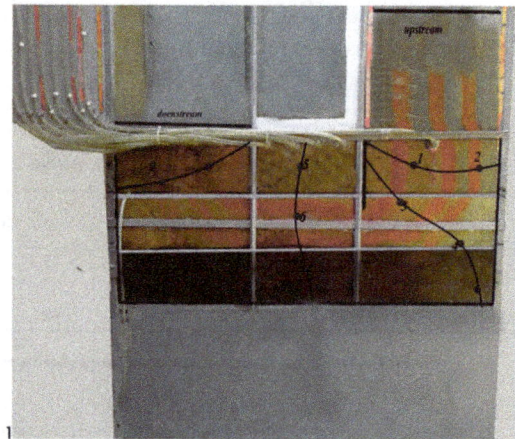

cutoff at the upstream side Θ=90

(cutoff at the downstream side Θ=90)

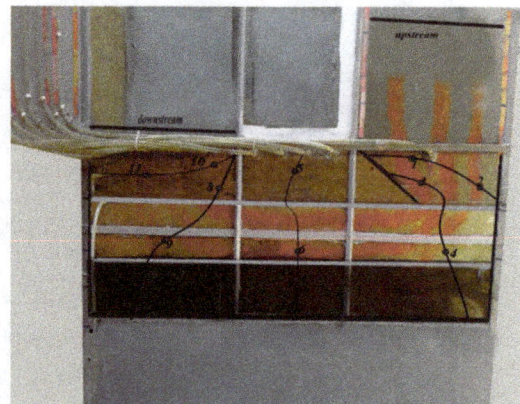

(cutoff at the upstream side Θ=45)

(cutoff at the upstream side Θ=120)   (cutoff at the downstream side Θ=120)

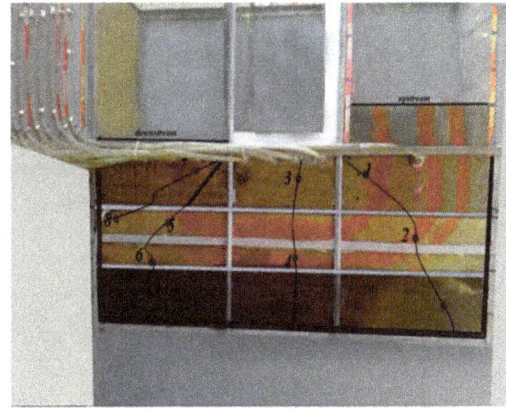

Figure (2). The model images for third experiment included hydraulic structure resting onmulti -layer soils.

7. Discussion of Results

Herein, the discussions of the results for multi layers soils according the following parameters:

7.1. Effect of Inclined Cutoff and its Position on the Uplift Head

As shown in figure (3). Whenthe Cutoff in upstream side of hydraulic structure was inclination with different angles, its noticed that the uplift pressure underneath the hydraulic structure decreases as(Θ) decreases towardU/S side for Θ(45° , 90°,120°) where the maximumredaction in uplift pressure according to the general case Θ=0(without any cutoff)was(-30% , -36% ,-23%) respectively ,so that the best angle is (45°).

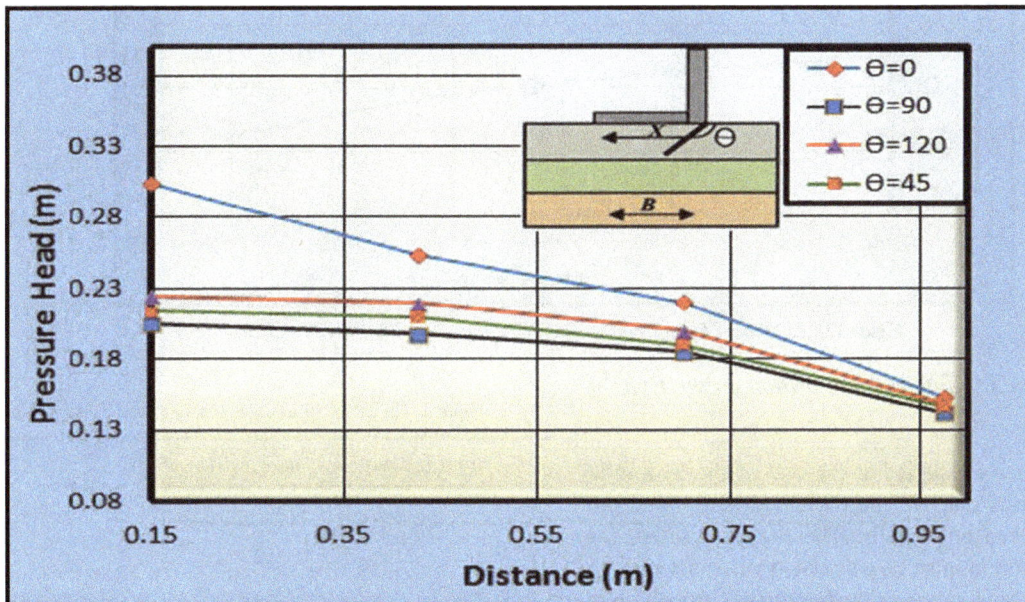

Figure (3). Uplift headfor different values of(Θ) values for cutoff in U/S.

From figure (4). When the Cutoff in downstream side of hydraulic structure is used, the uplift pressure obtained decreases as (Θ) decreases towardU/S side for Θ (90°and 120°) and the maximum redaction in uplift pressure according to the general case Θ=0 was (6%and5.4%). It is noticed that the redaction of the uplift pressure is small to the replacing ofthe cut-off, therefore, it is not recommended under any angle of inclination.

As cutoff positioned in U/S and D/S part of dam structure as shown figure(5), the uplift reduced strongly with maximum difference in value of uplift pressure was 43%.

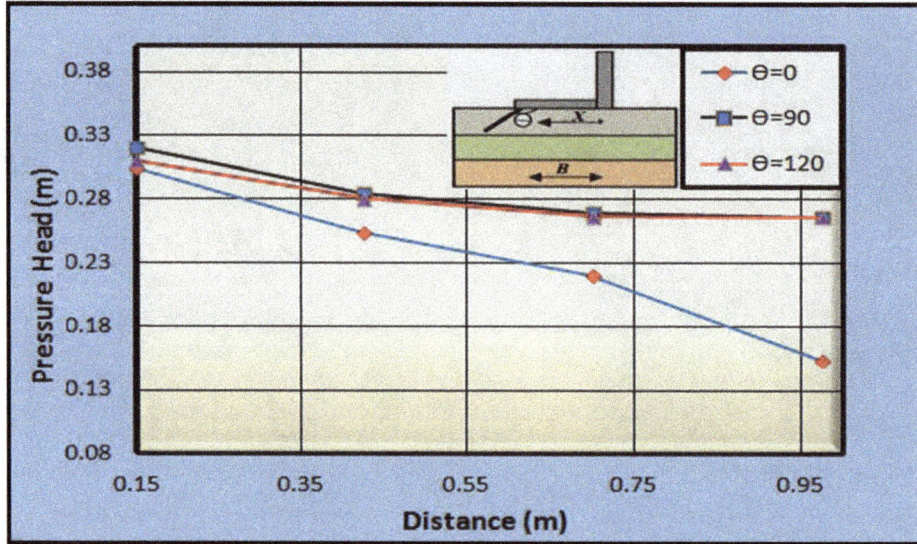

Figure (4). Uplift head for different values of(Θ) values for cutoff in D/S.

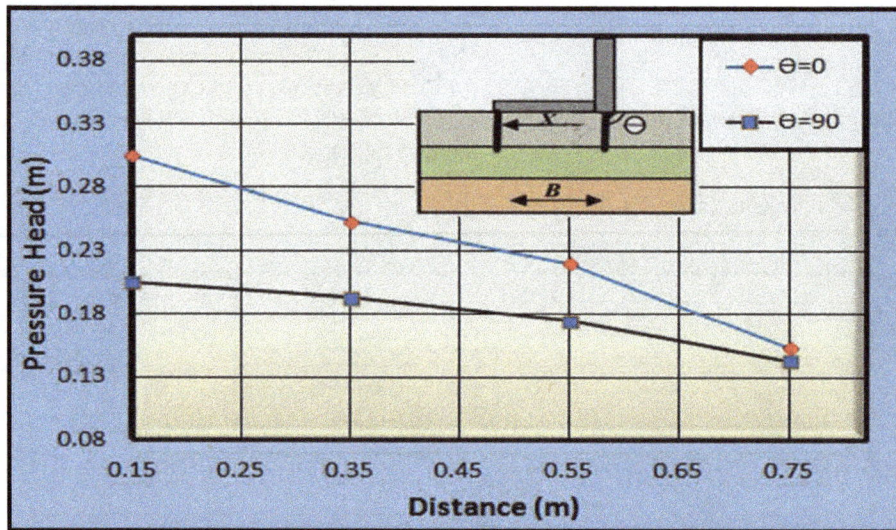

Figure (5). Uplift head for different values of(Θ) values for cutoff in U/S and D/S.

7.2. Effect of Inclined Cutoff and its Position on the Exit Gradient

The exit gradient was studied at the end of the hydraulic structure for the cases that will be discussed herein and the results are represented graphically. The factor of safety for each angle of inclination must be calculated where is equal to the division of exit gradient on critical gradient (Icr) which is dependent on the specific gravity (Gs) and void ratio (e) of the soil particles [Icr=(Gs-1)/(1+e)]. In this study andfor this type ofsoil(Gs =2.75 ,e =0.77, Icr =0.988).

In figure(6) whenthe cutoff put in upstream side of hydraulic structure it is found that the redaction in values of exit gradient were so small and as follows for Θ(45°,90°,120°) where the maximumredaction inaccording to the general case Θ=0(without any cutoff) (-9% , -9.3% ,-2.1%) respectively.

The result shows that the upstream cut-off inclination angle has no noticed effect on exit gradient. From the result the factor of safety against piping for this case were shown in table (2).

Table 2. The factor of safety against piping when cutoff atU/S.

U/S Cut-off Inclination	max (exist gradient)	Fs
0°	0.4	2.47
45°	0.382	2.586
90°	0.375	2.63
120°	0.362	2.729

In figure(7) whenthe cutoff is put it in downstream side for Θ (90°and 120°) the maximum redaction in exist gradientaccording to the general caseΘ=0 was (13.87% , 9%), respectively . So that the factor ofsafetyagainst piping phenomenonof this case can becalculated as shown in table (3).These results show thatusing cutoff in D/S sideinclination toward D/S for (Θ=120) increasing thefactor of safety against piping.

Table 3. *The factor of safety against piping when cutoff atD/S.*

U/S Cut-off Inclination	max (exist gradient)	Fs
0°	0.4	2.47
90°	0.35	2.82
120°	0.31	3.18

That exit gradientwhen the cutoff in the U/S andD/S part of dam structure decreases compare with general case as shown in figure (8) were maximum differencein value of exist gradient was 7%.

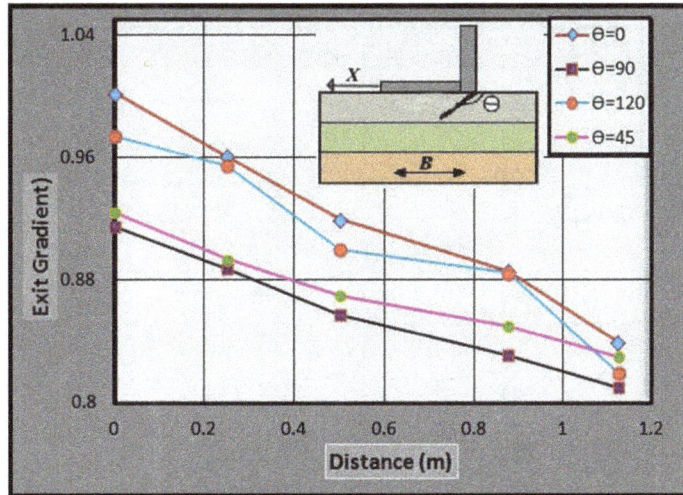

Figure (6). *Exit gradient values for different(Ө) values for cutoff in U/S.*

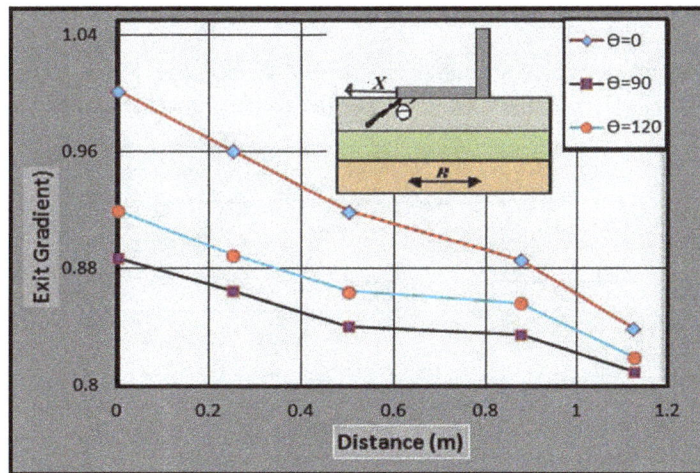

Figure (7). *The exit gradienta range of (Ө) values forcutoff in D/S.*

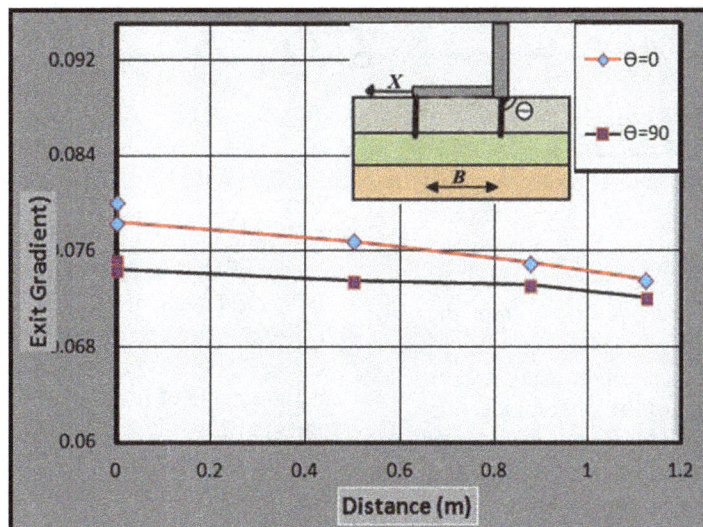

Figure (8). *The Exit gradienta range of (Ө) values for cutoff inU/S and D/S.*

7.3. *Effect of Cutoff Position and Inclination on the Seepage Quantity*

When theCutoff is in upstream side of hydraulic structure,as shown in figure(9), it's found that the seepage decreases while (Θ) decrease towardU/S , and the least quantity of seepage occurred when (Θ) value around (90).

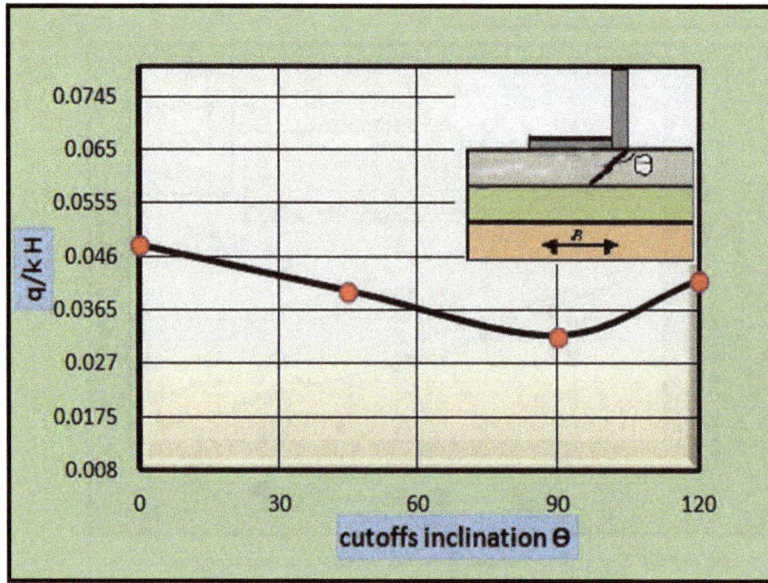

Figure (9). *The Seepage quantitya range of different values (Θ)for cutoff inU/S part of structure.*

In figure(10) Whenthe Cutoff is in upstream side of hydraulic structure the seepage decreases while (Θ) increases, and the least quantity of seepage occurred when (Θ) value around (120), then the seepage increases rapidly for (Θ ≤ 90).

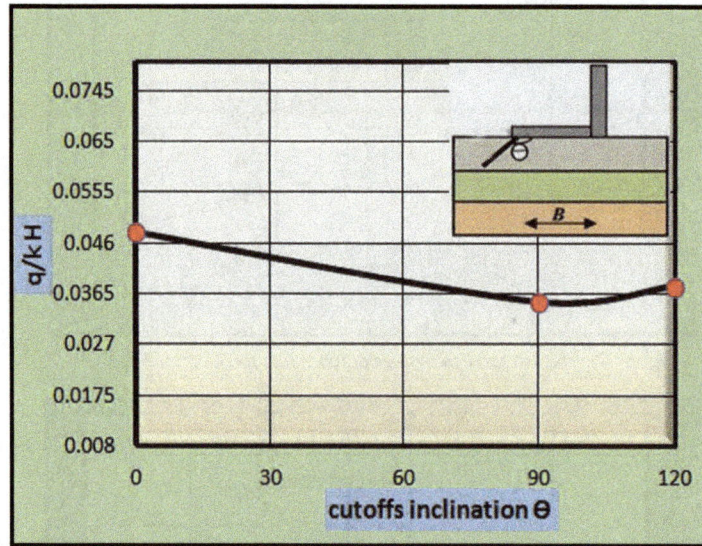

Figure (10). *The Seepage quantitya range of different values (Θ)for cutoff in D/S part of structure.*

8. Conclusions

The following main conclusions can be drawn from the results presented in this research:

1. The shapes of the flow net depend upon many factors such as diameter of the particle of the soil and location of cutoff.
2. The shape of the flow net obtained by experimental work (seepage tank model) and the result by slide model shows a good agreement.
3. The minimum value of uplift pressure is obtained when the cutoff used in upstream part of hydraulic structure is at the right angle of inclination(Θ= 90°).
4. The exit gradient in downstreamside and amount of seepage decreased to minimum when putting the cutoff at right angle of inclination for both up or downstream part cutoff.
5. Placing an inclined cut-off at the hydraulic structure heel and toe is not recommended under any angle of inclination.

6. Using double cutoff at the up and downstream part of dam at the right angle is useful to reduce the uplift pressure, exist gradient, and quantity of seepage at the same time.

References

[1] Abourohim, M. A., (1992) "Experimental study for the effect of seepage past Hydraulic Structures on the uplift pressure along the floo", Alexandria Engineering Journal, Volume 31, Issue 1.

[2] AL-Ganaini, M.A., (1984),"Hydraulic Structure", Beirut,pp47-60, (In Arabic).

[3] Selim, M.A., (1947) "Dams on Porous Media", Transaction ASCE, Vol. 1 Roy, S.K.(2010). "Experimental Study On Different Types Of Seepage Flow Under The Sheet Pile Through Indigenous Model.", M.Sc. Thesis Insoil Mechanics and Foundation Engineering, University Of Jadavpur.12, pp 488-526.

[4] Desai S.C. and Christian T.J. (1977). "Numerical methods in geotechnical engineering."McGraw-Hill Book Company, New York.

[5] Zheng-yi F. and Jonathan T.H.(2006). "The epsilon method: analysis of seepage beneath an impervious dam with sheet pile on a layered soil."NRC Research Press, Canada Geotechnical Journal Volume 43, P59–69.

[6] Aziz, L. J. (2008). "Lateral Resistance of Single Pile Embedded in Sand with Cavities."D.Ph Thesis, University of Technology, Iraq.

[7] EL-Fitiany, M. A. Abourohim, R. I. and El-Dakak, A. Y. (2003). "Three dimensional ground water seepage around a simplehydraulic structure." Alexandria Engineering Journal, Volume 42 ,Issue 5, September.

[8] Arslan, C. A. and Mohammad, S. A. (2011)."Experimental and Theoretical Study for Pizometric Head Distribution under Hydraulic Structures." Department of Civil engineering; College of engineering, University of kirkuk Volume 6, No.

[9] AL-Kubaisy,Y.K.Y.,(2004), " Effect of Soil Nonlinearity and Construction Sequence on the Behavior of Sheet Pile Wall", Department of Building and Construction Eng.,University of Technology ,Iraq.

[10] Das, B. M. (2008)."Advanced Soil Mechanics." Third Edition. Taylor and Francis, 270 Madison Ave, New York, NY 10016, USA.

11

Simulation analysis for schedule and performance evaluation of slip forming operations

Hesham Abdel Khalik[1], Shafik Khoury[1], Remon Aziz[1], Mohamed Abdel Hakam[2, *]

[1]Structural Engineering Department, Alexandria University, Alexandria, Egypt
[2]Construction Engineering and Management Department, Pharos University in Alexandria, Alexandria, Egypt

Email address:

heshamkhaleq@alexu.edu.eg (H. A. Khalik), Remon.aziz@alexu.edu.eg (R. F. Aziz), shafik.khoury@alexu.edu.eg (S. Khoury), Mohamed.hakam@Pua.edu.eg (M. A. Hakam)

Abstract: Slipforming operation's linearity is a source of planning complications, and operation is usually subjected to bottlenecks at any point, therefore, predicting construction duration is a difficult task due to the construction industry's uncertainty. Unfortunately, available planning tools do not carefully consider the variance and scope of the factors affecting Slipforming. Discrete-event simulation concepts can be applied to simulate and analyze construction operations and to efficiently support construction planning. The aim of this paper is to facilitate the adoption of DES and assist in determining most effective parameters that affect Slipform operation's duration in addition to better illustration of operation characteristics and overlapped parameters effects. To achieve this goal, a two-stage methodology for the development of an integrated simulation approach for Slipforming silo construction operations was proposed. Typical construction sequences in Slipforming construction were first identified, and then the statistical distributions of controlling activities on the sequences were surveyed. Subsequently, a DES model for predicting the duration of Slipforming construction was proposed, applied to a Slipform project and validated. The performance of the proposed model is validated by comparing simulation model results with a real case study showing average accuracy of 98.7%.Moreover research results defines the most effective factors arrangement that directly affects project schedule and to be taken in account by presenting the proportion of effectiveness of each value on research objectives. This research is considered beneficial for practitioners to estimate an overall construction schedule of building projects, especially in preconstruction phases.

Keywords: Simulation, Modeling, DES, Slipform, Fractional Factorial Design, Sensitivity Analysis, Temporary Structures

1. Introduction

Construction projects are usually delivered in an uncertain environment in which project resources and activities interact with each other in a complex manner [1] Therefore construction planning is the most challenging phase in the project development cycle due to the complicated, interactive, and dynamic nature of construction processes [2]. Modeling and simulation of construction process supports construction planning and can help in reducing the risks concerning budget, time and quality on a construction project [3].

Discrete-event simulation provides a promising alternative solution to designing and analyzing dynamic, complicated, and interactive construction systems [4]. Due to vertical Slipforming process's linear nature, it is considered a complicated process where it depends on efficient management of numerous parameters, moreover by considering the variability that always exists in construction operations, Slipform operations requires careful and thorough planning where Structure cross section; jacking rate; and concrete layer thickness can affect the Slipforming rate therefore project duration and so can the, pouring method, the site location, equipment location, and many other factors. Therefore, scheduling by coordinating the aforementioned parameters, resources of workers, machines and materials in a time-efficient way is required in order to realize the construction project within the anticipated time and budgeted costs[5].

Slipform process is considered a complicated construction process where it depends on efficient management of the five main parameters; the silo cross section (m), wall thickness (m), concrete layer thickness (cm), pouring method (Bucket, Pump or Hoist) and form jacking rate (cm/hr). Slipforming success key is to control the setting time of the fresh concrete so that the forms can be lifted at a predetermined speed and the concrete sets and hardens at the desired depth in the shallow forms [6]. Therefore, by considering the variability that always exists in construction operations, and in order to study the influence of the aforementioned parameters on productivity and duration of Slipform operations [7]. This research utilizes Simulation analysis using discrete event simulation for estimating and predicting the optimal and best combination of parameters of Slipform that affects the Slipform process. Simulation technique has been used widely in modeling construction operations in order to study different combinations of resources using different programming techniques and software [8], [9].Simulation analysis is difficult to implement, however, is very complex with respect to making simulation models [4].

Due to lack of research in Slipform application to concrete structures in Egypt, the presented research develops a realistic DES model for silo construction using Slipforms. To demonstrate the influence of various parameters those effect the Slipform operation's time and productivity, a full Silo construction operation model was developed in order to simulate the full project processes encountering all intermediate activities and resources. The model is generated and developed utilizing the "EZstrobe" Simulation Software. Simulation model is developed where the potential control units in a slip-form system are described for silos. Accordingly, this study can assist users and practitioners to develop a reliable schedule of Slipforming operation. The significance of this research can be described as, (1) Assists construction practitioners with a validated and adaptable simulation model for estimating and predicting the Slipforming projects duration; (2) Determine the most important factors that construction specialists should take in to consideration while preparation of project specifications and procedure which affect the vertical construction Slip forming operation and (3) Presents recommendations to the construction industries and the future development of vertical construction Slip forming operations.

2. Background

2.1. DES in Construction

To establish optimized construction operations, planners need to evaluate the productivity of each construction plan, including the sequence of various tasks and the efficient al location of resources based on decision support techniques. As decision support techniques, linear programming and construction simulation have been found to be remarkable tools. Linear programming tools have a limitation in the construction industry, however, since the results of linear programs do not include the sequence of tasks performed. Construction simulation is most beneficial during the preliminary phase when there are no data related to the

project, such as resource information, actual cycle times of each piece of equipment, actual costs, and productivity [10]. Discrete Event Simulation, referred to as simulation has proved to be an effective tool for complex processes analysis [11], besides being a well-established approach for analyzing, scheduling, and improving construction processes in the AEC arena [12]. The methodology of discrete-event simulation, which concerns "the modeling of a system as it evolves over time by a representation in which the state variables change only at a countable number of points in time" [13] provides a promising alternative solution to construction planning by predicting the future state of a real construction system following the creation of a computer model of the real system based on real life statistics and operations [14].

Many studies have introduced methods of evaluating time, and productivity for construction, such as a deterministic analysis, experimentation with the real system, mathematical modeling, simulation analysis, and various decision-making tools. Deterministic analysis, which has been utilized mostly because it is simple and easy to apply, is limited, however, in that it seldom resolves queuing or waiting-line problems [2]. Experimentation with the real system: On one extreme, is very realistic but is expensive, slow, lacks generality, and is sometimes impossible to do. Mathematical modeling on one extreme, is very precise but requires that important aspects of the process be disregarded, requires a high degree of mathematical ability, and becomes too complex for most real life construction situations. Simulation is the third technique it is very convenient because, while being realistic, it is also inexpensive, fast, and flexible. Simulation analysis has allowed construction planners and estimators to estimate and predict productivity and time and to evaluate construction operations by considering the variability that always exists in construction operations prior to the start of site work. Simulation analysis is difficult to implement, however, and is very complex with respect to making simulation models. Several simulation systems have been designed specifically for construction [16] and [17].. These systems use some form of network based on Activity Cycle Diagrams to represent the essentials of a model, and employ clock advance and event generation mechanisms based on Activity Scanning or Three-Phase Activity Scanning. These systems are designed for both simple (e.g., CYCLONE) and very advanced (e.g., STROBOSCOPE) modeling tasks but do not satisfy the need for a very easy to learn and simple tool capable of modeling moderately complex problems with little effort. EZstrobe is designed to fill this void in currently existing simulation tools and to facilitate the transition to more advanced tools (e.g. STROBOSCOPE) as the system is outgrown [18]. The basic modeling elements of EZstrobe are shown in Table 1, for the detailed understanding of EZstrobe program and its application, the reader can refer to "EZSTROBE General Purpose Simulation System Based on Activity Cycle Diagrams" [19]

Consequently, construction simulation allows experimentation on and evaluation of different scenarios in the phase of sensitivity analysis so that the user can get different system responses through various scenarios composed of changing resources and work task specifications. To identify any change in the productivity per unit of simulation models. This study adopted the EZstrobe modeling and programming technique to simulate this process. The elements of EZstrobe, originally developed by J. Martinez and Photios G. in 1994, are used to model and simulate slip-form operations. The program simplifies the simulation modeling process and makes it accessible to construction practitioners with limited simulation

For the modeling of construction operations, the determination of resource units associated with a construction operation, the basic work tasks with their related duration, and the resource unit flow routes over the

procedure, were acquired by data collection from numerous sources like Slipform previous researches and papers, real case studies, Construction productivity books, Sites Questionnaire, Slipform equipment used Data sheets and finally Online construction sites.

Table 1. EZstrobe Basic Modeling Elements

Element	Name	Description
NormalName Duration	Normal	Unconstrained in its starting logic and indicates active processing of (or by) resource entities.
CombiName Duration	Combi	Logically constrained in its starting logic, otherwise is similar to the NORMAL work task modeling element.
QueName	Queue	QUEUE Represents a queuing up or waiting for the use of passive state resources
	Fork	Probabilistic routing element. It typically follows an activity but can also follow another Fork.
>0 , 1	Draw Link	Connects a Queue to a Conditional Activity.
1	Release Link	Connects an Activity to any other node except a Conditional Activity.

2.2. Slipform Principles

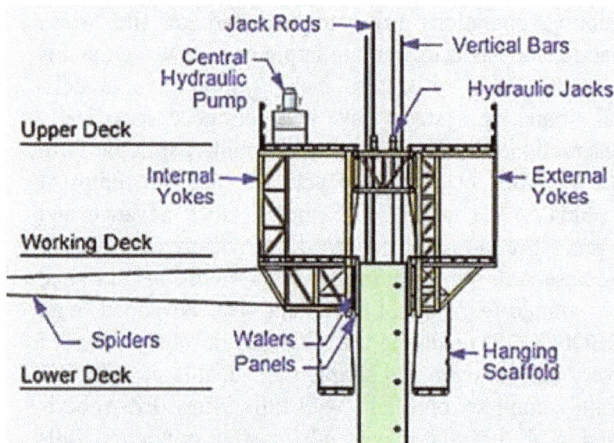

Figure 1. Slipform System Components

Slip-forming is a method of erecting silos by sliding up the whole form using an automated jacking device embedded in concrete and pouring continuously concrete, once concrete has developed early strength enabling it to stand by itself after placing. The essential elements of a Slipforming assembly are two parallel wall panels (about 1.2 m tall) supported by steel frames and horizontal yokes connected to hydraulic jacks as shown in Figure 1. After Slipform is

completely assembled on a concrete base, the forms are filled slowly with concrete. When the concrete in the bottom of the forms has gained sufficient rigidity, the upward movement of the forms is started and continued at a speed that is controlled by the rate at which the concrete sets. Many challenges face slip-form usage in the construction industry. The rate of movement of the forms is controlled and matches the initial setting of concrete so that the forms leave the concrete after it is strong enough to retain its shape while supporting its own weight. The forms move upward by mean of jacks climbing on smooth steel rods embedded in the hardened concrete and anchored at the concrete foundation base. These jacks may be hydraulic, electric, or pneumatic and operate at speeds up to 24 in. /h (609.6 mm/h). Lifting rates may vary from 2 or 3 in. per hr to in excess of 12 in. per hr, depending on the temperature and other properties of the concrete as shown in Figure 2. The Slipform rate is planned based on the concrete structure complexity, manning, the skills of the work force and limitation in the material supply. Modern Slipform technology enables a variety of shapes and forms to be produced to within strict geometrical tolerances. In general, the walls are vertical and of uniform thickness. If required, however, the shape and wall thickness can be varied in a seamless manner as the work progresses by means of screw type controls and overlapping wall panels. Slipform concrete construction was first used around the first of the century

(1904 - 1905) in Kansas City for a rectangular grain tank that was approximately 25 to 30 feet high. In the late 1920s, a number of concrete structures were cast using a system of formwork that was moving during the placing of concrete. Early application of Slipform was limited to storage bins and silos with a constant thickness all over the wall height. The slipping process has evolved from small hand screwed jacks with threaded rods (in which everyone turned one-quarter of a turn at the sound of the foreman's whistle) to pneumatic jacks developed sometime in the 1930's and then hydraulic jacks developed in the early 1940's.Since the late 1950s, Slipform construction has come a long way; locomotion is accomplished by jacks climbing on smooth steel rods or pipes anchored at the base of the structure. Accordingly, the list of recent application expanded to include silos, towers cores, bridge piers, power plant cooling, chimney shafts, pylons, and the legs of oil rig platforms. As discussed above, it is convenient to consider the Slipform operation as consisting of three main elements; the concrete, the batching, transporting, placing, compacting and curing of the concrete and the shape, size and speed of the Slipform. Therefore the basic criteria for selecting slip-form as a formwork method will be project time; required speed; cross section uniformity and height; number of openings; and necessary stoppages in the height. Various attempts have been introduced in order to illustrate a basic understanding for the process of Slipforming using different types of methods,[20], [21], , [7] ,[22], [23] and [24] these studies implemented different types of analysis and approaches. The risk of modifying operational information and having mistaken should be eliminated. Consequently, a jointly prepared and implemented Method Statement is an essential tool to achieve efficiency and quality. Resulting from the previous it can be clear that a common understanding of the process and efficient management are key factors for a successful operation.

3. Proposed Methodology

3.1. Overview

This research presents an integrated methodology for assisting in proper planning, estimating and prediction of construction duration generated by Slipform systems using EZstrobe software. To demonstrate the influence of Slipform technique implementations in construction industry, a Full Slip form Construction project was analyzed to estimate the proposed methodology's prediction of construction durations. The proposed system focuses on the structure part of Silo construction for the following elements "Earthworks, Foundations and Walls". A major benefit of using EZstrobe in simulation is that a simulation can be run for several times corresponding times of the duration of a project's completion will be automatically generated. The suggested model's major parameters under study for modeling are shown in Fig. 3. Traditionally, the simulation model data inputs are those of the real project parameters i.e. (structure geometric, material quantities, resources, costs and etc.). Simulation model's elements can be divided in to two main elements; Queues, that holds building quantities such as material quantities and resources amounts; and Combis that holds activity description and task durations. Queue's data quantities source is generated from material quantity surveying, while Combis durations are generated from a mathematical process conducted on quantities and productivity rates by user. The case study presented in Section 4 illustrates further details of the operations simulation that is applied in this work.

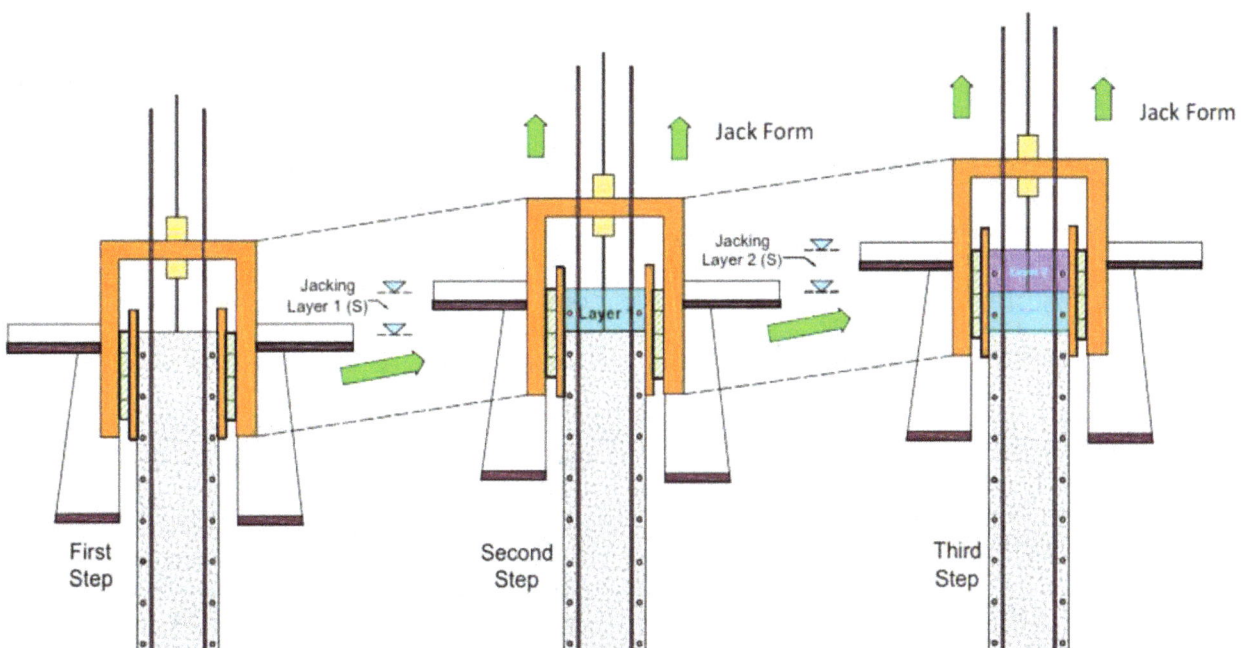

Figure 2. Silo Slipform Construction Process

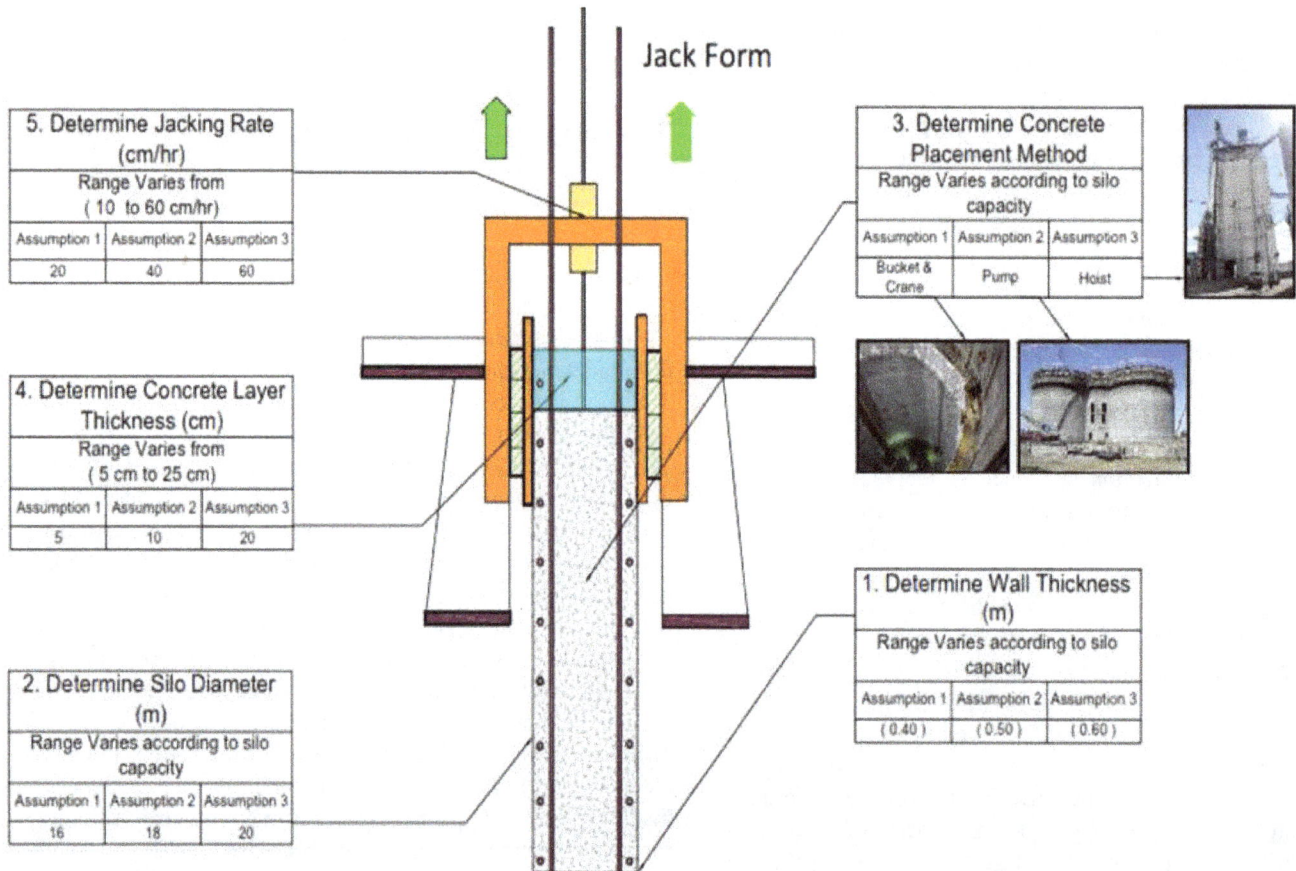

Figure 3. *Simulation Model Frame Work Inputs*

3.2. Modelling Assumptions

Based on the suggested project conditions the following assumptions are made for simulation modeling:

1. Model is based upon four stages of construction processes encountering Silo earthwork, Foundation work, Slipform Assembly and Slipforming works where each process covers the related activities as shown in Table 2.

2. In earthmoving operations, for general earthworks such as cut and dump operations, which were proposed in this study, the accuracy, degraded by multipath errors, was assumed to have been accepted and based upon a GPS system where surveying operations are not required [25].

3. Foundation work is excavated to a depth of 1.5 m below level for all instances and all scenario combinations as shown in sensitivity analysis.

4. Trucks used for hauling excavated earth to dumps are to have a capacity of 15 m3 per trip.

5. In foundation works, the concrete pouring is based upon a pump placement system with an average 3 to 4 m3/min.

6. For the Slipform assembly the total duration of assembly of Slipform is determined based upon 4 weeks to assemble.

7. The selection of placing system for the concrete has been studied based upon three different systems; conventional crane and bucket delivery with adequate crane capacity, pump lines into hoppers on the deck and hoist placement system.

8. For the Slipforming works the Concrete placement method average duration varies between (7.5 min/m³), (3 min/m³) and (22.5 min/m³) for crane and Bucket placement, Pump placement and hoist placement respectively.

9. The lifting rate is determined with regard to the supply of all the components and the availability of manpower, so that interdependent activities can proceed without hold ups.

10. Uniform Jacking stroke rates varies based upon the concrete setting time, the Slipform hydraulic pump capacity therefore

11. Jacking stroke rates is specified from 10 cm/ hr to 60 cm/hr.

12. The interval between placing increments is limited to a maximum of 60 min in all of the Slipform resources combinations.

13. Concrete is placed in increments that result in uniform horizontal bands that are not so deep as to prevent proper vibration of the concrete and rising of trapped air to the surface which varies from 5 cm to 25 cm.

14. The frame work for the model input data is furtherer illustrated in Figure 3 showing the previous main factors affecting the Slipform productivity and project completion date as factors from 1 to 5 as following

- Silo Diameter (SD)
- Wall Thickness (WT)
- Concrete Placing Method (CP)
- Layer Thickness (LT)
- Jacking Rate (JR)

15. Factorial design based on principle block is to be used for resources variation in sensitivity analysis with respect to 5 factors only which are (CP, LT, JR, SD and WT)

16. When estimating the daily productivity in the system, 60 min per hr was used for each 24 h during slip operation as no stoppage was allowed and work continued continuously.

Table 2. Simulation model sub models classification

No	Phase	Phase Model Circulation
1	Silo Earth Works	Excavating
		Hauling to Dump Areas
		Dumping
		Surface Finish
2	Silo Foundations Works	Foundation PC Work
		Foundation Rebar
		Foundation Form Work
		Foundation RC Placing
3	Silo Slipform Assembly	Slipform Assembly
		Raise and Connect Jacks
		Install Panels, Jacks
		Install Platform and Decks
4	Silo Slipforming Process	Slipform Jacking
		Concrete Placing
		Rebar and Steel Work
		Rebar Raise to Platform

4. Preliminary Prototype

As for this research case study, the preliminary prototype of the proposed integrated framework has been implemented and developed by using EZstrobe as the DES tool. Furtherer, to illustrate the framework model capabilities a Slipform case of study was chosen to apply the proposed software to a silo project located in Bandar Abbas, Iran for the Hormozghan cement factory project. The case of study chosen in current research is based upon Hormozghan cement factory project (Bandar Abbas, Iran). (Tarek Zayed 2008) It included Raw Meal Silos and towers with 6000-ton cement production per day. The silo was designed to store row material to feed a pre-heater tower that was used as a reserve for production line. Raw materials were transported from the Row Mill to the uppermost part of the silo. From there the row material flowed down a concrete cone, which distributed material to all outlets. All silos and towers of the cement factory were constructed using slip-form lifting system. Samarah Construction Co. (general civil contractor) performed the slip-forming part of this project. Silo has 16 meter- inner diameter, 50 cm thickness and 50 meters-height and a total Slipforming duration of 13.9 days. Concrete was poured using bucket and crane with a rate of 8 m³/hr. The data presented in Table 3 for time data collection is limited to the

Case study where for scenarios combination in sensitivity analysis the durations vary based on the scenario combination of resources

4.1. Simulation Model Input Data

Simulation model input data are inserted manually by the user in two main modeling elements as described earlier; first: Queue(s) elements input data that hold materials, labor and equipment quantities; and Second: Combi(s) elements input data that hold activities durations and attributes.

4.1.1. First; Material and Resources Quantities

Silo material quantities are estimated using traditional Quantity survey where QS estimates quantities from project drawings and provides the relevant parameters (such as length, width, height, area, and volume) needed to perform the quantity takeoffs of the weight (in tons) of steel, the area (m) of the forms, and the volume (m3) of concrete for each construction element (i.e., column, beam, wall, and slab). Table. 4 present a quantity table that summarizes the quantities of materials that will be needed for the tasks in the project. Moreover Resources can be categorized in to two main groups Labor and Equipment. Both labor and equipment amounts are shown in table 5 where it presents the resource type, resource name and quantity.

4.1.2. Second; Activity Durations

In order to create the durations input data, each project activity is assigned to duration with a certain distribution. Activities durations are shown in Table 6 where it presents the category of activity, activity name, type of duration and assigned duration. It is noted that due to the uncertainties in the industry both uniform and triangular distributions owns the majority of duration types as illustrated in the aforementioned table. Based on project case study and best practices the duration's types and distributions were chosen in the proposed model.

4.2. Simulation Model Development

An EZstrobe network for a full Slipform project has been created and developed. This network indicates the construction tasks, the logical links between tasks, and the resources required for the project. Specifically, the following sub models are encountered through building the simulation model the four models are connected through a fusion Queue for ease of studying each sub model. In addition, crews' formulation for concrete pouring, steel rebar and formwork crews are involved.

The methodology dramatically facilitates the generation of input data for a simulation. It is notable to mention that study considers the influence that uncertainties have on the productivity of construction tasks, therefore it is noted that durations have probabilistic durations of tasks. Moreover, there is competition among resources; for instance, concrete pumps are shared for the tasks of pouring concrete for PC beneath raft and for raft RC pouring as well as concrete crew workers. This competition is governed with a rule is that the first need receives the highest priority for being served. The following section illustrates the DES model phases.

4.2.1. Silo Earth Works Model

4.2.1.1. Model Description

Silo Earth works Model describes Earth works for the Slipform project, where it encounters the excavation works, hauling excavated soil to dumps and dumping, moreover the sub model shows compacting works for the excavated area. Silo earth work Sub model process is furtherer illustrated in Figure 4; amount of quantities and resources are shown in tables 4 and 5, while durations are shown in table 6.

4.2.1.2. Model Circulation

a. Silo ground area is excavated with required dimensions and to required level using a GPS machine guidance system where no surveying is required and accuracy is accepted.
b. Excavated material is hauled by truck to nearest dump, unloads and returns back to load excavated soil.
c. After excavation, the area is compacted using a compactor and level is ready for Foundation works.
d. Equipment used for earth work is (Truck, Loader, and Compactor) with the following activities durations Table 5.

4.2.2. Foundations Works Model

4.2.2.1. Model Description

Foundations Works Model describes typical foundation works for the Slipform project, where it encounters the plain concrete pouring under foundations, steel and rebar works for foundations, form work and Reinforced concrete pouring for foundations. Foundations Work Sub model process is furtherer illustrated in Figure 5, amount of quantities and resources are shown in tables 4 and 5, while durations are shown in table 6.

4.2.2.2. Model Circulation

a. Model starts with Pouring P.C for foundations on required level with thickness 0.15 m using the following resources (Pump placing System, Concrete Crew and PC Concrete Available). Placed concrete is left for 24 hours for final Hardening.
b. The Rebar and steel work starts after the finish of the PC hardening activity using the following resources (Steel crew and Rebar available)
c. After Rebar and steel work finish, the Formwork starts using the following resources (Foundation form work crew and Form work Available)
d. After Form work finish the RC pouring starts using the following resources (Pump placing System, Concrete Crew and RC Concrete Available)

4.2.3. Slipform Assembly Works Model

4.2.3.1. Model Description

Slipform Assembly Works Model describes the Slipform assembly process, where it encounters raising the jacks and connecting it using horizontal straps, installing panels, installing hydraulic jacks and platforms. Slipform assembly Sub model process is furtherer illustrated in Figure 6, amount of quantities and resources are shown in tables 4 and 5, while durations are shown in table 6.

4.2.3.2. Model Circulation

a. Slipform assembly starts with raising forms using Jacks and connecting it using Horizontal straps, after that Panels are installed using the Panels available, Then Hydraulic jacks are installed using hydraulic jacks available,
b. Finally Platforms Upper, Lower and working decks are installed using material available.
c. Activities for each activity are based upon average total duration for Slipform assembly divided by the number of activities. As Slipform assembly varies from 3 to 4 weeks based upon shape and geometry of silo structure

4.2.4. Slip Forming Works Model

4.2.4.1. Model Description

Slip forming Works Model describes the Slip forming operation sequence for Silo construction, the sub model encounters jacking forms based on jacking rates and layer thickness and concrete setting time, moreover concrete placement, rebar work and raising rebar to working decks. Slip forming Sub model process is furtherer illustrated in Figure 7; amount of quantities and resources are shown in tables 4 and 5, while durations are shown in table 6.

4.2.4.2. Model Circulation

a. The Slip forming model starts with the Jacking activity where jacking varies as shown in mentioned earlier from 10 cm/hr to 60 cm/hr; one jack height is almost 2 inch (5cm). Jacking activity duration depends mainly on the Slipform mechanical capacity, concrete setting time and concrete layer thickness. Jacking rate durations based on Layer thickness.
b. Then, Concrete is poured based on the pouring method which varies from Crane and buckets to Pumps and also depends on the silo cross section and jacking step.
c. When form is raised up to 20 cm, horizontal rebar is installed and Slipform gets ready for next jacking step.
d. Repeat steps 2 to 5 until the completion of concrete silo.
e. When form is raised up to 3 m the tower crane lifts reinforcements and embedded to working deck.

Table 3. Case of Study Durations

No	Activity	Duration (min)
1	Jacking Rate (min)	10
2	Concrete Placing (min)	Triangular [7,7.3,7.8] (Crane) Triangular[3.5,4,4.5] (Pump) Uniform[20,25] (Hoist)
3	Rebar Installation (min)	Triangular [7,8,9]
4	Material Lifting (min)	Triangular [5,6,7]

Table 4. Silo Project Material Quantities

ID	Phase	Material Type	Material Description	Unit	Quantity
1	Earthworks	Soil	Amount of Excavated Soil	m³	555
		Soil	Amount of Compacted Soil	m²	340
2	Foundation Works	Plain Concrete	Foundation PC Beneath Raft	m³	45
		Reinforced Concrete	Foundation RC Raft	m³	340
		Steel	Raft Rebar Amount	ton	50
		Formwork	Raft Formwork Amount	m²	60
3	Assembly Works	Slipform	Yokes Set for Slip Assembly	Set	1
		Slipform	Straps Set for Slip Assembly	Set	1
		Slipform	Panels Set for Slip Assembly	Set	1
		Slipform	Platforms Set for Slip Assembly	Set	1
4	Slipforming Works	Reinforced Concrete	Slipform Continuous Concrete Crew Number	m³	1300
		Steel	Slipform Continuous Steel Crew Number	ton	100

Table 5. Project Resources Amounts

ID	Phase	Resource Type	Resource Description	Unit	Amount
1	Earthworks	Equipment	Number of Excavators for Excavation	No.	1
		Equipment	Number of Trucks for Hauling	No.	1
		Equipment	Number of Compactors for Levelling	No.	1
2	Foundation Works	Labor	Foundation Formwork Crew Number	Crew No.	1
		Labor	Foundation Steel Rebar Crew Number	Crew No.	1
		Labor	Foundation Concrete Placing Crew Number	Crew No.	1
		Equipment	Number of Foundation Concrete Pumps	No.	1
3	Assembly Works	Labor	Slipform Techs Crew Number	Crew No.	1
		Equipment	Hydraulic Jacks Set Number	No.	1
4	Slipforming Works	Labor	Slipform Continuous Concrete Crew Number	Crew No.	1
		Labor	Slipform Continuous Steel Crew Number	Crew No.	1
		Labor	Slipform Finishing Surface Crew Number	Crew No.	1
		Equipment	Slipform Crane Available	No.	1

Table 6. Project Activities Durations

ID	Phase	Activity Name in DES	Activity Type	Duration Type	Duration [min]
1	Earth Work	LoadTruck	Combi	Uniform	[10,11]
		HaulTruck	Normal	Uniform	[5,7]
		DumpTruck	Normal	Uniform	[0.5,1]
		ReturnTruck	Normal	Uniform	[4.5,5]
		SitePrep	Combi	Triangular	[60,90,120]
		SurfaceFinish	Combi	Uniform	[60,90]
2	Foundation Work	PCFoundPour	Combi	Uniform	[3,4]
		WtConcHard	Combi	Uniform	[1440,1600]
		FoundRebarWrk	Combi	Uniform	[100,120]
		FoundFormWork	Combi	Uniform	[60,75]
		RCFoundPour	Combi	Uniform	[3,4]
3	Slipform Assembly Work	StartFormAss	Combi	Uniform	[5040,6720]
		RaiseForms	Combi	Uniform	[5040,6720]
		ConnectJacks	Combi	Uniform	[5040,6720]
		InstallPanels	Combi	Uniform	[5040,6720]
		HydJacksInstall	Combi	Uniform	[5040,6720]
		Install Platform	Combi	Uniform	[5040,6720]
4	Slip forming Model	JackForm	Combi	Deterministic	10
		PlaceConcrete	Combi	Triangular	[8.5,9.5,10.5]
		RebarandEmbd	Combi	Triangular	[7,8,9]
		RaiseSttoDeck	Combi	Triangular	[5,6,7]

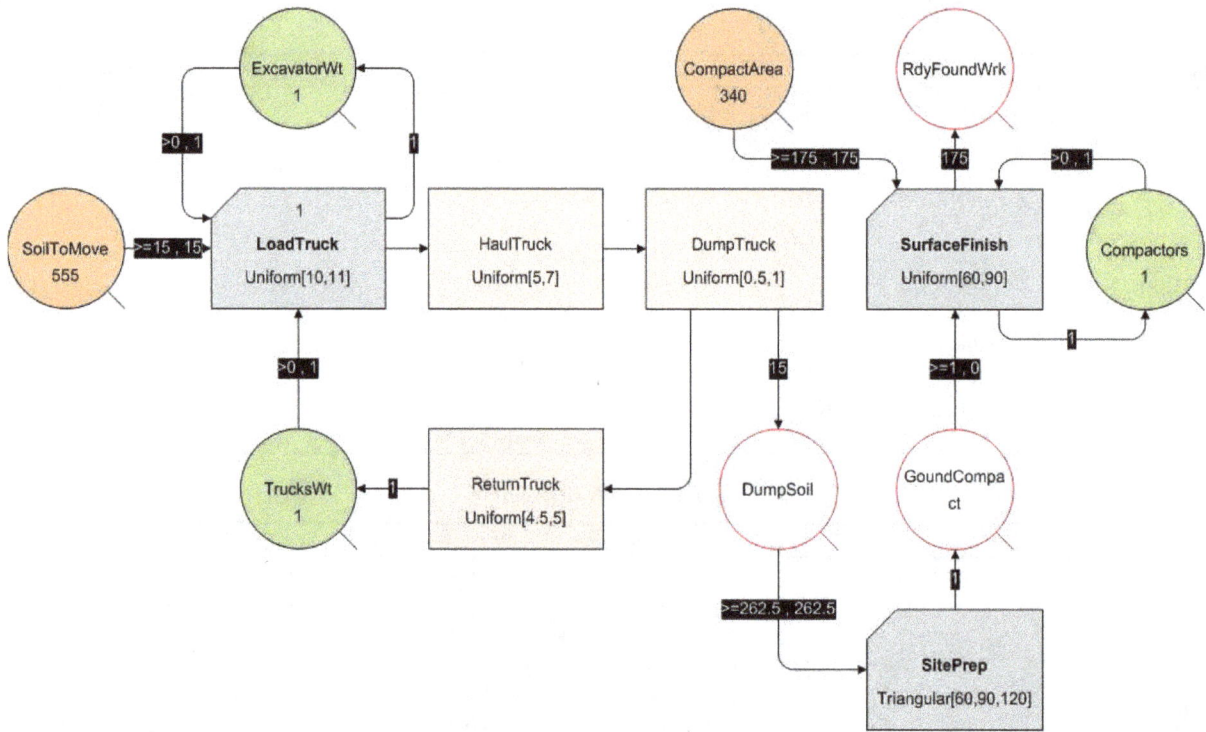

Figure 4. Slipform Earth Work Phase Model

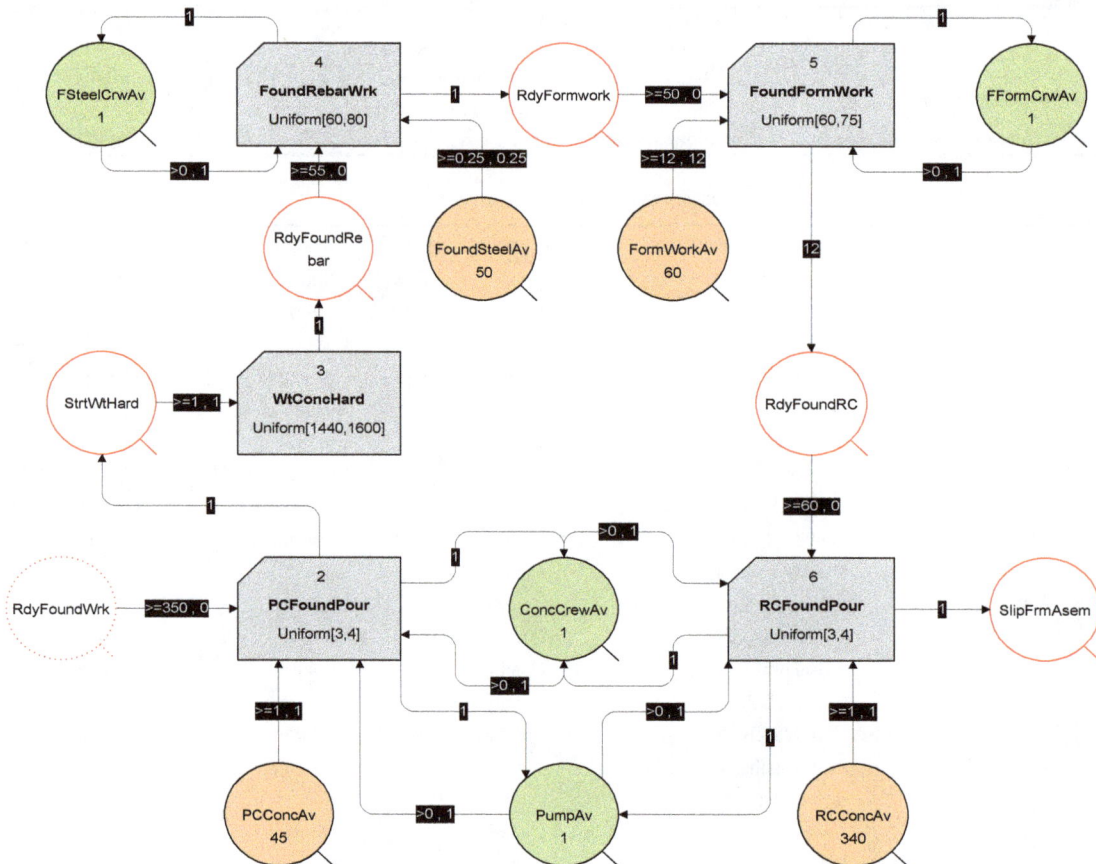

Figure 5. Slipform Foundations Works Phase Model

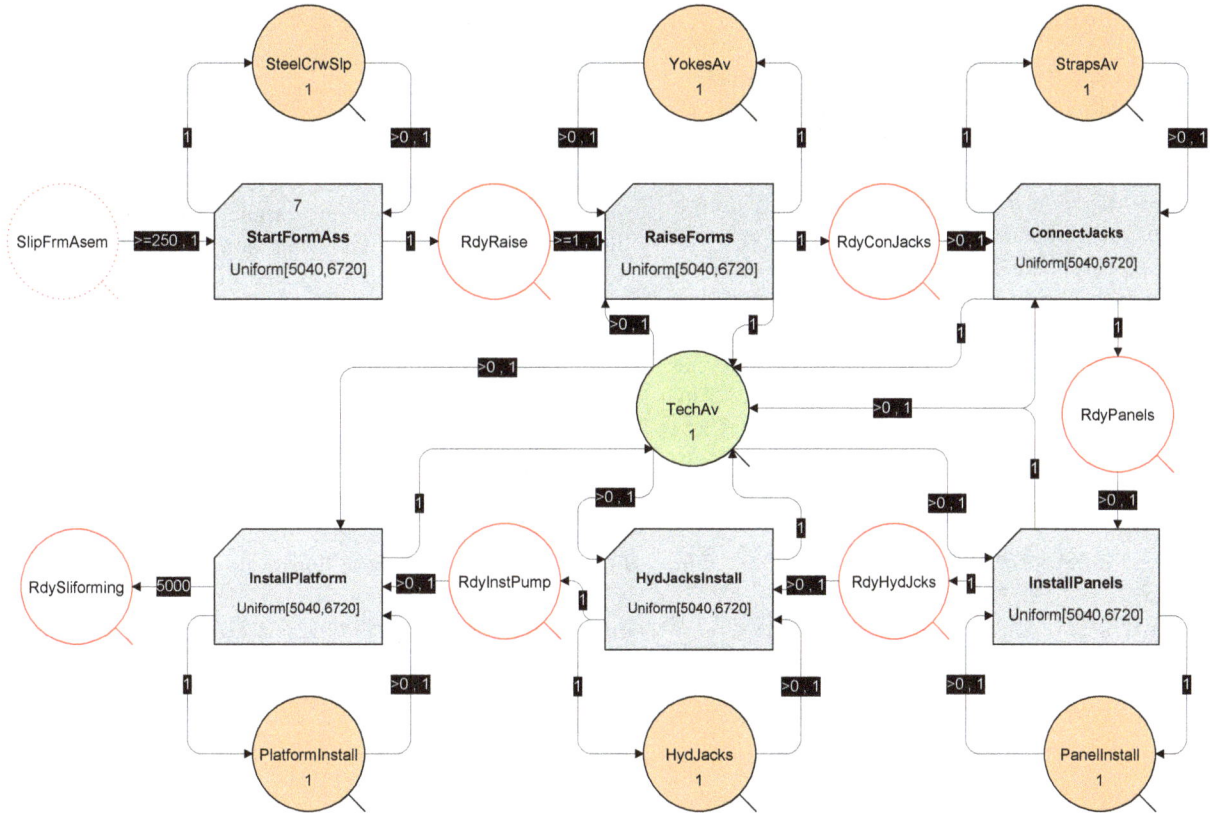

Figure 6. *Slipform Assembly Work Phase Model*

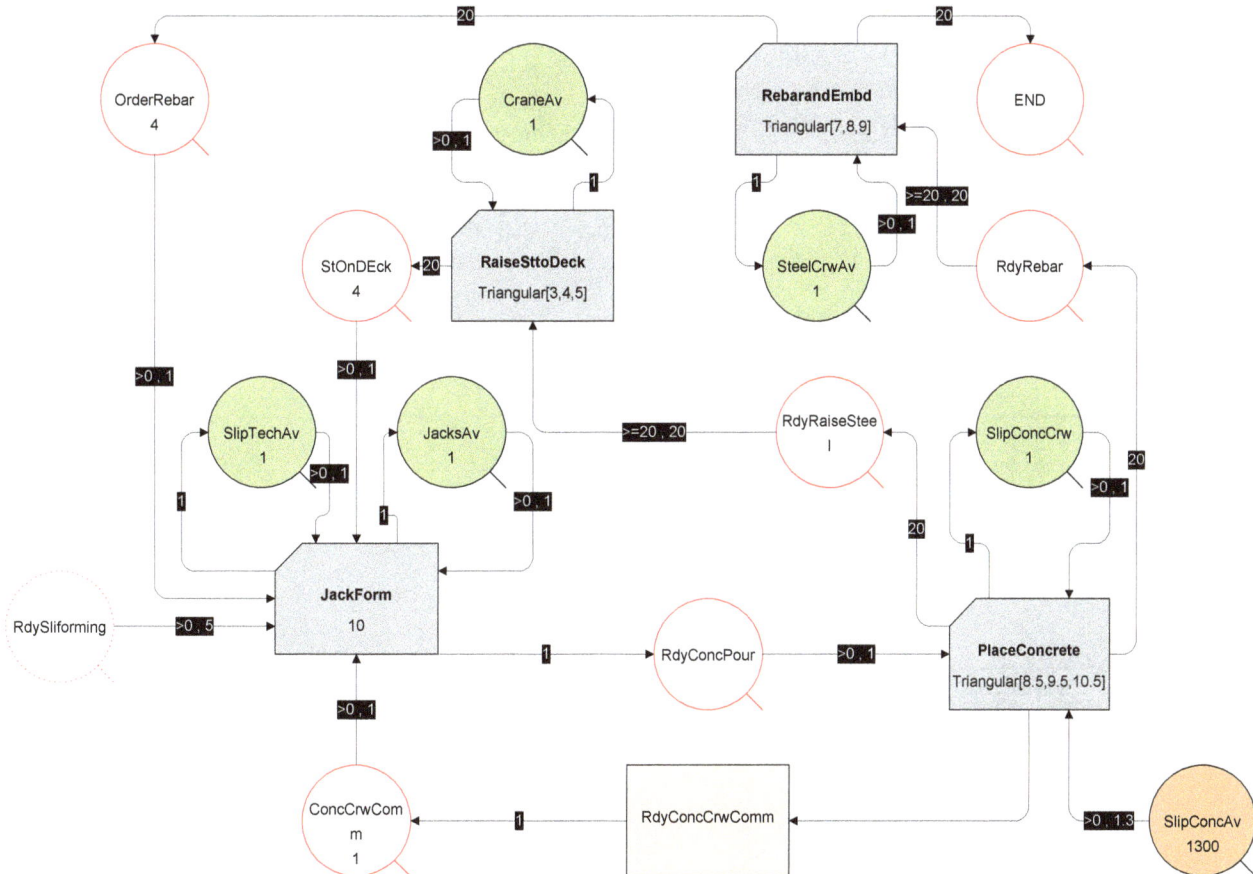

Figure 7. *Slipform Slipforming Phase Model*

4.3. Scenarios Factor's Configurations

Factors generally influence the Slipform construction duration includes; silo diameter, wall thickness, and concrete placing method, jacking rate and layer thickness. A series of experiments is performed on the 5 process factors to investigate their effect on the productivity rate and project duration, study the interaction among the factors and determine the most effective factor. A fractional factorial experiment approach using 5 factors each at three levels has been adopted. Table 7 presents the three levels of each factor under study as, three diameters of silo cross section (16, 18 and 20) m; three thicknesses of walls (0.4, 0.5 and 0.6) m; three types of concrete placing methods (Pump, Bucket and Hoist) m; three thicknesses of concrete layer pouring (5, 10 and 20) cm and three values of jacking rates (20, 30 and 40) cm/hr. Resulting in twenty seven experiment as 1/9 fractional factorial of 3^5 design based on the Principle block. The configurations of the twenty seven scenarios for the 5 factors selected for investigation together with the three levels for factorial design are given in Table 8. The levels of each factor were selected on the basis of good performance deduced from prior testing; production of practical and economic combinations.

5. Simulation Sensitivity Results

Based upon the aforementioned scenarios configurations, 27 simulation alternatives associated with various resource allocation strategies were produced. Table 9 presents the various values of model response to each scenario configuration as results (1) and (2) for total project duration and Slipforming duration. From this table it can be noted that all responses values showed Slipforming duration and total project duration comparable to that of case study results. In order to analyze the results and determine each parameter's significance, a set of steps are performed. First: calculate the average effect of factors levels on both results as shown in table 9. The average effect is calculated as the average of the 9 scenarios configuration with the same factor level for the required factor. I.e. for the average impact of level (0) of Factor (A) on Slipforming duration is calculated as following = (15.7+8.1+19.7+17+14+25.4+13.2+7.8+34.8)/9 = 17.3 days, this example is highlighted in table 10. Second: the degree of effectiveness of the three factors under study for total duration and Slipform duration are calculated and illustrated as shown in tables 11 and 12 respectively. The percentage of effectiveness is the result of dividing the difference between highest and lowest response for each factor by the sum of differences as shown in aforementioned. I.e. the effective percentage of factor A "Silo Cross Section" is due to dividing the difference between Highest project duration (58.2) and lowest project duration (56.6) for the same factor by the sum of differences "45.1" =(58.2-56.6)/45.1 = 3.6 % as highlighted in table 11.

Sensitivity analysis results are presented graphically in Figures 8 and 9 where these figures show the effective degree of each factor on both total project duration and Slipforming duration. Consequently, the highest effective degree on both results points to the most important factor that should be carefully considered when planning Slipforming operations. Duration reduction or increase is directly influenced by the effective degree of each factor as shown in aforementioned figures. Further figures and results analysis shows that the most effective factor among the five factors under study is Factor C: Concrete Placing Method, where in both figures factor C holds the highest proportion of the chart, followed by Factor E: Slipform Jacking Rate, followed by Factor B: Wall Thickness, then followed by Factor A: Structure Cross Section, Lastly Factor D: Concrete Layer Thickness. Although the arrangement of factor's effective degree on both results follows the same pattern, it is notable to mention that the factors effect is highly observed on the Slipforming duration rather than the total project duration. Generally, concrete placing method has the most significant effect on both results, while the concrete layer thickness has the lowest effect on both results.

Table 7. Simulation Configuration Levels

Factor Level	Factor Description				
	A: Silo Cross Section (m)	B: Wall Thickness (m)	C: Concrete Placing Method	D: Concrete Layer Thickness (cm)	E: Slipform Jacking Rate (cm/hr)
0	≡ 16	≡ 0.4	≡ Pump	≡ 5	≡ 20
1	≡ 18	≡ 0.5	≡ Bucket	≡ 10	≡ 30
2	≡ 20	≡ 0.6	≡ Hoist	≡ 20	≡ 40

Table 8. Factors levels and values

Scenario	Factor Level				
	A Silo Diameter	B Wall Thick.	C Placing Method	D Layer Thickness	E Jacking Rate
1	0	0	0	0	0
2	0	0	1	2	1
3	0	0	2	1	2
4	0	1	0	2	2
5	0	1	1	1	0

Scenario	Factor Level				
	A Silo Diameter	B Wall Thick.	C Placing Method	D Layer Thickness	E Jacking Rate
6	0	1	2	0	1
7	0	2	0	1	1
8	0	2	1	0	2
9	0	2	2	2	0
10	1	0	0	1	1
11	1	0	1	0	2
12	1	0	2	2	0
13	1	1	0	0	0
14	1	1	1	2	1
15	1	1	2	1	2
16	1	2	0	2	2
17	1	2	1	1	0
18	1	2	2	0	1
19	2	0	0	2	2
20	2	0	1	1	0
21	2	0	2	0	1
22	2	1	0	1	1
23	2	1	1	0	2
24	2	1	2	2	0
25	2	2	0	0	0
26	2	2	1	2	1
27	2	2	2	1	2

Table 8. (Continue).

Scenario	Factor Description				
	A Silo Diameter	B Wall Thick.	C Placing Method	D Layer Thick.	E Jacking Rate
1			Bucket	5	20
2		0.4	Pump	20	40
3			Hoist	10	60
4			Bucket	20	20
5	16	0.5	Pump	10	20
6			Hoist	5	40
7			Bucket	10	40
8		0.6	Pump	5	60
9			Hoist	20	20
10			Bucket	10	40
11		0.4	Pump	5	60
12			Hoist	20	20
13			Bucket	5	20
14	18	0.5	Pump	20	40
15			Hoist	10	60
16			Bucket	20	60
17		0.6	Pump	10	20
18			Hoist	5	40
19			Bucket	20	60
20		0.4	Pump	10	20
21			Hoist	5	40
22			Bucket	10	40
23	20	0.5	Pump	5	60
24			Hoist	20	20
25			Bucket	5	20
26		0.6	Pump	20	40
27			Hoist	10	60

Table 9. *Simulation Results Values.*

Scenario	Factor Description					Simulation Results	
	A Silo Diameter	B Wall Thickness	C Placing Method	D Layer Thickness	E Jacking Rate	(1) Total Project Duration (Day)	(2) Slipforming Duration (Day)
1			Bucket	5	20	56.3	15.7
2		0.4	Pump	20	40	47.7	8.1
3			Hoist	10	60	58	19.7
4			Bucket	20	20	55	17
5	16	0.5	Pump	10	20	52.6	14
6			Hoist	5	40	65.6	25.4
7			Bucket	10	40	52.3	13.2
8		0.6	Pump	5	60	47.4	7.8
9			Hoist	20	20	75.6	34.8
10			Bucket	10	40	50.7	11.1
11		0.7	Pump	5	60	45.6	6.7
12			Hoist	20	20	67.5	28.4
13			Bucket	5	20	56.2	17.8
14	18	0.8	Pump	20	40	48.8	9.3
15			Hoist	10	60	64.6	26.2
16			Bucket	20	60	51.2	12.4
17		0.9	Pump	10	20	53.4	15.3
18			Hoist	5	40	71.1	32.5
19			Bucket	20	60	48.1	10
20		1	Pump	10	20	53.2	14
21			Hoist	5	40	65.8	25.2
22			Bucket	10	*40*	52.3	13.4
23	20	1.1	Pump	5	60	45.7	8
24			Hoist	20	20	74.7	36.7
25			Bucket	5	20	60	20.3
26		1.2	Pump	20	40	49.7	10.6
27			Hoist	10	60	74.1	33.8

5.1. Significance and Discussions

While this research results can directly define the most important factors that should be carefully considered when a new project is under study in the preliminary phase, the interaction between factors based upon the factors levels are presented to furtherer illustrate the factors significance from the obtained results as shown in figures 10 and 11 for total project duration and Slipforming duration. Results show that among the five factors under study, the most significant factor that effects both Slipforming duration and total project duration is the concrete placing method, followed by the jacking rate, then the structure wall thickness, structure cross section and finally the concrete layer thickness.

Therefore this research's significance can be described in the following; (A) This research is beneficial for practitioners to estimate an overall construction schedule of Slipforming building projects, especially in preconstruction phases by applying the required project parameters and values in proposed model; (B) Determine the most effective facto in Slipforming operations that directly affects the duration of such projects, (C) better understanding of relevant projects processes characteristics and circulation. Although the direct application of the proposed construction time forecast model is limited to Slipform silo buildings, the approaches proposed in this research can be adopted to forecast project time for other Slipform project types and/or in other locations. The current study focuses on the aspect of duration. Cost criteria may also be added to evaluate various construction strategies.

Table 10. *Average effect of factors on sensitivity results*

Factor	Factor Level	Factor Value	Average Factor Effect on Simulation Results	
			Total Project Duration (Days)	**Slipforming Duration (Days)**
A: Silo Cross Section (m)	0	16	56.7	17.3
	1	18	56.6	17.7
	2	20	58.2	19.1
B: Wall Thickness (m)	0	0.4	54.8	15.4
	1	0.5	57.3	18.6
	2	0.6	59.4	20.1
C: Concrete Placing Method	0	Bucket	53.6	14.5
	1	Pump	49.3	10.4
	2	Hoist	68.6	29.2
D: Concrete Layer Thickness (cm)	0	5	57.1	17.7
	1	10	56.8	17.9
	2	20	57.6	18.6
E: Slipform Jacking Rate (cm/hr)	0	20	67.2	23.8
	1	40	56.0	16.5
	2	60	48.3	13.8

Table 11. *Effectiveness percentage of factors on total project duration*

Factor	Factor Level	Factor Value	Total Project Duration (Days)	Difference (High PD - Low PD)	Effectiveness Percentage (%)
A: Silo Cross Section (m)	0	16	56.7	1.6	3.6%
	1	18	56.6		
	2	20	58.2	*hint = 58.2-56.6*	*Illustration = 1.6/45.1*
B: Wall Thickness (m)	0	0.4	54.8	4.7	10.3%
	1	0.5	57.3		
	2	0.6	59.4		
C: Concrete Placing Method	0	Bucket	53.6	19.2	42.6%
	1	Pump	49.3		
	2	Hoist	68.6		
D: Concrete Layer Thickness (cm)	0	5	57.1	0.8	1.7%
	1	10	56.8		
	2	20	57.6		
E: Slipform Jacking Rate (cm/hr)	0	20	67.2	18.9	41.8%
	1	40	56.0		
	2	60	48.3		
			Sum	*45.1*	100%

Table 12. *Effectiveness percentage of factors on Slipforming duration*

Factor	Factor Level	Factor Value	Total Project Duration (Days)	Difference (High PD - Low PD)	Effectiveness Percentage (%)
A: Silo Cross Section (m)	0	16	17.3	1.8	5.0%
	1	18	17.7		
	2	20	19.1		
B: Wall Thickness (m)	0	0.4	15.4	4.6	12.9%
	1	0.5	18.6		
	2	0.6	20.1		
C: Concrete Placing Method	0	Bucket	14.5	18.8	52.1%
	1	Pump	10.4		
	2	Hoist	29.2		
D: Concrete Layer Thickness (cm)	0	5	17.7	0.9	2.4%
	1	10	17.9		
	2	20	18.6		
E: Slipform Jacking Rate (cm/hr)	0	20	23.8	9.9	27.6%
	1	40	16.5		
	2	60	13.8		
			Sum	36	100%

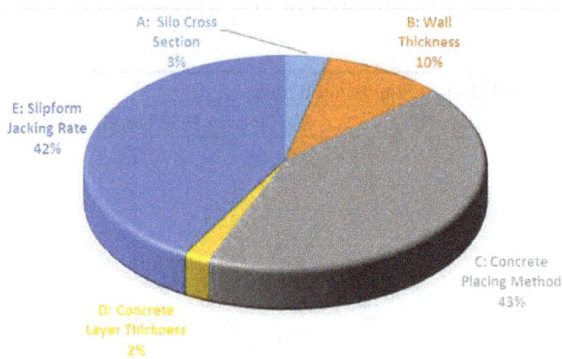

Figure 8. *Effective percentage of each factor on total project duration.*

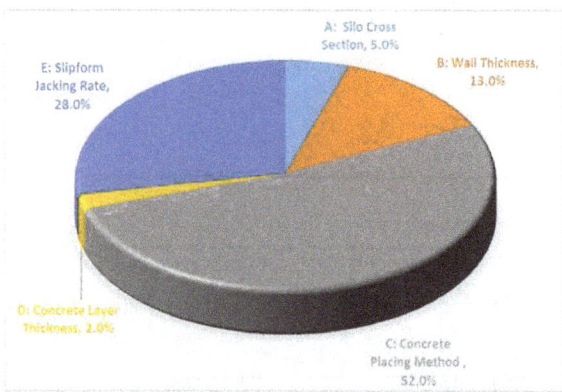

Figure 9. *Effective percentage of each factor on Slipforming duration.*

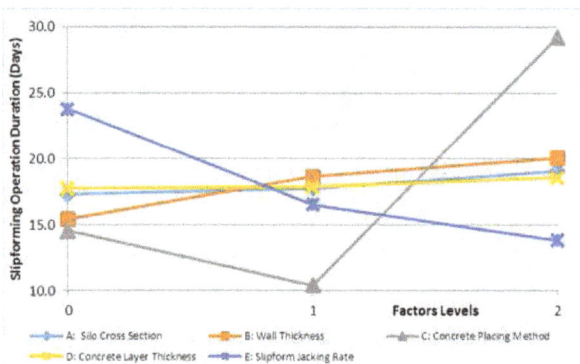

Figure 10. *Interaction of Factors Values and Levels on Slipforming Duration*

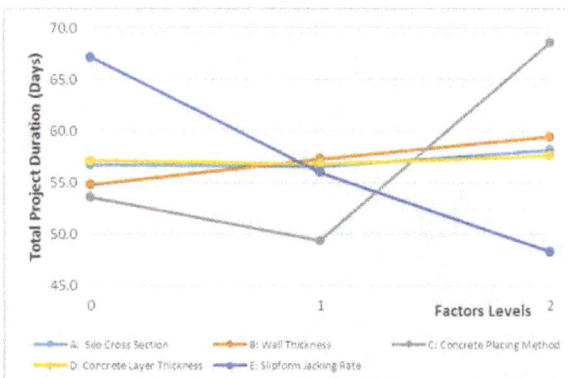

Figure 11. *Interaction of Factors Values and Levels on total project Duration*

6. Conclusions

The responsive use of discrete event simulation model for estimating and predicting preliminary schedule duration of Slipform projects application to concrete silos is presented in this research. The EZstrobe simulation software was used to develop the model. This model is considered a full slip form project and operation model with the aforementioned capabilities. The proposed model is divided into four model phases for the Slipform operation, Silo earthworks, Foundation works, Slipform Assembly and Slipforming process. Simulation is an ideal decision–support tool because it allows quick and data-rich exploration of "what if" scenarios at any given point in time. Several factors have been considered through the model for predicting productivity such as, Silo diameter, wall thickness, placing method, concrete layer thickness and Jacking rate of forms. The models are tested and show high accuracy in predicting slip-form productivity as by validating proposed Slipform model with case study it showed a robust of 98.7%. The output of this research can helpfully assist practitioners and researchers in the field of construction simulation and Slipforming, because they provide a planning and scheduling tool for slip-form operation in silo construction. Moreover computational results show that the use of this approach can significantly help improving the efficiency of the production system. In addition to simulation models that are flexible enough to modify and add more features and develop a reliable schedule of Slipforming operation suitable for a specific project by optimizing various operation parameters.

Acknowledgements

I would like to extend my appreciation to BEC Slipform System Company particularly Mr. Eddie Williams (Managing Director), for his great help and professional advice in current research.

References

[1]   Weizhuo Lu, Thomas Olofsson. "Building information modeling and discrete event simulation: Towards an integrated framework." Automation in Construction, 2014: 73-83.

[2]   Halpin, D., and Riggs. Planning and analysis of construction operations. New York. Wiley, 1992.

[3]   Jürgen Melzner, Sebastian Hollermann, Hans-Joachim Bargstädt. "Detailed Input data Source for Construction Process Simulation." The Third International Conference on Advances in System Simulation. Bauhaus-University Weimar, Germany, 2011.

[4]   Lu, Ming. "Simplified Discrete-Event Simulation Approach for Construction Simulation." Journal of Construction Engineering and Management, 2003.

[5] I-Chen Wu, André Borrmann, Ulrike Beißert, Markus König, Ernst Rank. "Bridge construction schedule generation with pattern-based construction methods and constraint-based simulation." Advanced Engineering Informatics, 2010: 379-388.

[6] K.T. Fossa, A. Kreiner, J. Moksnes. "Slipforming of advanced concrete structures." Tailor Made Concrete Structures – Walraven & Stoelhorst , 2008.

[7] M.R. Sharifi, S. Baciu and T. Zayed. "Slip-Form Productivity Analysis for Concrete Silos." International Construction Specialty Conference. Calgary, Alberta, Canada, 2006.

[8] S. Abourizk, D. W. Halpin, and J. D. Lutz,. "State of the art in construction simulation." Proceedings of the 1992 Winter Simulation Conference. Arlington, VA, 1992. 1271-1277.

[9] Zayed, T. and Halpin, D. "Simulation Of Concrete Batch Plant Production." Journal of Construction Engineering and Management, 2001: 132-141.

[10] Kannan, G., et al . "A framework for incorporating dynamic strategies in earthmoving simula-tions. In WSC'97: Proceedings of the 1997 Winter Simulation Conference, Atlanta, Ga., 7–10 December 1997. ." 1997.

[11] Ericsson, U. Diffusion of Discrete Event Simulation in Swedish Industry, One way to an Increased Understanding, PhD Thesis. Gothenburg, Sweden: Chalmers University of Technology, 2005.

[12] A. Skoogh, T. Perera, B. Johansson. "Input data management in simulation – Industrial practice and future trends." Simulation Modelling Practice and Theory, 2012: 181-192.

[13] Law, A. and Kelton, D. 1982. Simulation modeling and analysis, McGraw-Hill, New York.

[14] Jahangirian et al. "Simulation in manufacturing and business." European Journal of Operational, 2010: 1-13.

[15] Wang, Halpin. "Simulation experiment for improving construction processes." Proceedings of the 2004 Winter Simulation Conference, Washington, D.C. 2004.

[16] AbouRizk, S. "Role of simulation in construction engineering and management." Journal of Construction Engineering and management, 2010: 1140-1153.

[17] Martinez, J.C. STROBOSCOPE: State and Resource Based Simulation of Construction Process, Doctoral Dissertation. Ann Arbour, 1996.

[18] Martinez, J.C. "EZstrobe -- general-purpose simulation system based on activity cycle diagrams." Proceedings of 1998 Winter Simulation Conference. 1998. 341 – 348.

[19] Fossa, K. T. "Slipforming of Vertical Concrete Structures. PhD Thesis." Norway, 2001.

[20] Anagnostopoulos, Christos. Application of the maturity method in Slipforming operations. 2003.

[21] T. Zayed "Slip-Form Application to Concrete Structures" Journal of Construction Engineering and management, 2008

[22] Kim, H. Mechanical Characteristics of GFRP Slip Form. Ph.D. Thesis, Hanyang University, 2012.

[23] Hyejin Yoon, Won Jong Chin, Hee Seok Kim, Young Jin Kim. "A Study on the Quality Control of Concrete during the Slip Form Erection of Pylon." Engineering, 2013: 647-655.

[24] Hanna. Concrete Formwork Systems. New York: Marcel Dekker, 1998.

[25] H.Khalik et al. "Simulation Analysis For Productivity And Unit Cost By Implementing (Gps) Machine Guidance In Road Construction Operations In Egypt." 2012.

12

Uncertainty and Risk Factors Assessment for Cross-Country Pipelines Projects Using AHP

Hesham Abd El Khalek[1], Remon Fayek Aziz[2], Hamada Kamel[3, *]

[1]Construction Engineering and Management, Faculty of Engineering, Alexandria University, Alexandria, Egypt
[2]Construction Engineering and Management, Faculty of Engineering, Alexandria University, Egypt
[3]Faculty of Engineering, Alexandria University, Alexandria, Egypt

Email address:
heshamkhaleq@gmail.com (H. A. El Khalek), remon_fayek@hotmail.com (R. F. Aziz), hamadakamel1974@gmail.com (H. Kamel)

Abstract: Infrastructure cross-country pipelines projects carry out higher risk than traditional because they entail high capital outlays and intricate site conditions. The high-risk exposure associated with infrastructure cross-country pipelines projects needs special attention from contractors to analyze and manage their risks. They cannot be eliminated but can be minimized or transferred from one project stakeholder to another. Therefore, current research aims for identifying the risk factors that affect infrastructure cross-country pipelines projects based on experts experience and company's point of view which participated in similar projects. The risk factors classified under two categories to company level risks and project level risks. The risk factors were assessed using risk assessment models that facilitate this assessment procedure, prioritize these projects based upon its risk indexes and evaluate risk contingency value. Analytical hierarchy process (AHP) used to evaluate risk factors weights (likelihood) and FUZZY LOGIC approach to evaluate risk factors impact (Risk consequences) using software aids such as EXCEL and MATLAB software, accordingly risk indexes for both company level and project level evaluated and overall project risk index determined. Five case studies in different countries were selected to determine the highest risk factors and to implement the designed models and test its results. Results show that project no 3 in Iraq conquer the highest risk index (39.75%); however, project 5 in Egypt has the lowest risk index (5.24%). Results of risk factors in other countries are (32.81%) in Emirates, (17.27%) in Saudi Arabia and (11.67%) in Libya. Therefore, the developed model can be used to sort projects based upon risk, which facilitate company's decision of which project can be pursued.

Keywords: Risk Management, International Construction, Risk Factors, Optimization Model, Analytic Hierarchy Process (*AHP*), *FUZZY LOGIC* Approach, *MATLAB* Software, Validation Process

1. Introduction

The business of construction has changed a great deal resulting from the effects of growing globalization and competition [22]. The fast-growing international trade and developments, such as the World Trade Organization agreements [32] and the Asia-Pacific Economic Cooperation forum (APEC, 2003) have provided new opportunities to the construction industry. Facilitated by sophisticated communication technologies, advanced project management, and by profits attraction, large-scale projects are no longer local events but international affairs involving parties of different nations [13, 21, 22].

Infrastructure such pipelines project is by its nature the corner stone of our society. It lays the foundation for a healthy economy and civilization. Such projects carry out higher risk than traditional because they spend high capital outlays and have complicated site conditions. Generally, projects that are implemented in the infrastructure field as cross-country pipeline projects are considered as investments, due to the high initial cost and the project's long time horizon.

Risks cause cost overrun and schedule delay in many projects. The effectiveness of risk management becomes an important issue in project management. To make risk management more efficient and effective, all parties must understand risk responsibilities, risk event conditions, risk preference, and risk management capabilities. There are many types of potential sources of risk and uncertainty that affect infrastructure cross-country pipelines projects. These sources of risk and uncertainty include political, economical, cultural, market, and technical risks that might reduce the contractor(s) and/or subcontractor(s) profit. It is essential that contractors and subcontractors conquer these sources of risk and uncertainty. [1, 13, 30, 3, 9, 16].

2. Background

A cross-country pipeline construction projects are exposed to an uncertain environment due to its enormous size (physical, manpower requirement and financial value), complexity in design technology and involvement of external factors. These uncertainties can lead to several changes in project scope during the process of project execution. Unless the changes are properly controlled, the time, cost and quality goals of the project may never be achieved [23].

The cross-country petroleum pipelines are sensitive for risks because they traverse through varied terrain covering crop fields, forests, rivers, populated areas, desert, Hills, sea bed and offshore [13, 4, 6, 23]. Pipelines represent critical infrastructure can create significant social and environmental impacts, such pipeline exposed to natural disasters (such as landslides, earthquakes). [3].

Large-scale construction projects are exposed to an uncertain environment because of such factors as planning and design complexity, presence of various interest groups (project owner, owner's project group, consultants, contractors, vendors etc.), resources (materials, equipment, funds, etc.) availability, climatic environment, the economic and political environment and statutory regulations [30, 31, 23].

Sources of risk and uncertainty always exist in construction projects and often cause schedule delay or cost overrun [5, 33, and 36]. Project risk is defined as "the exposure to loss/gain", "the probability of occurrence of loss/gain multiplied by its respective magnitude" [15], Cooper and Chapman [7] define it as "exposure to the possibility of economic or financial loss or gain, physical damage or injury, or delay, as a consequence of the uncertainty associated with pursuing a particular course of action." Al-Bahar [2] defines it as "the exposure to the chance of occurrences of events adversely or favorably affecting project objectives as a consequence of uncertainty.

Project management considers risk management as one of the key knowledge areas for managers [33, 37].

2.1. Risk Management Process

Risks do not exist in isolation but evolve in the context of a project. In order to reduce the potentially disastrous consequences of risks, project managers seek to understand them and deal with them appropriately. Project managers have given this process the name Project Risk Analysis and Management (PRAM) [5]. This process can be broken up into a number of components, identification, assessment, allocation, mitigation and management.

2.2. Risk Identification

The first stage of the risk management process is to actually identify the relevant risks to the project. Dias A., [10] stated the aims of this phase are to: 1. Identify all the significant types and sources of risk and uncertainty associated with each of the investment objectives. 2. Determining Key parameters relating to these objectives ascertain the causes of each risk.3. Assess how risks are related to other risks and how risks should be classified and grouped for evaluation.

There are a number of methods available to project planners when seeking to identify the relevant risks to their project. Chapman [5] has assessed the benefits of three of the most commonly used methods, Brain storming and nominal group technique and Delphi technique.

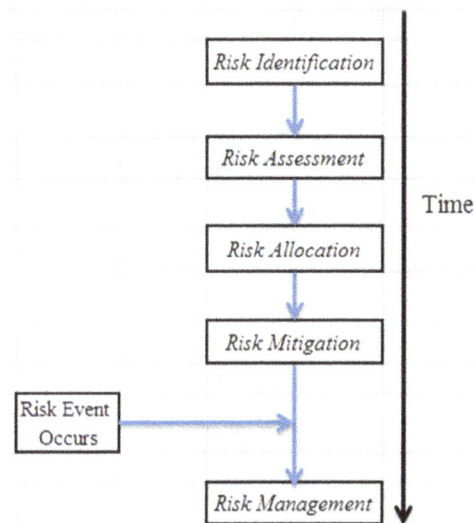

Figure 1. *The sequence of risk management process.*

2.3. Risk Assessment

Risk assessment is the process of reviewing and understanding risks in order to determine their significance for the project and its outcomes. This process include determining the relative importance of all risks which can impact the project and estimating the probability of the risk occurring and the likely size of the risk. Risk assessment is important as it help project developers to concentrate theirre sources (in terms of both time and money) in the areas where they can make the most significant contribution to the eventual project outcome. It also allows project developers to understand which aspects of the project are the most sensitive to risk events. There are two types of method which may be used for risk assessment. [29].

2.4. Risk Allocation

Once project risks have been identified and assessed, they are usually allocated to one of the parties involved. The party that becomes responsible for the risk must bear any costs associated with its crystallization.

2.5. Risk Mitigation

Risk mitigation is the process of understanding the risks to which a project is exposed and attempting to reduce the significance of those risks prior to their occurrence. There are many types of risk mitigation measures available. Suitable methods vary depending on the type of risk and the characteristics of the organization that is mitigating that risk.[10].

2.6. Risk Management

Risk management is the process of understanding how a risk has affected a particular project and putting in place measures to reduce the effect of that risk event. The aim of risk management is to restore the project to its 'pre-risk' state as quickly as possible and with the smallest possible cost [29].

Risk and uncertainty of cross country pipelines construction projects did not receive sufficient attention from researchers. Therefore, current research is trying to open this area by studying several case studies in cross country pipelines construction projects. It only considers the first two steps of risk management: identification and assessment. The following sections will explain risk identification and assessment model building for cross country pipelines construction projects.

3. Research Objectives

The objectives of current research are:
1. Identify main risk and uncertainty factors and their sub-factors that affect projects for the company level and the project level in cross country pipelines projects.
2. Evaluate the most risky factors that affect infrastructure cross-country pipelines projects using software aids.
3. Building risk assessment model and apply the proposed model on real cases.

4. Study Methodology

The first stage in this research methodology, is to specify the several variables (numerical and linguistic), that would affect the project. This will be done by gathering all the related variables from database of previous projects, the project environment (host country conditions, project's characteristics and location). The project risk decision factors selection based upon evaluation of a wide range of risk decision factors and their sub factors gathered from the literature. The second stage is to identify those variables, remove the redundant variables, and classifies them, Then, grouping these decision factors under main categories according to their relativeness.

The third stage, Questionnaire was designed to rate the significant level of project risk factors with in each category of risks by using five –point scale (1-5) to build the risk factors model in both company and project level in order to assign the most important factors and remove ineffective factors. The fourth stage, two risk index (R) models in both

company and project levels will be designed to assess the effect of sources of risk and uncertainty on construction project based on the equation (1) which is adapted from (Dias, 1996) [10]..

RISK=LIKELIHOOD X CONSEQUENCE

$$R = \sum_{i=1}^{n} W(xi) * E(xi) \qquad (1)$$

R   :   Risk index for a construction projects.
Wi (xi) :   Weight for each risk area i using Eigen value method.
Ei (xi) :   Effect score for each risk factor (xi).
Xi   :   Different risk factor (i).
I   :   1, 2, 3,............., n.
n   :   Number of risk factors.

The risk model consists of two parts: risk factors weights (W) and their worth score (E). Risk factors weights will be determined using the AHP; while the worth scores (Risk effectiveness) will be assessed using four approaches, Dias approach [10], Value curve approach according to Zayed [34, 35], New approach according to Salman M. [29] and new approach according Fuzzy logic approach. Finally five case studies have been employed to demonstrate the application of proposed model.

5. Data Collection

To identify the risk factors and sub factors in international projects, a questionnaire survey in the form of face-to-face interview was conducted with 93 practitioners, who are experts in the field. The selection of the experts was based on that they work in cross country pipelines projects, participate in international projects or tend to go in new markets. The positions of the participants vary among project managers, project planners, proposals, quality control, estimators, safety, site and cost control engineers from all disciplines from the participated. There are two phases of data collection which are implemented through research displayed in table (1).

Table 1. Study Questionnaires.

Questionnaire No	Description	Objectives
Over all Data		
A. Questionnaire 1	Criteria Development	Building Risk model
Focused Data		
B. Questionnaire 2	AHP, Risk Performance surveys for five projects	Model application

6. Identification of Cross Country Pipelines Projects Risk Factors

Zayed and Chang [35] proposed one model combining the company and project risks in one model. BU Qammaz [4] identified risk associated with international construction projects (ICPR model) which risk sources were categorized under 5 main categories which are country, inter-country, construction, project team, and contractual issues; and these criteria were believed to best reflect the nature of the considered risk sources, The hierarchical representation of risk sources is known as a hierarchal risk breakdown structure (HRBS). Many authors proposed different risk breakdown structures to classify risk in these projects in two categories, in the company management level and in the project management level. Company level risks contain risk factors which connect with the characteristics of the host country. It concerns the political situation, economic conditions, unethical practices, legal system maturity, and the stability and level of security in the country. The factors considered under this category are: bribery, government instability, tension/conflicts/terrorism, bureaucratic difficulties, immaturity/unreliability of legal system, change of regulations/laws (government interventions), and instability of economic conditions (inflation/currency fluctuation) [11, 34, 38].

Different risk breakdown structures to classify risks in project level were proposed reflecting different experts opinions [7], In addition Zayed T [34] proposed the more importance risk factors in the project level concerning emerging technology usage, contracts and legal issues, resources, design stage, construction stage, quality, and other areas, such as weather, natural causes of delay in addition to physical damages. Based upon literature and cross-country pipelines projects expert's opinions in questionnaire forms,

Risk factors classified under to main classes company and project levels risk factors as displayed in Fig (2).

Twenty two factors have been selected in the company risk level and thirty three factors have been selected in the project risk as the most significant factors affecting the project and forming the risk factors model. Figures 2, 3 and 4 display the risk hierarchy models in both company and project levels.The factors were identified and classified under five main categories in the company risk and nine main categories in the project risk according to their relativeness.

7. Risk Assessment Model Development

According to the aforementioned factors, a risk index(R) model is designed to assess the effect of sources of risk and uncertainty on a construction project from contractor (company) prospective the model displayed in figures 2, 3, 4 respectively. It provides a logical, reliable, and consistent method of evaluating potential projects, prioritizing them, and facilitating company's decision in the promotion. The risk index (R) model based upon equation 1 characterizes the various sources of risk and uncertainty in a project and assesses their effect on such project. The R-index consists of two parts: weights of risk factors and sub-factors and their effect score.

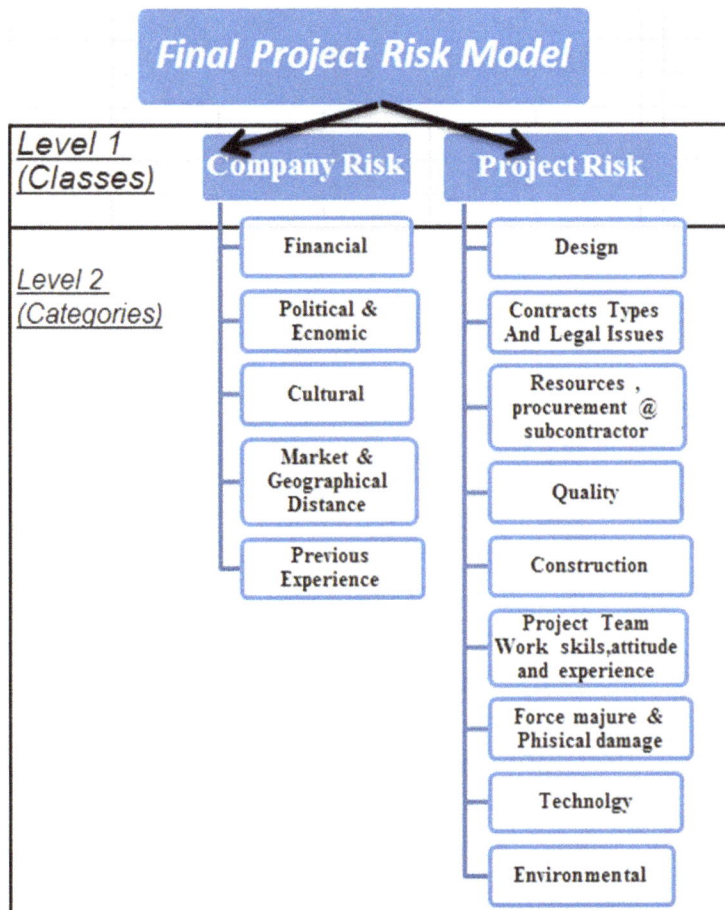

Figure 2. Developed Risk hierarchy model in company and project Levels.

Risk Factors in the company Level

Financial
- *Financial Type*
  - * puplic financial
  - * PPP financial
- Tax or capital movements restriction
- Currency exchange rate
- Currency Exchange difficulty
- *payment risk*
  - -Low advance payment
  - -Shotage of client financial res.
  - - Payment delay

Political & Ecnomic
- Dependece on or importance of majour power
- Relation and Houspitality with neighboring country
- Bribery & Bureaucratic difficulties
- Change of regulation/law s)
- Government instability
- Immaturity/unreli ability of legal system
- Instability of economical conditions
- Tension/conflict s /terrorism

Cultural
- Interaction of management with local contracts
- Cultural Differences
- Poor Attitude of the Host Country towards Foreign Companies

Market & Geographica l Distance
- Current market volume and competitors
- Future market volume and competitors
- Geographical Distance

Previous Experience
- in internationl projects
- in similar projects
- in host country

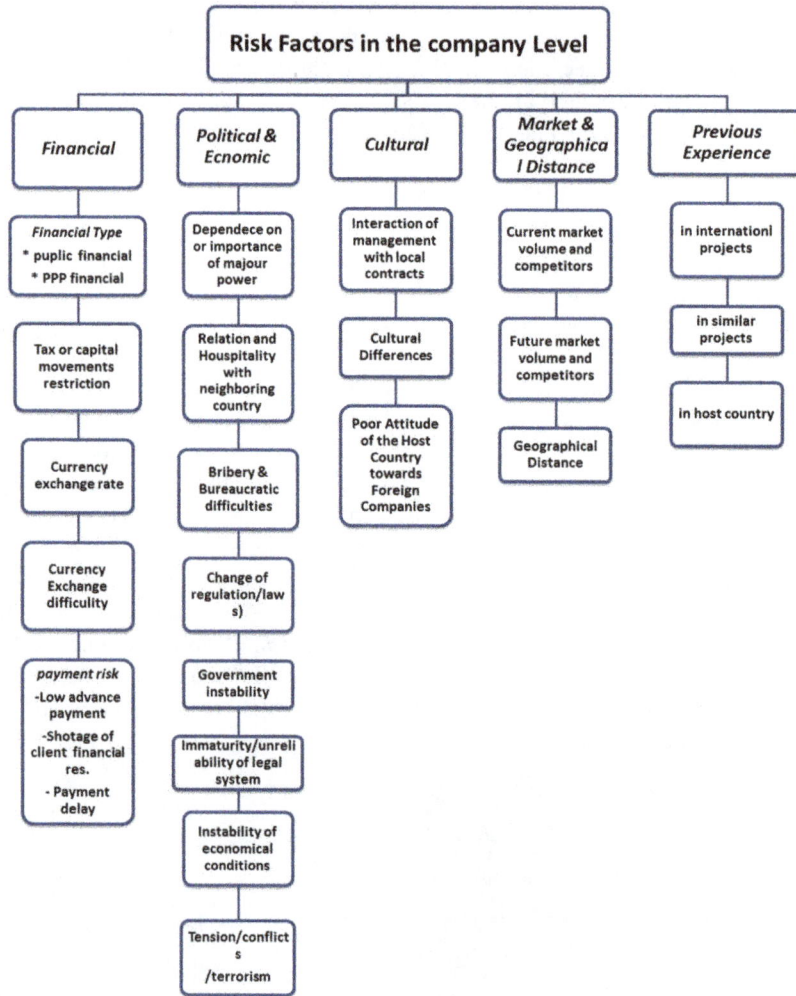

Figure 3. *Risk hierarchy model in company Level.*

Risk Factors in Project Level

Design
- Delay in Design and regularty Approval
- Defective design errors and rework
- Unsuitable Design

Contracts Types And Legal Issues
- type of the contract
  - - Fixed Price Contract
  - - Unit Price Contract/ BOQ
  - - Cost + Contract
  - - EPC Turn key Contract
- Potential of contracts disputes and claims
- Vagueness of Contract Conditions about Risk Allocations
- Problem in disputes setteIment due to international Law

Resources , procurement and subcontractor
- Shortage of skilled workers
- Availability of special Equipment
- Delay in Material supply
- subcontractor unavailabilty or poor performance

Quality
- Strict Quality Requirem ents
- Bad quality of material
- Bad quality of work execution

Construction
- Managerial Complexities
- Project prouductivity Risk (schedule delay/Tight Schedule/High Liquidated Damages)
- Third party delays
- Strict Safety and Health Requirements
- Cost over run
- work change order

Project Team Work skils,attitude and experience
- Contractor
- Client
- Consultant
- Designer
- Joint Venture/ Partner

Force majure & Phisical damage
- Weather and natural Causes of delay
- Phisical damage
- Unforeseen adverse ground conditions

Technology
- Technolgy Transfere
- Retention of Technology

Environmental
- Conformance to laws Land regulations/ Strict Environment Regulations
- Project risk manageme nt system
- public acceptance to the project

Figure 4. *Risk hierarchy model in project Level.*

8. Model Application

8.1. Program Verification for Five Projects

Five projects in different countries were selected to verify model application, the projects are as follows:
1. Project 1: Nuayyim Field ASL Pipelines project (Saudi Arabia).
2. Project 2: Habshan Saiem Plant and Pipelines development (Arab United of Emirates).
3. Project 3: Nasria Pipeline /16 "/ 200km/Oil Pipeline Company (OPC) (Iraq).
4. Project 4: Sareer Plant / Entisar Field Pipeline 195 KM (Libya).
5. Project 5: Desouq Fields Development pipelines - 132 Km (Egypt).

8.2. Part 1: AHP Survey

It was required form the participants to make pairwise comparison among risk factors and risk sub factors represent the relative importance between them based on the numerical scale (1-9) using Analytical Hierarchy Process (AHP) Figure 5 provides an example to explain the pair wise process. The assignment of weights requires logical and analytical thinking, so it is preferred to focusing on the participants who have good experience and knowledge under each case study to participate in the AHP survey questionnaire to ensure that only valid and good quality data are acquired. It is assumed that the group members will carry out necessary brainstorming sessions and reach to a consensus for the required tasks. In other words, rather than asking the same questions to individual members separately, only one response is received from the group and it is believed to represent the democratic majority point of view of the group. [25, 26, 27, 28, 18]

	absolutely more Important	substantially more Important	moderately more Important	slightly more Important	Equal Important	slightly more Important	moderately more Important	substantially more Important	absolutely more Important	
	9	8	7	6	5	4	3	2	1	2	3	4	5	6	7	8	9	
Factor A					_5_													Factor B
											_3_							Factor c
If (A) is more Important than (B or C) use this side								If (B or C) is more Important than (A) use this side		

Figure 5. *An example was provided to explain the pairwise process.*

Factors Category	Performance Limits	Much Lower			Lower		Same Risk as previous projects	Higher		Much Higher	
		Extremely Ineffective	substantially Ineffective	moderately Ineffective	slightly Ineffective	Neither Effective Nor Ineffective	slightly Effective	moderately Effective	substantially Effective	absolutely Effective
		1	2	3	4	5	6	7	8	9
A	Normal						6			
	P1 P2		2						8	

Maximum Ineffective Performance
Normal Effective Performance
Expected Risk Performance
Maximum Effective Performance

Figure 6. *An example was provided to explain Performance scale.*

8.3. Part 2: Assigning Risk Performance

It was required form the participants to assign 3 point represent the low risk performance (P1), the high point of risk performance (P2) and the Expected risk performance ($P_{Expected}$) for all sub factors in both company and project risk factors based on the numerical scale (1-9). Figure 6 provides an example to explain Assigning Risk Performance for each risk factor. The main points in the performance scale are:

Minimum Risk Performance (P1): this point represents maximum Ineffective risk performance. It indicates the risk factor impact if things go well (optimistic Impact).

Maximum Risk Performance (P2): this point represents maximum effective risk performance. It refers to the risk factor impact if things do not go well (pessimistic Impact).

Expected Risk Performance (P Expected): This point represents best estimate of the risk impact (most likely impact).

Neither effective nor ineffective point: This point represents normal risk performance which means the same risk as previous projects.

Extremely Ineffective: The lowest risk point in the performance scale. It is means there is extremely no risk.

Absolutely Effective: The highest risk point in the performance scale. It is means there is extremely high risk.

Excel spread sheet software was designed to solve the weights, impacts and receive the results obtained from fuzzy program *(Expected Risk Performance (P Expected),* hence the overall risk can be determined based on equation 1 [29, 34] for four risk evaluation methods.

DIAS approach [10], P2=100 approach [34, 35], P2 only [29] and new model based on FUZZY LOGIC approach (not scope of this paper due to limited space reason).

8.4. Model Results

Results of risk factors in company level based on the new model of fuzzy approach displayed in table 2 and figures 7, 8 reveal that Current market volume and competitors, previous experience in host country, have the highest risk value in project no (1) in Saudi Arabia, On the other hand Shortage of skilled workers, Subcontractor unavailability or poor performance and Strict Quality Requirements have the highest risk value in project level.

Table 2. *Low and high risk factors in each project in different countries under current study in both Company and project level based on new model of fuzzy approach.*

Project	Country	Project	Level	High Risk Factors	High Risk value	Low Risk Factors	Low Risk value
Project (1)	Nuayyim Field ASL Pipelines	Nuayyim Field ASL Pipelines	Company risk	Current market volume and competitors	0.0941	Tension/conflicts/terrorism	0.0009
				PR EXP in host country	0.0889	Change of regulation/laws)	0.0008
				Future market volume and competitors	0.0397	Government instability	0.0007
						Instability of economical conditions	0.0007
			Project Risk	Shortage of skilled workers	0.0742	Project risk management system	0.0011
				subcontractor unavailabilty or poor performance	0.0740	Unforeseen adverse ground conditions	0.0009
				Strict Quality Requirements	0.0535	Physical damage	0.0007
						Joint Venture Team Work skils, attitude and experience	0.0003
Project (2)	United arab of Emarates	plant and Pipelines of Habshan Saiem	Company risk	Change of regulation/laws)	0.1089	Currency exchange rate	0.0018
				Dependence on or importance of major power	0.0929	Cultural Differences	0.0015
				Future market volume and competitors	0.0717	Poor Attitude of the Host Country towards Foreign Companies	0.0015
						Tax or capital movements restriction	0.0012
			Project Risk	Shortage of skilled workers	0.0972	Potential of contracts disputes and claims	0.0023
				Delay in Material supply	0.0617	Bad quality of work execution	0.0023
				Cost over run	0.0551	Bad quality of material	0.0023
						public acceptance to the project	0.0018
Project (3)	IRAQ	Nasria Pipeline 16 _ 200kmOil	Company risk	Tension/conflicts/terrorism	0.1045	Immaturity/unreliability of legal system	0.0044
				Dependece on or importance of majour power	0.0720	Relation and Hospitality with neighboring country	0.0032
				Previous Experience in host country	0.0702	Cultural Differences	0.0024
				Geographical Distance	0.0508	Poor Attitude of the Host Country towards Foreign Companies	0.0008
			Project Risk	subcontractor unavailabilty or poor performance	0.0753	Contractor Team Work skils, attitude and experience	0.0026
				Defective design errors and rework	0.0511	public acceptance to the project	0.0024
				Managerial Complexities	0.0464	Potential of contracts disputes and claims	0.0017
						Type of the contract	0.0013
Project (4)	Lebia	Sareer Plant /	Company	Previous Experience in host	0.1058	Instability of economical	0.0014

Project	Country	Project	Level	High Risk Factors	High Risk value	Low Risk Factors	Low Risk value
		Entisar field Pipeline 195 km	risk	country		conditions	
				Current market volume and competitors	0.0677	Tension/conflicts/terrorism	0.0013
				Future market volume and competitors	0.0478	Relation and hospitality with neighboring country	0.0007
						Government instability	0.0004
				Cost over run	0.0353	Type of the contract	0.0026
			Project Risk	Unsuitable Design	0.0291	Conformance to laws Land regulations/Strict Environment Regulations	0.0017
				Weather and natural Causes of delay	0.0263	public acceptance to the project	0.0008
				payment risk	0.0571	Relation and hospitality with neighboring country	0.0009
			Company risk	Instability of economical conditions	0.0460	Dependence on or importance of majour power	0.0009
		Desouq Fields development pipelines - 132 Km		Previous Experience in zoon area	0.0158	Interaction of management with local contracts	0.0007
Project (5)	Egypt			Delay in Material supply	0.0680	Conformance to laws Land regulations/Strict Environment Regulations	0.0016
			Project Risk	Delay in Design and regularty Approval	0.0322	Physical damage	0.0014
						work change order	0.0010
						public acceptance to the project	0.0007

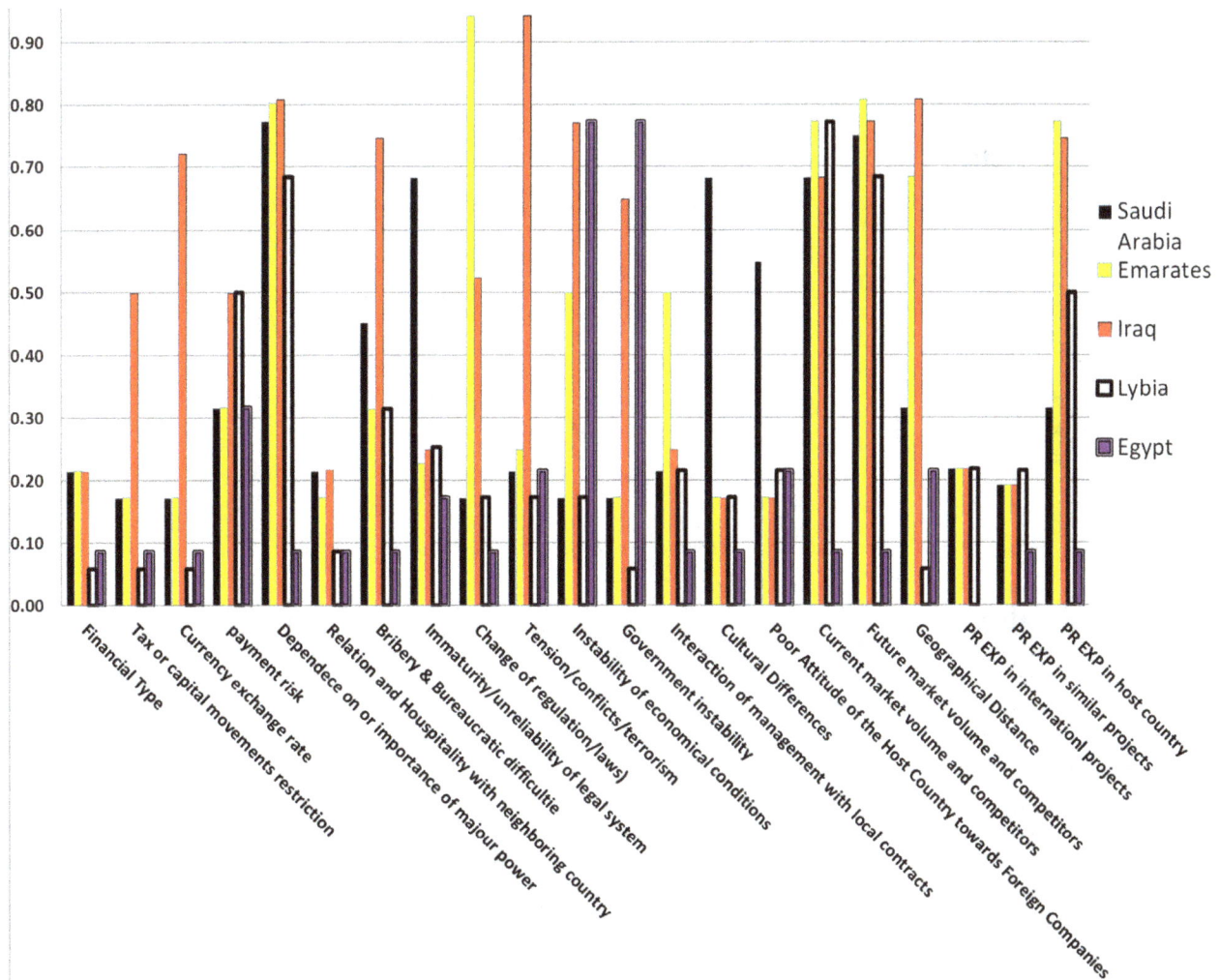

Figure 7. Risk attributes values in company level for each project (Model based on fuzzy approach).

The Change of regulation/laws, dependence or importance of major power, Future market volume and competitors, current market volume and competitors and geographical distance have the highest risk value in company level in project no (2) in Emirates, in addition Shortage of skilled workers and Delay in Material supply have the highest risk value in project level.

Tension/conflicts/terrorism, dependence on or importance of major power and previous experience in host country have the highest risk value in company level in project no (3) in Iraq, On the other hand subcontractor unavailability or poor performance and defective design errors and rework have the highest risk value in project level. previous experience in host country and Current market volume and competitors have the highest risk value in company level in project no (4) in Libya, On the other hand Cost overrun, unsuitable design and weather and natural Causes of delay have the highest risk value in project level.

Payment risk and Instability of economic conditions have the highest risk value in company level in project no (5) in Egypt, on the other hand delay in material supply and Delay

in Design and regularity Approval have the highest risk value in project level.

The above analysis indicates that previous experience in host country attribute, Current market volume and competitors, The Change of regulation/laws, dependence or importance of major power and payment risk and Instability are considered high risk in the five existed profile projects that mean, the decision makers should concentrate well on such attributes to decrease their risk before proceeding with similar to their project. Also The above analysis indicates that availability of resources factors is considered high risk in the most existed profile projects that mean, the decision makers should concentrate well on such attributes to decrease their risk before proceeding with their project by making sure that the project local resources are available when needed and the required imported resources with their paper works (type, cost, import licenses, taxes, delivery time, etc.,) will be settled in the project feasibility study stage. Moreover, from figures (7, 8) it is interesting to note that some factors have low risk value and in another project have high risk value based on each project conditions.

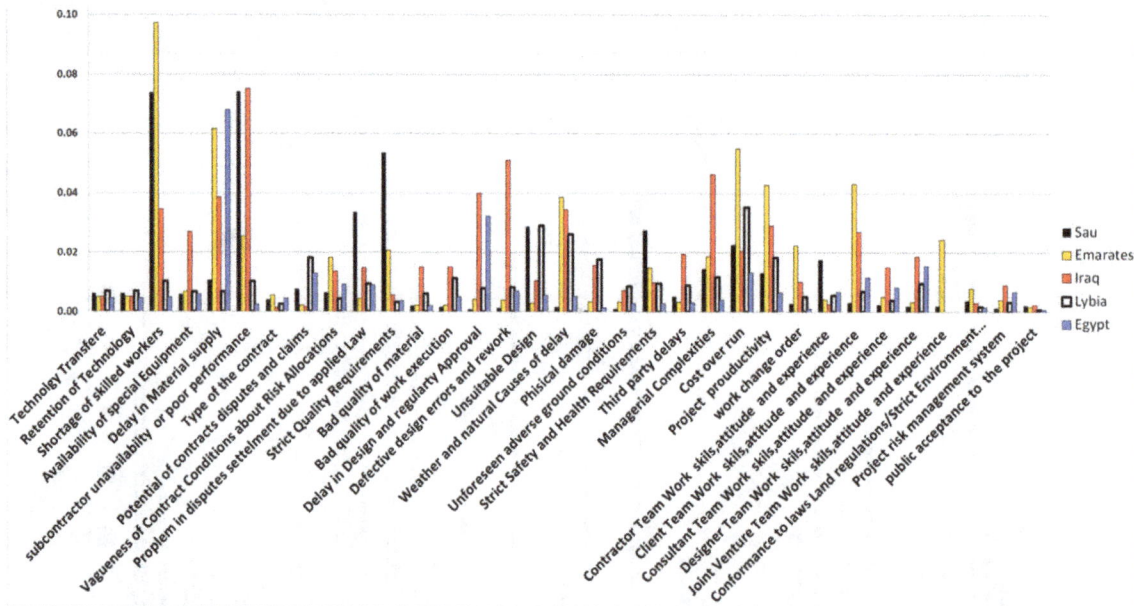

Figure 8. Risk factors values in project level for each project (Model based on fuzzy approach).

9. Project Risk Index

The results of Company and project risk indexes for each project conjunction with each approach, gathered from excel program are tabulated in table 3. The purpose of calculating the risk index for a project is to answer the question "is the

project viable enough to be successfully promoted by the contractor and for what extent?" The project risk factors should be revised by decision makers and risky factors possible provisions may be suggested and reevaluated to decrease its risk to catch the accepted limits.

Table 3. Company and project risk indexes and contingency value for each project conjunction with each approach.

Project/Location	Project 1: Nuayyim Field ASL Pipelines/Suadi				Project 2: Habshan Saiem plant and Pipelines /Emarates			
	Suadi				Emerates			
Evaluation	Dias	P2=100	P2_Only	Fzy/Sgm	Dias	P2=100	P2_Only	Fzy/Sgm
Comp Risk index	0.42	0.47	0.70	0.40	0.43	0.58	0.75	0.57
Proj Risk index	0.49	0.48	0.72	0.44	0.49	0.58	0.78	0.58
Final Risk index	20.9%	22.6%	50.1%	17.3%	21.1%	33.8%	58.2%	32.8%
Fuzzy Risk Value				22.6%				28.4%

Table 3. *Continued.*

Project / Location	Project 3: Nasria Pipeline /16 "/ 200km/Oil Pipeline Company(OPC) /IRAQ				Project 4: Sareer Plant / Entisar field Pipeline 195 km (Libya)				Project 5: Desouq Fields development pipelines - 132 Km/Egypt			
	IRAQ				Libya				Egypt			
Evaluation	Dias	P2=100	P2_Only	Fzy/Sgm	Dias	P2=100	P2_Only	Fzy/Sgm	Dias	P2=100	P2_Only	Fzy/Sgm
Comp Risk index	0.44	0.59	0.78	0.61	0.36	0.39	0.64	0.36	0.31	0.28	0.55	0.19
Proj Risk index	0.52	0.62	0.80	0.65	0.35	0.34	0.53	0.32	0.46	0.37	0.64	0.27
Final Risk index	22.8%	36.5%	62.0%	39.8%	12.7%	13.4%	34.5%	11.7%	14.3%	10.5%	34.6%	5.2%
Fuzzy Risk Value				33.6%				19.8%				16.2%

High risk projects may be accepted by decision maker as it is if the project quantitative factors are feasible enough to overcome the deficiencies in high qualitative attributes. Table 3 and figures 9, 10, 11 provide the results of risk index for projects under the study for company and project level and overall project risk index.

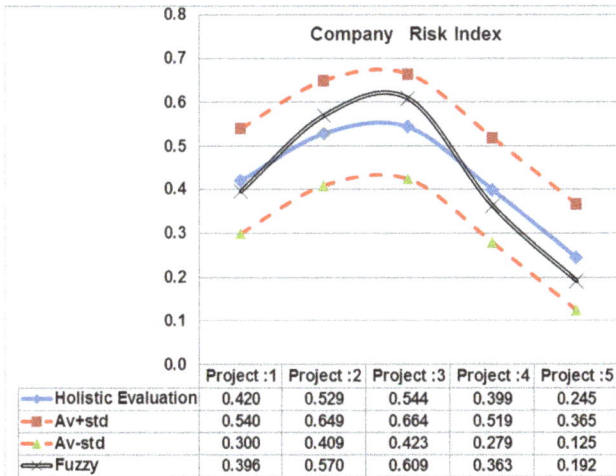

Figure 9. *Company risk index in addition to the holistic evaluation according to developed fuzzy model results.*

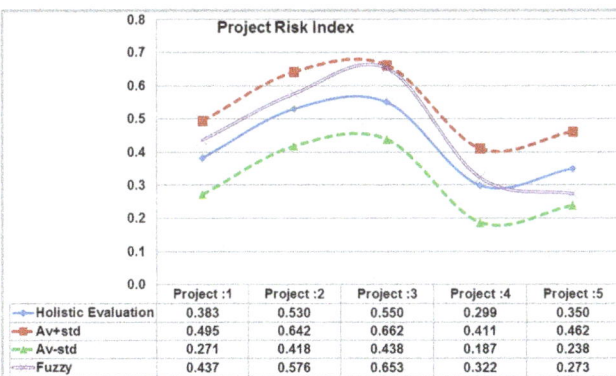

Figure 10. *Project risk index in addition to the holistic evaluation according to developed fuzzy model results.*

From figure 11, Results of final risk index in the overall project based on fuzzy approach show that project 3 in Iraq conquer the highest risk (39.75%); however, project 5 in Egypt has the lowest risk index (5.24%). Results of risk factors in other countries are (32.81%) in Emirates, (17.27%) in Saudi Arabia and (11.67%) in Libya. (Final Risk Index =

Company Risk Index * Project Risk Index). Therefore, the developed model can be used to sort projects based upon risk, which facilitate company's decision of which project can be pursued.

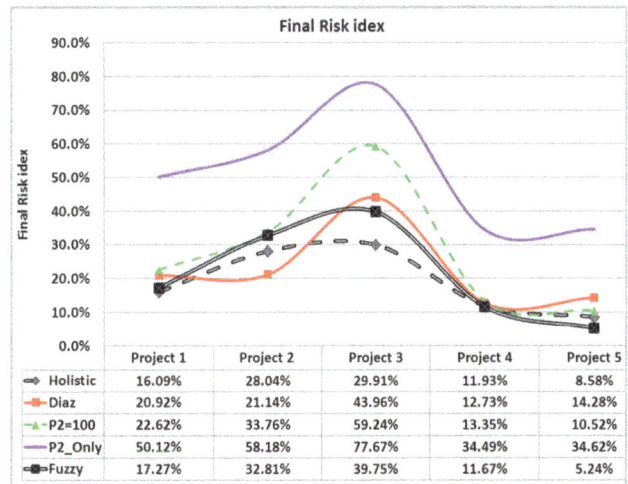

Figure 11. *Overall Project risk index for decomposed approach for each project.*

10. Using Risk Index in the Project Feasibility Study

The purpose of developing the risk model was to help the decision maker in evaluating the feasibility and risk of their project in its preliminary stages and before going forward with the project. The first step on using the risk model consists of assigning the factors importance weights and their performance (quality) levels, developing attributes value curves, and computing the project risk index. Once the factors indexes that forming the total project risk index have been determined, the high risk attributes that affect the total project risk will be known and the decision maker can put the possible strategies to improve their performance level and reevaluate them to increase the project viability or to reject the project if it was not satisfied with the resulting viability.

11. Conclusion

Most construction companies are willing to enter into international markets in order to maximize their revenues and growth potential benefits. A construction company's decision to expand into international markets must be based on a good understanding of the opportunities and threats associated with international business, as well as the development of company strengths relative to international activities.

This study provided the main factors of risk and uncertainty and their sub-factors in infrastructure cross country pipelines projects; it identified and analyzed these risk factors in direction, company and project levels.

In addition this study proposes Risk hierarchy model in both company and project Levels that performs three functions: identify risk factors in both company and project levels, used to evaluate sources of risk and uncertainty using software aids, accordingly prioritize infrastructure cross-country pipelines projects according their risk.

Results of risk factors in company level using software aids (fuzzy Logic approach model) show that Current market volume and competitors, previous experience in host country, have the highest risk value in project no (1) in Saudi Arabia, on the other hand Shortage of skilled workers, subcontractor unavailability or poor performance and Strict Quality Requirements have the highest risk value in project level.

The Change of regulation/laws, dependence or importance of major power, Future market volume and competitors, current market volume and competitors and geographical distance have the highest risk value in company level in project no (2) in Emirates, in addition Shortage of skilled workers and Delay in Material supply have the highest risk value in project level.

Tension/conflicts/terrorism, dependence on or importance of major power and previous experience in host country have the highest risk value in company level in project no (3) in Iraq, however subcontractor unavailability or poor performance and defective design errors and rework have the highest risk value in project level.

previous experience in host country and Current market volume and competitors have the highest risk value in company level in project no (4) in Libya, while Cost overrun, unsuitable design and weather and natural Causes of delay have the highest risk value in project level.

Payment risk and Instability of economic conditions have the highest risk value in company level in project no (5) in Egypt, as delay in material supply and Delay in Design and regularity Approval have the highest risk value in project level.

Results show that project no 3 in Iraq conquer the highest risk index (39.75%); however, project 5 in Egypt has the lowest risk index (5.24%). Results of risk factors in other countries are (32.81%) in Emirates, (17.27%) in Saudi Arabia and (11.67%) in Libya. For companies that pursues infrastructure cross-country pipelines projects have to consider seriously the previous risks when bidding in international markets. Therefore, the developed model can be used to sort projects based upon risk, which facilitate company's decision of which project can be pursued.

This study developed database of risk information of the most significance risk factors in the international projects in both company and project levels for future references. It is an essential tool to assess the level of risk associated with construction projects under study in the bidding phase in order to take preventive actions.

Risk model and developed decision support tool are company specific. Each company has its own risk knowledge leading to different rules and may have different risk attitudes. Thus, the cases given in this paper should be treated as an example of how the proposed risk model can be utilized in practice rather than a universally accepted evaluation for the risk factors assessment in international construction projects.

Acknowledgements

The authors express their gratitude to Pr. Hesham Abdel Khalek and Dr Remon Aziz for their generous contribution of time, energy, and knowledge in this study. They also extend their gratitude to all companies that facilitate the authors' research and participate in this research.

References

[1]   Antonio J., Monroy A, Gema S., R., and Lopez A., (2011). "Financial Risks in Construction Projects". African Journal Of Business Management Vol. 5(31), Pp. 12325-12328, 7 December, 2011.

[2]   Al-Bahar, J. F. and Crandall, K. C., 1990. "Systematic risk management approach for construction projects", Journal of Construction Engineering and Management, 116(3), 533-546.

[3]   Babatunde A, Damian M., Dan Van Horst., Lee Chapman. (20120. "Attacks On Oil Transport Pipelines In Nigeria: a quantitative exploration and possible explanation of observed patterns". applied geography 32 (2012) 636e651.

[4]   Bu-Qammaz, A. S. (2007), "risk assessment of international construction projects using the analytic network process", master of science thesis, middle east technical university.

[5]   Chapman, C. and Ward, S., (1997). "Project Risk Management Processes, Techniques and Insights", John Wiley, UK.

[6]   Cooper D, Grey S, Raymond G and Walker P, 2007, "Project Risk Management Guidelines: Managing Risk in Large Projects and Complex Procurements", John Wiley & Sons, Ltd, ISBN 0-470-02281-7.

[7]   Cooper, D. and Chapman C., (1987), "Risk Analysis for Large Projects- Models, Methods and Cases", John Wiley, UK.

[8]   Carreño, M. L., Cardona, O. D. and Barbat, A. H. "Evaluation of the Risk Management Performance", 250th Anniversary of the 1755 Lisbon Earthquake, technical University Of Catalonia, Barcelona, Spain.

[9]   Xiaoping D., And Pheng L., (2012). "Understanding the Critical Variables Affecting the Level of Political Risks In International Construction Projects". KSCE Journal of Civil Engineering (2013) 17(5): 895-907.

[10] Dias A, Ioannou P., (1996). "Company and Project Evaluation Model for Privately Promoted Infrastructure Projects. Journal of Construction Engineering and Management, ASCE 1996; 122(1): 71–82. March.

[11] Dikmen I, Birgonul T and Han S. (2007). "Using Fuzzy Risk Assessment to Rate Cost Overrun Risk in International Construction Projects". International Journal of Project Management 25 (2007) 494–505.

[12] Enrique Jr., Ricardo C., Vicent E. and Aznar J., (2011). "Analytical Hierarchical Process (AHP) as a Decision Support Tool In Water Resources Management". Journal of Water Supply: Research and Technology—Aqua | 60.6 | 2011.

[13] Gunhan S, Arditi D.(2005). "International expansion decision for construction companies". ASCE, j constreng manage 2005; 131(8): 928–37.

[14] Hyun C., Hyo C. and Seo J. W., (2004). "Risk Assessment Methodology for Underground Construction Projects", ASCE Journal of Construction Engineering and Management, 130, 258-272.

[15] Jaafari A. (2001). "Management of risks: uncertainties and opportunities on projects: time for a fundamental shift". international journal of project management 19 (2001) 89±101

[16] John, W., G. and Edward Jr. (2003). "International Project Risk Assessment: Methods, Procedures, and Critical Factors"., A Report of The Center Construction Industry Studies The University Of Texas At Austin.

[17] Liu, A., Wang B, Ma Qingguo B. (2011). The Effects of Project Uncertainty and Risk Management on IS Development Project Performance. A Vendor Perspective International Journal of Project Management 29 (2011) 923–933.

[18] Ludovic V, Marle F, Bocquet J, C. (2011). "Measuring Project Complexity Using the Analytic Hierarchy Process". International Journal of Project Management 29 (2011) 718–727.

[19] Ming W. and Hui C. (2003). "Risk Allocation and Risk Handling of Highway Projects In Taiwan". Journal of Management in Engineering, Asce / April 2003.

[20] Mohamed A. and Aminah F., (2010). "Risk management in the construction industry using combined fuzzy FMEA and fuzzy AHP". journal of construction engineering and management, ASCE / September 2010.

[21] Ofori, G. (2000). "Globalization and construction industry development: research opportunities". Construction Management and Economics, 18, 257-262.

[22] Ofori, G and Chan, S L." (2000) Factors influencing growth of construction enterprises in Singapore". Construction Management and Economics (in press).

[23] Prasanta D., (2002). "An Integrated Assessment Model For Cross Country Pipelines". Environmental Impact Assessment Review, 22, (2002) 703–721.

[24] Prasanta D., (2010). "Managing Project Risk Using Combined Analytic Hierarchy Process and Risk Map". Applied Soft Computing 10 (2010) 990–1000.

[25] Saaty TL. The Analytic Hierarchy Process. 1980. New York: Mc graw- Hill, 1980.

[26] Saaty TL. Decision Making For Leaders. Belmont, California: Life Time Leaning Publications, 1985.

[27] Saaty TL. (1990). "How to Make a Decision: The Analytic Hierarchy Process". European Journal of Operational Research, North-Holland 1990; 48: 9±26.

[28] Saaty TL, Kearns KP. (1991). "Analytical Planning: The Organization of Systems". The Analytic Hierarchy Process Series 1991; Vol. 4RWS.

[29] Salman, A. (2003). "Study Of Applying Build Operate And Transfer Bot Contractual System On Infrastructure Projects In Egypt". PHD Thesis, Zagazig University, Faculty of Eng.

[30] Seung H., James E. and Jong H., (2005). "Contractor's risk attitudes in the selection of international construction projects". journal of construction engineering and management, ASCE / march 2005 / 283.

[31] Suat, G and David A. (2005). "Factors affecting international construction". journal of construction engineering and management, ASCE / march 2005 / 273.

[32] WTO Annual Report 2004 - World Trade Organization.

[33] Wang M, Chou H. Risk allocation and risk handling of highway projects in Taiwan. J Manage Eng, ASCE 2003; 19(2): 60–8. April.

[34] Zayed T, Mohamed A, Jiayin P. (2008). Assessing Risk And Uncertainty Inherent In Chinese Highway Projects Using AHP "Internal Journal Of Project Management" 26 (2008) 408–419.

[35] Zayed, T, and Chang, L. (2002). Prototype Model for Build-Operate-Transfer Risk Assessment. Journal of Management In Engineering / January 2002 / 7.

[36] Zayed TM, Halpin DW. Quantitative assessment for piles productivity factors. J Constr Eng Manage, ASCE 2004; 130(3): 405–14. May/ June.

[37] PMI (Project Management Institute). A guide to project management body of knowledge: PMBOK (Project Management Book of Knowledge) Guide. 2nd ed. Upper Darby, PA, USA; 2000.

[38] Baghdadi, A, Kishk, M., (2015). "Saudi Arabian aviation construction projects: Identification of risks and their consequences". Procedia Engineering 123 (2015) 32–40.

[39] Khodeir, L., Mohamed, A., (2014). "Identifying the latest risk probabilities affecting construction projects in Egypt according to political and economic variables". Housing and Building National Research Center, HBRC Journal (2015) 11, 129–135.

[40] Rafindadi, A., Miki, M., Kovai, I. and Ceki, Z., (2014). "Global Perception of Sustainable Construction Project Risks", 27 th IPMA World Congress, Procedia- Social and Behavioral Sciences 119 (2014) 456–465.

[41] Tanaka, H., (2014). "Toward project and program management paradigm in the space of complexity: a case study of mega and complex oil and gas development and infrastructure projects". 27 th IPMA World Congress, Procedia- Social and Behavioral Sciences 119 (2014) 65-74.

[42] Toh, T., Ting, C., Ali, A., Aliagha, G. and Munir, O., (2012). "Critical cost factors of building construction projects in Malaysia". International Conference on Asia Pacific Business Innovation and Technology Management, Procedia - Social and Behavioral Sciences 57 (2012) 360–367.

13

Effect of Using Waste Material as Filler in Bituminous Mix Design

Dipu Sutradhar[*], **Mintu Miah, Golam Jilany Chowdhury, Mohd. Abdus Sobhan**

Department of Civil Engineering, Rajshahi University of Engineering & Technology, Rajshahi, Bangladesh

Email address:

dipu.ruet.civil@gmail.com (D. Sutradhar), mintu@ruet.ac.bd (M. Miah), zilanichoudhuri@gmail.com (G. J. Chowdhury), msobhan@yahoo.com (Mohd. A. Sobhan)

Abstract: Bituminous concrete or asphaltic concrete is one of the highest and costliest types of flexible pavement layers used in surface course. Being of high cost specifications, the bituminous mixes are properly designed to satisfy the design requirements of the stability and durability. The mixture contains dense grading of coarse aggregate, fine aggregate and mineral filler coated with bitumen binder. The mineral filler passing through 0.075 mm sieve performs some important roles in bituminous mixes. Marshall Stability of bituminous mix increases as the amount of filler increases. The Asphalt Institute recommends the use of 4 to 8% filler in asphalt concrete. The common filler materials like cement, lime stone, granite powder etc. is not easily and economically available in Bangladesh. Waste concrete dust and brick dust are considered to be cheaper and in abundant supply in Bangladesh. In this study an attempt is made to find the effect of types of filler on the behavior of bituminous mixes. According the properties of bituminous mixes containing filler like waste concrete dust and brick dust is studied and compared with the mixes containing filler like fine sand and stone dust mixture generally used. The Marshall method of mix design was used for the comparison. The Marshall stabilities of mix types containing filler fine sand and stone dust mixture, waste concrete dust and brick dust were found 9.8 KN, 11.1 KN and 11.3 KN respectively which satisfy the limiting value of 5.33 KN according to Marshall Design criteria. The study indicates the possibility of using waste concrete dust and brick dust as filler in bituminous mix.

Keywords: Bituminous Paving Mixes, Brick Dust, Waste Concrete Dust, Stone Dust, Filler, Marshall Mix Design

1. Introduction

Bituminous roads are defined as the roads in the construction of which bitumen is used as binder. It consists of an intimate mixture of aggregates, mineral filler and bitumen. The quality and durability of bituminous road is influenced by the type and amount of filler material is used. [1] The filler tends to stiffen the asphaltic cement by getting finely dispersed in it. Various materials such as cement, lime, granite powder, stone dust and fine sand are normally used as filler in bituminous mixes. Cement, lime and granite powder are expensive and used for other purposes more effectively. Fine sand, ash, waste concrete dust and brick dust finer than 0.075 mm sieve size appear to be suitable as filler material. The use of waste powder as filler in asphalt mixture has been the focus of several research efforts over the past few years. Phosphate waste filler [2], Jordanian oil shale fly ash [3], bag house fines [4], recycled waste lime [5], municipal solid waste incineration ash [6] and waste ceramic materials [7] have been investigated as filler. It was proved that these types of recycled filler could be used in asphalt mixture and gave improved performance. So the present study has been taken in order to investigate the behavior of bituminous mixes with different types of filler materials locally available.

If filler is mixed with less bitumen than it is required to fill its voids, a stiff dry product is obtained which is practically not workable. Overfilling with bitumen, on the contrary, imparts a fluid character to the mixture. [8]The filler has the ability to increase the resistance of particle to move within the mix matrix and/or works as an active material when it interacts with the asphalt cement to change the properties of the mastic. [9] Elastic modulus of asphalt concrete mixture can increases by the addition of mineral filler. But excessive amount of filler may weaken the mixture by increasing the amount of asphalt needed to cover the aggregates [10, 11].The effects of these fillers are also dependent on gradations.

The objective of this study is to produce an

aggregate-asphalt mix with a controlled void. If the void of the mixed aggregate is too low, the mix will be unable to carry sufficient asphalt and therefore will be difficult to compact due to insufficient lubrication. It will not be sufficiently durable as the film on the aggregate particles will be too thin. On the other hand, if the void is too high, it is probable that the mix will be lacking in stability. Because each aggregate will receive less support from those surrounding it. [12, 13] Fillers could improve the temperature susceptibility and durability of the asphalt binder and asphalt-concrete mixture. The effects of these fillers are also dependent on gradations. To have a good mixture, aggregates and filler should bind properly. [14] A strong backbone for the mixture can provided by the good packing of the coarse, fine, and filler aggregates.

The performance of bituminous mix also depends on amount of filler in the mix. [15] The workability of a mix depends, to some extent, on the amount and type of the filler present in the mix. [16, 17] The mixture performance also affected by the interactions between asphalt and filler because of the larger surface area, filler may absorb more asphalt and its interaction with asphalt may lead to different performance of asphalt-concrete mixture. [18] The size distribution, particle shape, surface area, surface texture, voids content, mineral composition, and other physiochemical properties vary for several fillers. Therefore, their effect on the properties of asphalt-concrete mixture also varies.

Conventionally in Bangladesh, fine sand with stone dust is used as filler material in bituminous mix. In this study an attempt is made to find the effect of types of cheap & non-conventional filler on the behavior of bituminous mixes. For this purpose, Waste concrete dust and brick dust were used as non-conventional fillers. The performance characteristics of the mixture containing different types of filler were evaluated by examining fundamental material properties and by performing various laboratory tests. Then results obtained for mix type containing non-conventional fillers were compared with the results obtained for mix type containing conventional filler. The possibility of using non conventional filler (e.g. waste concrete dust and brick dust) are also investigated.

2. Materials

A bituminous mixture is normally composed of aggregate and bitumen. The aggregates are generally divided into coarse, fine and filler fractions according to the size of the particles. The following sections include the description of the coarse aggregate, fine aggregate, mineral fillers and bitumen used in this study.

2.1. Coarse Aggregate

Coarse aggregate for bituminous mix has been defined as that portion of the mixture which is retained on 2.36 mm (No. 08) sieve according to the Asphalt Institute. Basalt rock was used as coarse aggregate. It was crushed manually and brought to the sizes 25.0 mm or less. The aggregates were then sieved using U.S. standard sieves and separated out in different fractions.

Figure 1. Appearance of coarse aggregate.

2.2. Fine Aggregate

Aggregate passing through 2.36 mm sieve and retained on 0.075 mm sieve was selected as fine aggregate. Domar sand was the source of fine aggregates.

Figure 2. Appearance of fine aggregate.

2.3. Filler

Fine sand and stone dust mix, waste concrete dust and brick dust finer than 0.075 mm size sieve were used as filler in the bituminous mixes for comparison and economical point of view.

Figure 3. Appearance of fine sand and stone dust mix.

Figure 4. Appearance of waste concrete dust.

Figure 5. Appearance of brick dust.

2.4. Bitumen

In this study 85-100 grade bitumen was used. Same bitumen was used for all the mixes so the type and grade of binder would be constant.

Figure 6. Appearance of bitumen.

3. Methods

3.1. Laboratory Tests for the Properties of Materials

3.1.1. Properties of Aggregates

Tests were performed to determine the Aggregate Crushing Value, Aggregate Impact Value, Specific gravity, L.A Abrasion value and Water absorption according to the procedures specified by AASHTO and BS standards and results are summarized in Table 1.

Table 1. Physical Properties of aggregates.

Properties	AASHTO/BS Designation	Aggregates		Standard values (AASHTO)
		Coarse	Fine	
Aggregate Crushing Value (%)	BS812:Part3	12	-	< 30%
Aggregate Impact Value (%)	BS812:Part3	20	-	< 30%
Specific gravity	T85	2.84	2.89	2.60 – 2.90
L.A Abrasion value(Grade A)	T96	15	-	< 40

3.1.2. Properties of Filler

Tests were performed to determine the specific gravity of filler fine sand with stone dust, waste concrete dust and brick dust according to the procedures specified by AASHTO and given in Table 2.

Table 2. Physical Properties of fillers.

Filler type	AASHTO Designation	Specific gravity
Fine sand and stone dust mix	T85	2.71
Waste concrete dust	T85	2.43
Brick dust	T85	2.34

3.1.3. Properties of Bitumen

Penetration, specific gravity, ductility, softening point, flash and fire point of bitumen were determined according to the procedure specified by AASHTO standards. Properties of bitumen used in bituminous mix are given in Table 3

Table 3. Properties of bitumen.

Properties	AASHTO Designation	Test Value	Standard Values(AASHTO)
Penetration(1/10 th mm)	T49	96	85-100
Specific Gravity	T229	1.03	1.01-1.05
Ductility(mm)	T51	100+	Min. 100
Softening Point(°C)	T53	48	45°C -52°C
Flash Point(°C)	T48	295	280 °C-300°C
Fire Point(°C)	T48	315	300°C-320°C

3.2. Marshall Mix Design

In this research work Marshall Stability testing setups was used. Tests were performed to determine the Marshall stability, flow value, optimum bitumen content and amount of bitumen required for mix types containing different filler. To investigate the Marshall stability of bituminous mixes with different fillers 45 specimens of 101.6 mm diameter and approximately 63.5 mm thickness were prepared. It was observed from preliminary trails that about 1200 grams of aggregates were required to prepare one specimen of 101.6 mm (4 inch) diameter and 63.5 mm (2.5 inch) thick. Three specimens were prepared for each bitumen contents and 5 bitumen contents (4% to 6%) were used with an increment of 0.5% by weight of total mix. Preparation of specimen, compaction and testing were performed according ASTM D1559 (Marshall Mix Design Method). The aggregates and bitumen were rapidly mixed to yield a mixture having a uniform distribution of bitumen throughout. The bulk specific gravity and density, theoretical maximum specific gravity was determined according to ASTM D2726, ASTM D2041 respectively. After determination of specific gravities of the compacted specimens were immersed in the thermostatically controlled water bath maintained at a temperature of 60°C for 30 minutes. Marshall Stability and flow test were performed afterwards for each specimen by testing machine. Marshall testing machine is an electrically powered compression testing device. Load was applied to the test specimens through cylindrical segment testing heads at a constant rate of vertical strain of 51 mm (2inch) per minute until the maximum load was reached. The maximum load resistance and respective flow value were recorded and percent air voids were determined according to ASTM D3203.

4. Results and Discussions

Marshall Test results of compacted mix types containing filler fine sand with stone dust, waste concrete dust and brick dust are tabulated in table 4, 5 and 6 respectively. The tables contains the value of bitumen content, unit weight, Marshall Stability, flow value, percent air voids (%V_a), percent voids in mineral aggregate (%VMA) and percent voids filled with bitumen (%VFB). The relationship between unit weight, Marshall Stability, flow value, percent voids in total mix (%V_a), percent voids in mineral aggregates (%VMA) and percent voids filled with bitumen (%VFB) with percentage of bitumen content are shown in figure 7 to figure 12.

Table 4. Marshall Properties of specimens with filler fine sand and stone dust mix.

% BC	Unit weight(kg/m³)	Stability(kN)	Flow value(mm)	Va(%)	VMA(%)	VFB(%)
4.0	2348	8.0	2.50	6.80	20.1	66.17
4.5	2375	8.8	2.65	5.10	19.3	73.58
5.0	2391	9.5	2.83	3.90	18.8	79.26
5.5	2395	9.8	3.03	3.00	19	84.21
6.0	2390	9.3	3.25	2.50	19.8	87.37

Table 5. Marshall Properties of specimens with filler waste concrete dust.

% BC	Unit weight(kg/m³)	Stability(kN)	Flow value(mm)	Va(%)	VMA(%)	VFB(%)
4.0	2358	9.3	2.57	7.09	17.8	60.17
4.5	2380	10.1	2.75	5.30	16.5	67.88
5.0	2395	10.8	2.93	4.00	16.2	75.31
5.5	2398	11.1	3.13	3.10	16.5	81.21
6.0	2393	10.7	3.38	2.70	17.5	84.57

Table 6. Marshall Properties of specimens with filler brick dust.

% BC	Unit weight(kg/m³)	Stability(kN)	Flow value(mm)	Va(%)	VMA(%)	VFB(%)
4.0	2341	9.5	2.67	7.28	21.5	66.14
4.5	2367	10.3	2.85	5.5	20.41	73.05
5.0	2385	11	3.03	4.2	20	79
5.5	2390	11.3	3.23	3.2	20.5	84.39
6.0	2386	11.1	3.50	2.8	20.8	86.54

Key: BC= Bitumen content.

For fine sand with stone dust, waste concrete dust and brick dust filler samples, the bitumen content at maximum unit weight and stability were determined from figures 7 & 8 respectively and bitumen content at 4 % air voids in the total mix were determined from figure 10. The average of these three bitumen content is taken as optimum bitumen content.

The amount of bitumen to be added in a mixture cannot be too much or too little. An optimum amount of bitumen should use so that the aggregate are fully coated with bitumen and the voids within the bituminous material are sealed up. As such, the durability of the pavement can be enhanced by the impermeability achieved. Moreover, a minimum amount of binder is essential to prevent the aggregates from being pulled out by the abrasive actions of moving vehicles on the carriageway.

The optimum bitumen content for waste concrete dust and brick dust were found 5.33% and 5.37% which are almost the same as the optimum value obtain for conventional filler (e.g. fine sand with stone dust). It means the filler waste concrete dust and brick dust will provide the same surface area to absorb bitumen as filler fine sand with stone dust. The Marshall properties obtained for these three types of fillers reveal that due to having slightly higher bitumen content, nonconventional filler specimens are found to exhibit higher stability value (11kN& 11.18kN) compared to conventional filler specimens (9.68kN).

Figure 7 showing the relationship between unit weight and bitumen content indicates that the unit weight of compacted specimens for the mix increases initially with an increase in bitumen content, reaches a maximum value and then decreases. This is because while bitumen content increases in the mix, it fills the voids hence increase unit weight.

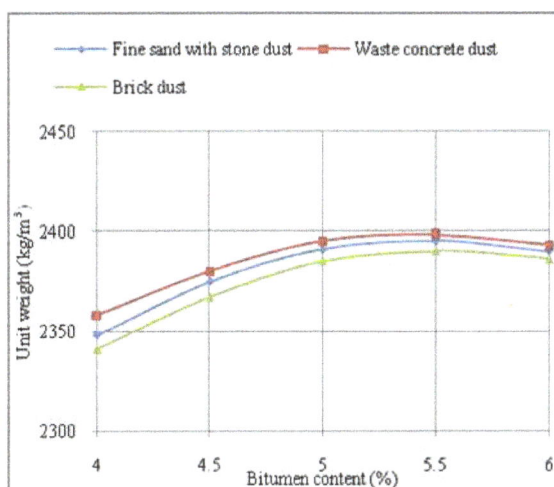

Figure 7. Variation of unit weight with bitumen content (%).

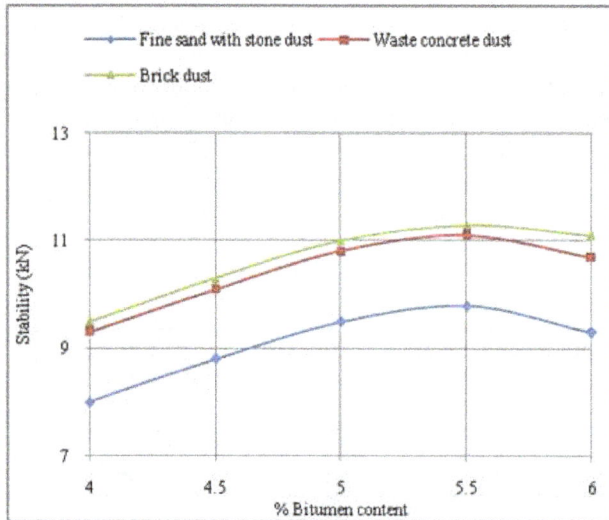

Figure 8. *Variation of stability with bitumen content (%).*

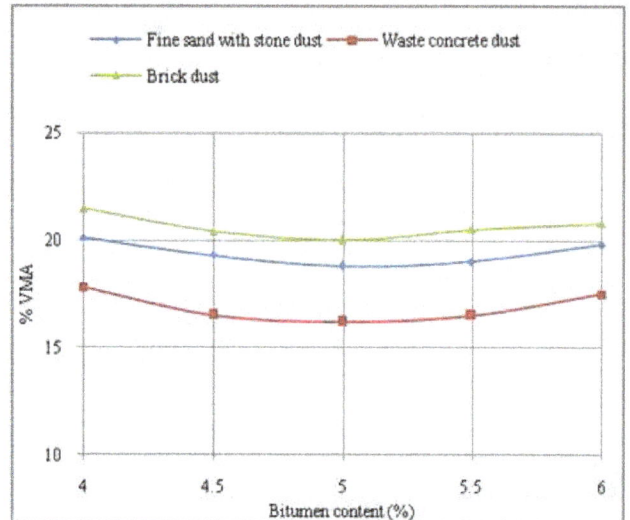

Figure 11. *Variation of % VMA with bitumen content (%).*

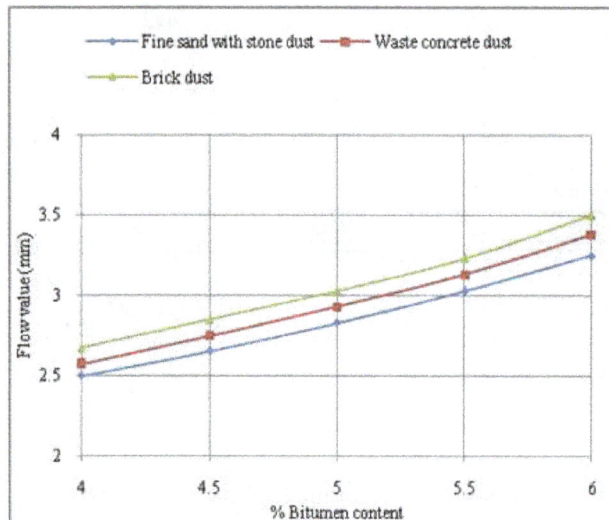

Figure 9. *Variation of flow value with bitumen content (%).*

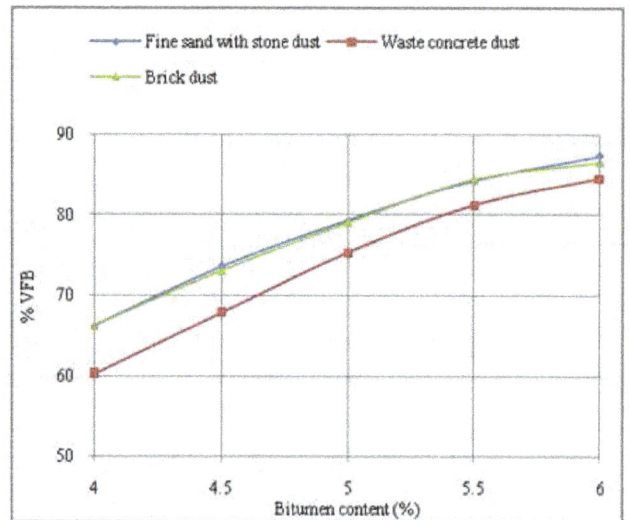

Figure 12. *Variation of % VFB with bitumen content (%).*

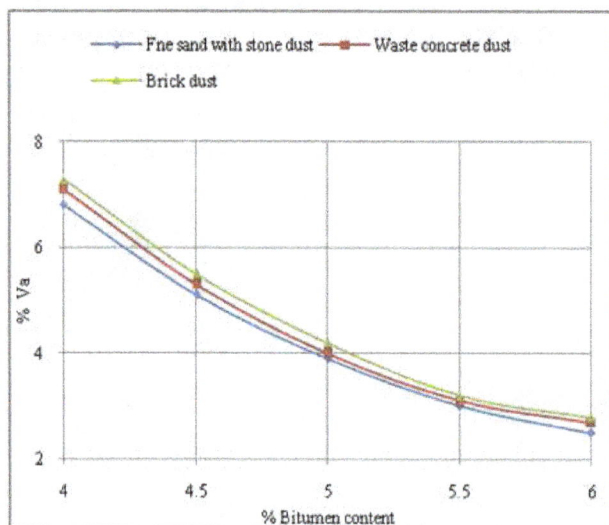

Figure 10. *Variation of % V_a with bitumen content (%).*

Figure 8 showing the relationship between Marshall Stability and bitumen content which is similar in nature to that of unit weight. At optimum bitumen content bituminous mixes containing brick dust and waste concrete dust as filler displayed stability of 11.18 KN and 11 KN respectively which are higher than the value 9.68 KN obtained for mix containing fine sand with stone dust as filler. Marshall Stability increases with the increase of viscosity of the asphalt cement. So the addition of brick dust and waste concrete dust filler in the mixture may produce a more viscous asphalt cement mixture binder thus increasing the stability.

Figure 9 showing the relationship between flow value and bitumen content, which indicates that the flow value of the specimens increase with increase in bitumen content. The rate of increase is being higher for higher proportions of bitumen. The value obtained for mixes with brick dust and waste concrete dust as filler are higher than the mix containing fine sand with stone dust as filler which indicates that brick dust and waste concrete dust are coarser than fine sand with stone dust mix. This means the more fine filler (e.g. fine sand with stone dust) tempers the bitumen mixtures by extending the bitumen binder, hence gives lower stability.

Figure 10 showing the relationship between percent air voids in total mix and bitumen content. It indicates higher percentages of voids in mineral aggregate were obtained from mixes prepared by brick dust filler. This may be due to the fact that brick dust filler is coarser than waste concrete dust and fine sand with stone dust filler.

Figure 11 showing the relationship between percent voids in mineral aggregate and bitumen content, which shows that the percentage of voids in mineral aggregates decreases initially with an increase in bitumen content, reaches a minimum value and then increases. It indicates higher percentages of voids in mineral aggregate were obtained from mixes prepared by brick dust filler. This is because brick dust filler is coarser than fine sand with stone dust filler. A minimum percentage of VMA is required in mixtures to allow adequate bitumen content to coat aggregate particles with sufficient bitumen

Figure 12 showing the relationship between percent voids filled with bitumen and bitumen content. It indicates that the %VFB of compacted specimens for the mixes increases with an increase in bitumen content.

The test results obtain for the three types of filler specimens satisfies the standard value specified by Roads & Highways Department of Bangladesh, only the exception is that the %VMA obtains for brick dust filler specimen is slightly more than the standard value. Also the value of % VFB calculated for three types of filler exceed the standard limit.

At optimum bitumen content, the values of unit weight, Marshall Stability, flow value, percent air voids (%V_a), percent voids in mineral aggregate (%VMA) and percent voids filled with bitumen (%VFB) are shown in table 7. It indicate that bituminous mixes with non-conventional filler (e.g. waste concrete dust & brick dust) are found to have satisfactory Marshall properties, which are almost same as those of conventional fillers (e.g. .fine sand stone dust mix) .

Table 7. Comparison between three types of mineral filler.

Properties	Fine sand and stone dust mix	Waste concrete dust	Brick dust	Standard values (RHD,Bangladesh)
OBC	5.3	5.33	5.37	4.9-6.5
Unit weight(kg/m³)	2393.4	2396.98	2388	-
Stability(kN)	9.68	11	11.18	Min.5.33 kN
Flow value(mm)	2.95	3.06	3.15	2-4
%V_a	3.36	3.41	3.6	3-5
%VMA	18.92	16.40	20.30	15-20
%VFB	82.23	79.20	82.23	70-80

Key: OBC= Optimum bitumen content.
RHD= Roads & Highway Department.

5. Conclusion

(1) It is found that bituminous mixes containing waste concrete dust and brick dust as fillers have almost same Marshall Properties as those of conventional (fine sand with stone dust) fillers.

(2) It is observed that with the increase of bitumen content, the Marshall Stability of the mixture also increased. Bituminous mixes containing brick dust as filler showed maximum stability of 11.3 KN at 5.5 % bitumen content. The maximum stability value obtained for the mixes containing waste concrete dust and fine sand with stone dust as filler were 11.1KN and 9.8 KN respectively.

(3) From the considerations of economy and availability, waste concrete dust and brick dust are suitable as filler as compared with mineral filler generally used in bituminous mix.

(4) With some modification in design mixes, can result in utilization of waste concrete dust and brick dust as fillers in bituminous pavement, thus save considerable investment in construction and partially solving the disposal of wastes.

Acknowledgments

For successful completion of the project authors tenders their best regard to Md. Abdul Alim, Professor and Head of the Department of Civil Engineering of Rajshahi university of Engineering & Technology (RUET). Also thanks are credited in favors of all the laboratory assistant of the Transportation Engineering of Rajshahi university of Engineering & Technology (RUET).

References

[1] Kadeyali: Principles and practice of Highway Engineering.3rd edition (1997).

[2] Katamine NM: Phosphate waste in mixtures to improve their deformation. J Transport Eng 2000; 126:382–9.

[3] Asi Ibrahim, Assa'ad Abdullah: Effect of Jordanian oil shale fly ash on asphalt mixes. J Mater CivEng 2005; 17:553–9.

[4] Lin Deng-Fong, Lin Jyh-Dong, Chen Shun-Hsing: The application of baghouse fines in Taiwan. Resour Conserv Recycle 2006; 46:281–301.

[5] Sung Do Hwang, Hee Mun Park, Suk keun Rhee: A study on engineering characteristics of asphalt concrete using filler with recycled waste lime. Waste Manage 2008; 28:191–9.

[6] Xue Yongjie, Hou Haobo, Zhu Shujing, Zha Jin: Utilization of municipal solid waste incineration ash in stone mastic asphalt mixture: pavement performance and environmental impact. Constr Build Mater 2009; 23:989–96.

[7]   Huang Baoshan, Dong Qiao, Burdette Edwin G: Laboratory evaluation of incorporating waste ceramic materials into Portland cement and asphaltic concrete. Constr Build Mater 2009; 23:3451–6.

[8]   Kalkattawi, H.R: Effect of Filler on the Engineering Properties of Asphalt Mixes, M.S. Thesis, King Abdul Aziz University, Jeddah, Saudi Arabia. (1993)

[9]   Anderson, D. A.: Guidelines for use of dust in hot mix asphalt concrete mixtures."Proc. Association of Asphalt Paving Technologists, 56, Association of Asphalt Paving Technologists, St. Paul, MN, 492–516, 1987

[10] Elliot, R.P., Ford, M.C., Ghanim, M., and Tu, Y.F. :Effect of aggregate gradation variation on asphalt concrete mix properties, Transportation Research Record, 1317, National Research Council, Washington, D.C., 1991

[11] Kandhal, P.S., Lynn, C.Y., and Parker, F.: Characterization tests for mineral fillers related to performance of asphalt paving mixtures, NCAT Rep. No. 98-2, 1998

[12] Bahia, H.U., Zhai, H., Bonnetti, K.,and Kose, S.: Non-linear visco-elastic and fatigue properties of asphalt binders, Journal of Association of Asphalt Paving Technology, 68, 1-34,1999

[13] Geber, R. and Gomze, L.A.: Characterization of mineral materials as asphalt fillers, Material Science Forum,659, 471-476, 2010

[14] Vavrik, W.R., Pine, W.J., Carpenter, S.H., and Bailey, R.: Bailey method for gradation selection in hot-mix asphaltmixturedesign,TransportationResearchBoard,NationalR esearchCouncil,Washington,D.C., USA., 2002

[15] Zulkati, A., Diew, W. Y. and Delai, D.S. :Effects of Fillers on properties of Asphalt-Concrete Mixture, Journal of Transportation Engineering, ASCE, Vol. 138, No. 7, 902-910.,2012

[16] Taylor, R.: Surface interactions between bitumen and mineral fillers and their effects on the rheology of bitumen-filler mastics. Ph.D. thesis, Univ. of Nottingham, UK., 2007

[17] Lesueur, D.: The colloidal structure of bitumen: consequences on the rheology and on the mechanisms of bitumen modification. Adv.Colloid Interface Sci., 145(1–2), 42–82., 2009

[18] Bahia, H. U., Faheem, A., Hintz, C., Al-Qadi, I., Reinke, G., and Dukatz, E.: Test methods and specification criteria for mineral filler used in HMA."NCHRP Research Results Digest, 357, Transportation Research Board, Washington, DC., 2011

[19] American Society for Testing and Materials (ASTM): Road and Paving Materials; Vehicle Pavement System, Annual Book of ASTM Standards, Section 4, Volume 04.03,2003.

14

Dynamic Analysis of High Rise Seismically Isolated Buildings

Mohammed Naguib, Fikry A. Salem, Khloud El-Bayoumi

Civil Engineering Dep., Faculty of Engineering, Mansoura University, Egypt

Email address:
engkhloud@yahoo.com (K. El-Bayoumi)

Abstract: The purpose of this paper is to offer a relative understanding of the seismic performance enhancements that a typical 40-story steel office building can achieve through the implementation of base isolation technology. To reach this understanding, the structures of a fixed-base office building and a base-isolated office building of similar size and layout were designed; their seismic performance was compared in both response spectrum analysis and time history analysis. As a result of this paper, building owners and construction industry professionals can recognize the benefits of implementing base isolation on a wider range of projects, thereby creating the potential for a significant increase in the technology's use.

Keywords: Triple Friction Pendulum Bearing, Structure Control, Seismic Isolation, Base Isolation, High Rise Buildings

1. Introduction

A critical aspect in the design of civil engineering structures is the reduction of response quantities such as velocities, deflections and forces induced by environmental dynamic loadings (i.e., wind and earthquake). Structural control methods are the most recent strategies for this purpose, which can be classified as active, semi-active, passive, and hybrid control methods [1]. Control methods have been slow in their acceptance in the structural design community because the systems are often prohibitively complicated, large and expensive. Over time, however, their utility is becoming more recognized and improvements in the technology are making them more viable options in new construction and retrofits.

In the last three decades or so, the reduction of structural response, caused by dynamic effects, has become a subject of research, and many structural control concepts have been implemented in practice [2].

Base isolation systems are one of the most successful and widely-applied methods of mitigating structural vibration and damage during seismic events. Base isolation systems have been installed in numerous full-scale structures [3]. Sliding isolator works on principle of friction. This approach is based on the premise that the lower the friction coefficient, the less the shear transmitted [4]. The type of base isolation technology that is used in this study is the Triple Friction Pendulum (TFP) bearing. The Triple Friction Pendulum (TFP) bearing differs from the single Friction Pendulum (FP) bearing in that there are 3 friction pendulum mechanisms existing in each bearing instead of just 1 mechanism. These mechanisms are activated at different stages as the seismic demand gets stronger. The 3 mechanisms are achieved by using 4 concave surfaces in a single bearing, with sliding occurring on two of the surfaces at a given time [5]. An image of the TFP's disassembled parts and a cross section of a TFP bearing are shown in Figures A, B below, respectively [6], [7].

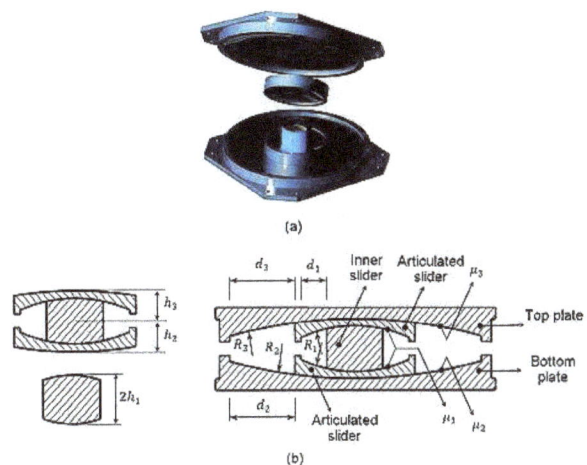

Figure A. Triple friction pendulum bearing, (a) Three-dimensional view; (b) Section view and basic parameters.

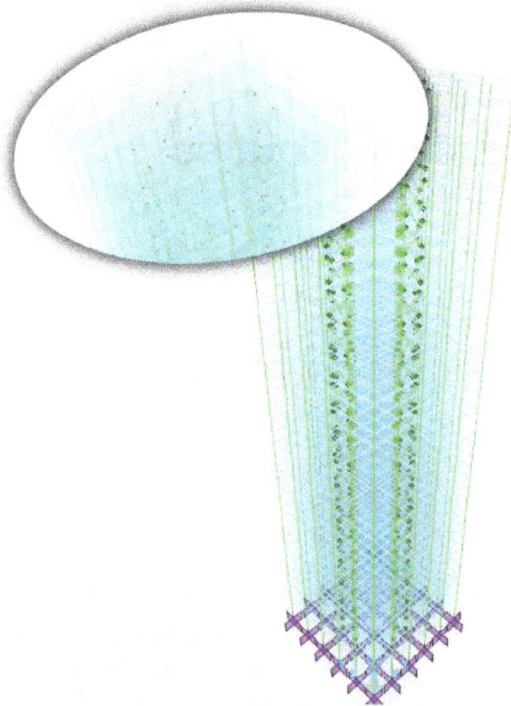

Figure B. 3D view of the SAP2000 model.

2. Structural Model

A model of (30*30) m 40-story building was created with steel columns, beams and sections of 0.2m width concrete slab. The steel superstructure had a lateral system of special concentrically braced frames (SCBF) in both the transverse and longitudinal directions, and that structural system was used for both of the fixed-base and isolated-base buildings designed for this study. Building place was assumed to be in Cairo and according to USGS worldwide seismic design tool [8] it was found that response spectrum parameters were 0.5815g and 0.3395g for S_{DS} and S_{D1} respectively, 0.509g and 0.872g for S_{M1} and S_{MS} respectively.

3. TFP Bearings

To create the isolated-base model, three TFP bearings were assumed with the following properties:

3.1. Calculating D_D (Upper Bound Analysis)

Table A. basic parameters for calculating D_D

	μ	μ_1	D_y	F_{d1}	W	#B	ΣW
TFP1	0.082	0.085	0.00693	0.37711	324	4	1296
TFP2	0.092	0.094	0.00789	0.277243	648	16	10368
TFP3	0.112	0.112	0.00107	0.202702	1296	16	20736
				Σ 7566.413			32400

1) Let the displacement be D_D =0.63

2) Effective stiffness: $Q_d = \mu . \Sigma W$ =3382.6

$$K_D = \Sigma F_D / D_D = 12010$$

$$K_{eff} = K_{D} + Q_d / D_D = 17379$$

3) Effective period: (Eq.17.5-2, ASCE 7-10) [9]

$$T_{eff} = 2\pi \sqrt{\frac{\Sigma W}{K_{eff}. g}} = 2.7402$$

4) Effective damping: (Eq.17.8-7, ASCE 7-10) [9]

$$\beta_D = \frac{E}{2\pi K_{eff} D_D^2} = \frac{4\mu \Sigma(D_D - D_y)}{2\pi K_{eff} D_D^2} = 0.1956$$

5) Damping reduction factor:

$$\beta = (\frac{\beta_{eff}}{0.05})^{0.3} = 1.5056$$

6) Check D_D':

$$D_D' = \frac{S_{D1}. T_{eff}^2}{4\pi^2. \beta} g = 0.6302$$

Table B. Summary of Isolation Bearing Properties.

Property	TFP1	TFP2	TFP3
Place	Corner columns	Outer columns	Inner columns
Vertical load ton	324	648	1296
$R_{1eff} = R_{4eff}$ mm	2133	3395	6934
$R_{2eff} = R_{3eff}$ mm	330	526	1074
$d_1^* = d_4^*$ mm	339.8	540.4	1103.48
$d_2^* = d_3^*$ mm	41.5	65.9	30.85
$\mu_1 = \mu_4$ Lower bound	0.071	0.078	0.093
$\mu_2 = \mu_3$ Lower bound	0.053	0.066	0.093
μ Lower bound	0.068	0.076	0.093
$\mu_1 = \mu_4$ Upper bound	0.085	0.094	0.112
$\mu_2 = \mu_3$ Upper bound	0.064	0.079	0.111
μ Upper r bound	0.082	0.092	0.112

3.2. Sap2000 Link/Support Property Data Input (Upper Bound)

SAP 2000 version 16.0 and later versions has a direct link property that simulates the actual behavior of triple friction pendulum bearing [10], and then the 3 bearings input data are shown in figures below:

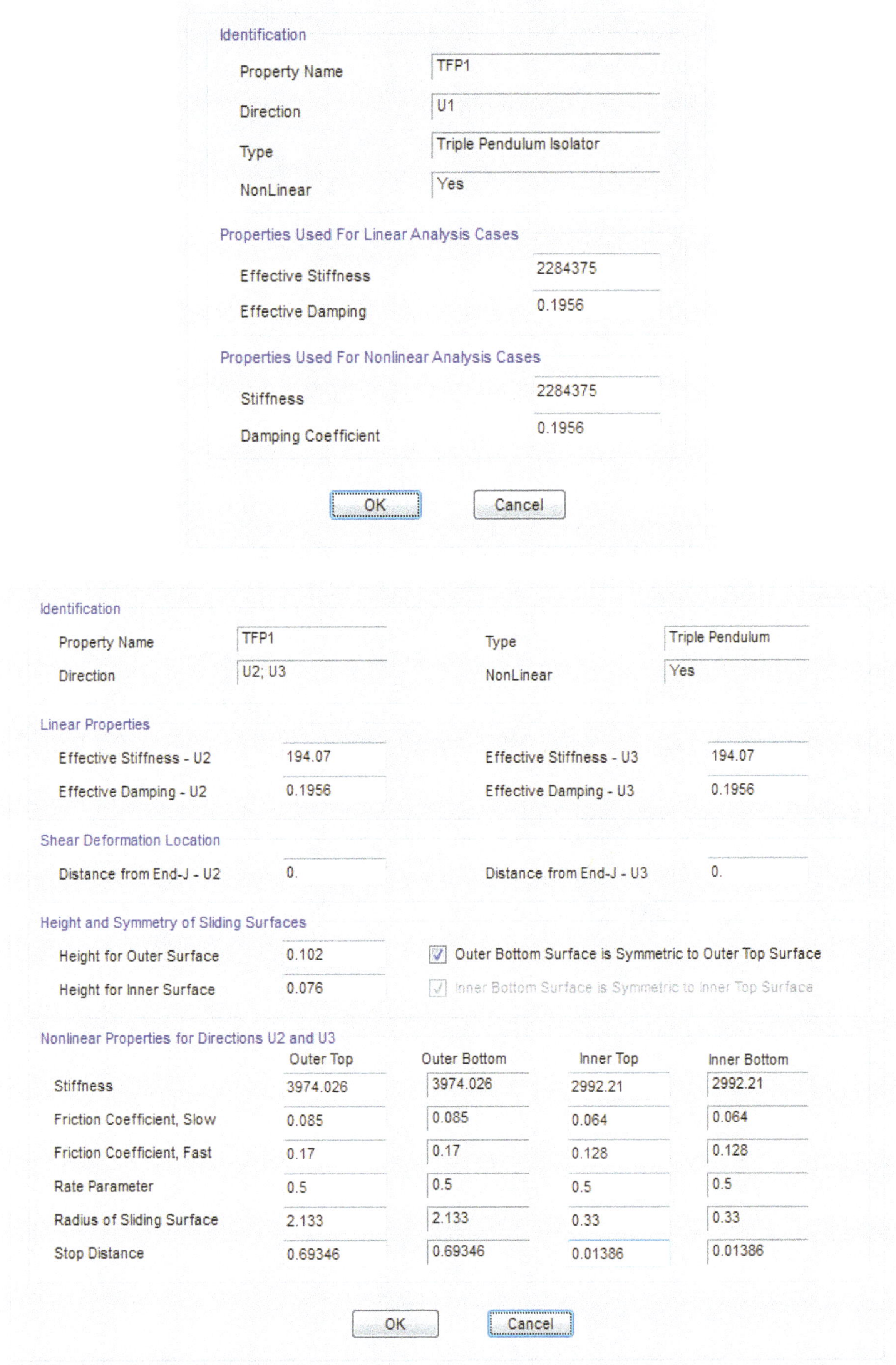

Identification

Property Name   TFP1

Direction   U1

Type   Triple Pendulum Isolator

NonLinear   Yes

Properties Used For Linear Analysis Cases

Effective Stiffness   2284375

Effective Damping   0.1956

Properties Used For Nonlinear Analysis Cases

Stiffness   2284375

Damping Coefficient   0.1956

OK   Cancel

Identification

Property Name	TFP1	Type	Triple Pendulum
Direction	U2; U3	NonLinear	Yes

Linear Properties

Effective Stiffness - U2	194.07	Effective Stiffness - U3	194.07
Effective Damping - U2	0.1956	Effective Damping - U3	0.1956

Shear Deformation Location

Distance from End-J - U2	0.	Distance from End-J - U3	0.

Height and Symmetry of Sliding Surfaces

Height for Outer Surface	0.102	☑ Outer Bottom Surface is Symmetric to Outer Top Surface
Height for Inner Surface	0.076	☑ Inner Bottom Surface is Symmetric to Inner Top Surface

Nonlinear Properties for Directions U2 and U3

	Outer Top	Outer Bottom	Inner Top	Inner Bottom
Stiffness	3974.026	3974.026	2992.21	2992.21
Friction Coefficient, Slow	0.085	0.085	0.064	0.064
Friction Coefficient, Fast	0.17	0.17	0.128	0.128
Rate Parameter	0.5	0.5	0.5	0.5
Radius of Sliding Surface	2.133	2.133	0.33	0.33
Stop Distance	0.69346	0.69346	0.01386	0.01386

OK   Cancel

Figure C. SAP2000 Friction Pendulum Bearing Properties for TFP1, a. Vertical direction U1; b. Lateral direction U2, U3.

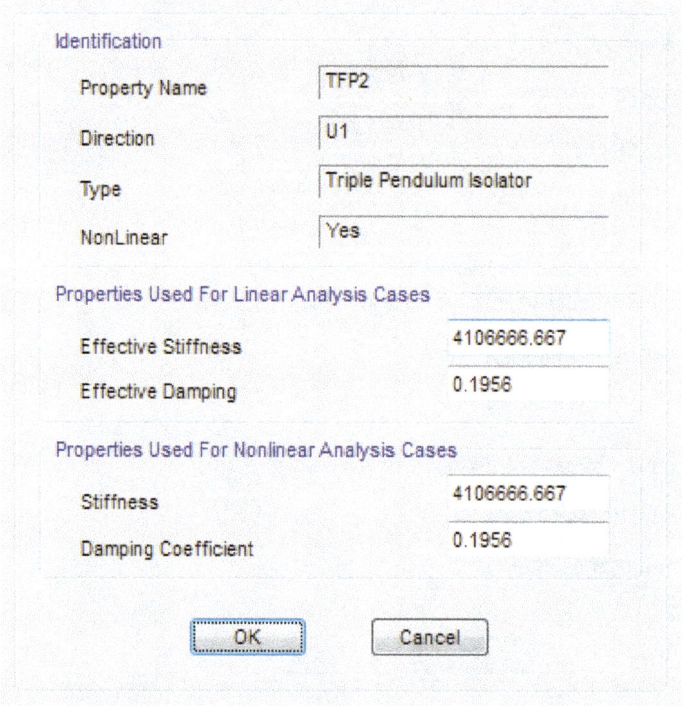

Identification

Property Name	TFP2
Direction	U1
Type	Triple Pendulum Isolator
NonLinear	Yes

Properties Used For Linear Analysis Cases

Effective Stiffness	4106666.667
Effective Damping	0.1956

Properties Used For Nonlinear Analysis Cases

Stiffness	4106666.667
Damping Coefficient	0.1956

OK   Cancel

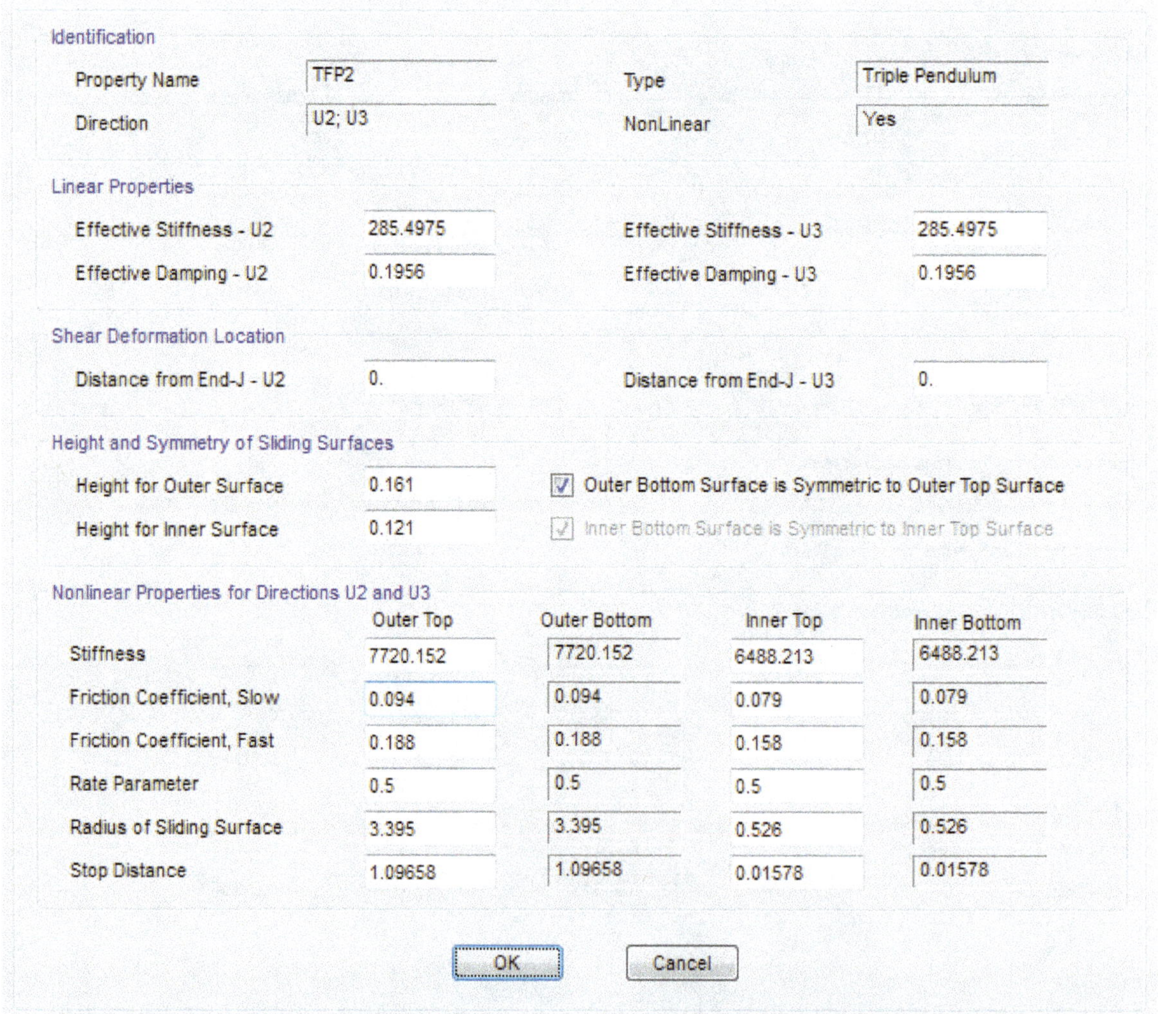

Identification

Property Name	TFP2	Type	Triple Pendulum
Direction	U2; U3	NonLinear	Yes

Linear Properties

Effective Stiffness - U2	285.4975	Effective Stiffness - U3	285.4975
Effective Damping - U2	0.1956	Effective Damping - U3	0.1956

Shear Deformation Location

Distance from End-J - U2	0.	Distance from End-J - U3	0.

Height and Symmetry of Sliding Surfaces

Height for Outer Surface	0.161	☑ Outer Bottom Surface is Symmetric to Outer Top Surface
Height for Inner Surface	0.121	☑ Inner Bottom Surface is Symmetric to Inner Top Surface

Nonlinear Properties for Directions U2 and U3

	Outer Top	Outer Bottom	Inner Top	Inner Bottom
Stiffness	7720.152	7720.152	6488.213	6488.213
Friction Coefficient, Slow	0.094	0.094	0.079	0.079
Friction Coefficient, Fast	0.188	0.188	0.158	0.158
Rate Parameter	0.5	0.5	0.5	0.5
Radius of Sliding Surface	3.395	3.395	0.526	0.526
Stop Distance	1.09658	1.09658	0.01578	0.01578

OK   Cancel

Figure D. *SAP2000 Friction Pendulum Bearing Properties for TFP2, a. Vertical direction U1; b. Lateral direction U2, U3.*

Identification

Property Name	TFP3
Direction	U1
Type	Triple Pendulum Isolator
NonLinear	Yes

Properties Used For Linear Analysis Cases

Effective Stiffness	9645375
Effective Damping	0.1956

Properties Used For Nonlinear Analysis Cases

Stiffness	9645375
Damping Coefficient	0.1956

OK   Cancel

Identification

Property Name	TFP3	Type	Triple Pendulum
Direction	U2; U3	NonLinear	Yes

Linear Properties

Effective Stiffness - U2	417.305	Effective Stiffness - U3	417.305
Effective Damping - U2	0.1956	Effective Damping - U3	0.1956

Shear Deformation Location

Distance from End-J - U2	0.	Distance from End-J - U3	0.

Height and Symmetry of Sliding Surfaces

Height for Outer Surface	0.33	☑ Outer Bottom Surface is Symmetric to Outer Top Surface
Height for Inner Surface	0.247	☑ Inner Bottom Surface is Symmetric to Inner Top Surface

Nonlinear Properties for Directions U2 and U3

	Outer Top	Outer Bottom	Inner Top	Inner Bottom
Stiffness	135150.838	135150.838	133944.1341	133944.1341
Friction Coefficient, Slow	0.112	0.112	0.111	0.111
Friction Coefficient, Fast	0.224	0.224	0.222	0.222
Rate Parameter	0.5	0.5	0.5	0.5
Radius of Sliding Surface	6.934	6.934	1.074	1.074
Stop Distance	2.209108	2.209108	0.002148	0.002148

OK   Cancel

Figure E. SAP2000 Friction Pendulum Bearing Properties for TFP3, a. Vertical direction U1; Lateral direction U2, U3.

4. Time History Data Input

In order to account for the variation of the building's response throughout the duration of each earthquake ground motion, a time history analysis was required. Ground motions representative of different hazard levels have been assembled for this research. All these ground motions are assembled from The Pacific Earthquake Engineering Research Center ground motion database [11].

Table C. *Time history EQ Ground Motions.*

NGA#	EQ Name	Year	Station	Magnitude
182	"Imperial Valley-06"	1979	"El Centro Array #7"	6.53
183	"Imperial Valley-06"	1979	"El Centro Array #8"	6.53
1605	"Duzce Turkey"	1999	"Duzce"	7.14
1158	"Kocaeli Turkey"	1999	"Duzce"	7.51

5. Analysis

5.1. Modal Analysis

The figure G below illustrates the modal periods resulted from the response spectrum modal analysis. And it was found that the average modal period increased by about 9.11%.

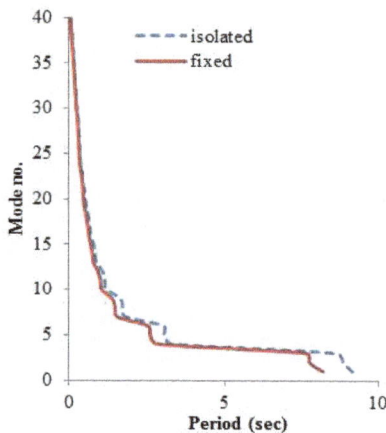

Figure F. *Modal periods for fixed and isolated models.*

5.2. In-Story Drift

For the response spectrum analysis, the design drift for the fixed-base was and isolated-base models were 0.00637, 0.00246 respectively, which met the design drift limit of 0.0150For the time history analysis; the design drifts of motion 182 were 0.016123, 0.000855 for fixed and isolated models, for motion 183 they were 0.016095, 0.000668 for motion 1158 they were 0.025387, 0.000216 and for motion 1605 they were 0.028296, 0.001506.

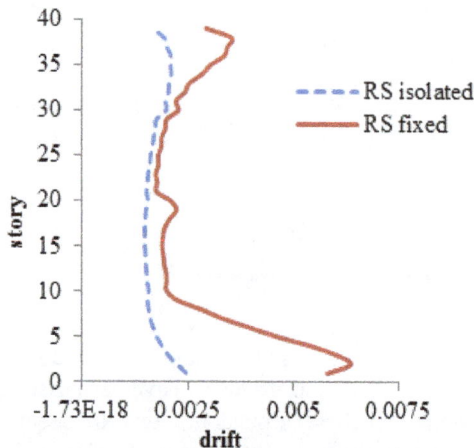

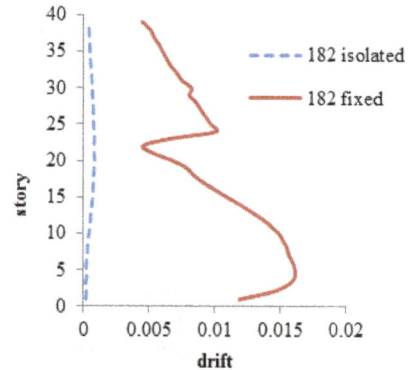

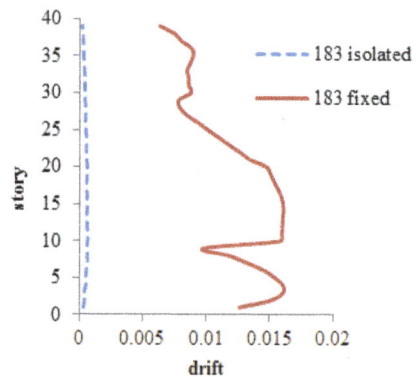

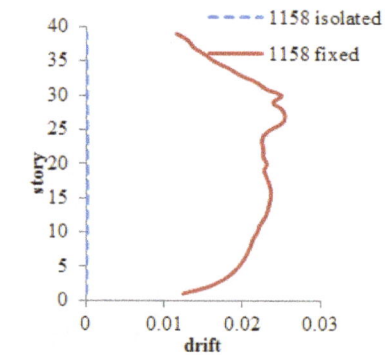

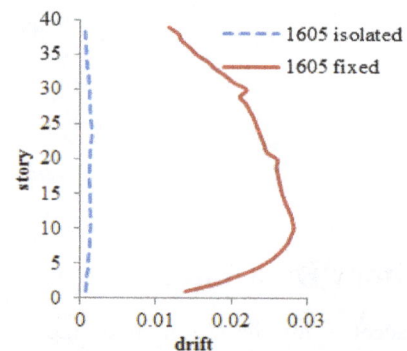

Figure G. *In story drift due to response spectrum and time history analysis.*

It was foundthat the in-story drift for the isolated model was 29.61% less than the fixed model in case of response spectrum analysis and 93.23%, 95.63%, 99.25%, 95.09% in time history analysis motions 182, 183, 1158, and 1605 respectively

5.3. Floor Acceleration

In case of response spectrum analysis; the resulted maximum story acceleration was 0.337 g for fixed model and 0.0612 g for isolated model. In case of time history analysis it were 0.583 g , 0.889 g ,0.519 g and 0.669 g for fixed model and 0.341 g , 0.614 g , 0.312 g and 0.405 g for isolated model in motions 182 , 183 , 1158 and 1605 respectively.

It was found that story acceleration for the fixed model was 72.87% higher than the isolated model in case of response spectrum analysis and 16.56%, 15.69%, 22.81% and 24.71% in time history analysis motions 182, 183, 1158, and 1605 respectively, Then it was noted that isolator system efficiency in decreasing story acceleration was directly proportional to motion intensity and reversely proportional to motion ground acceleration.

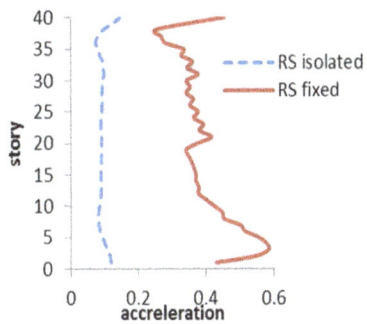

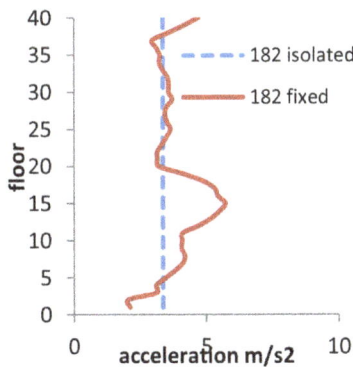

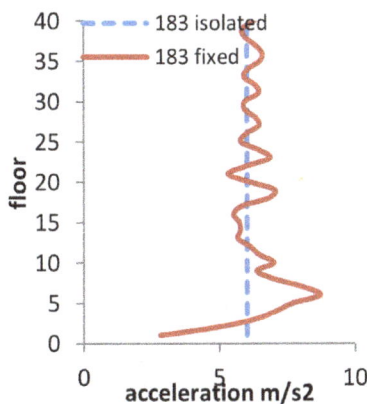

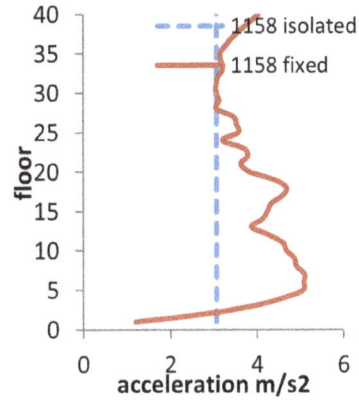

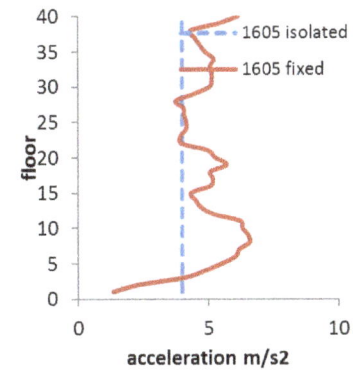

Figure H. Floor acceleration due to response spectrum and time history analysis.

5.4. Story Displacement

In case of response spectrum analysis; the resulted maximum story displacement was 0.333 m for fixed model and 0.287 m for isolated model. In case of time history analysis it were 1.19 , 1.44 , 2.44 and 2.64 for fixed model and 0.43 , 0.51 , 0.32 and 0.69 for isolated model in motions 182 , 183 , 1158 and 1605 respectively. Figure H illustrates the story displacements resulted from the response spectrum and time history analysis. And it was found that maximum story displacement for the isolated model was 13.79% less than the fixed model in case of response spectrum analysis and 63.92%, 64.41%, 87.08%, 74.1% in time history analysis motions 182, 183, 1158, and 1605 respectively.

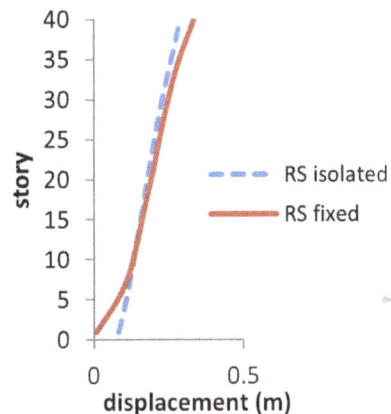

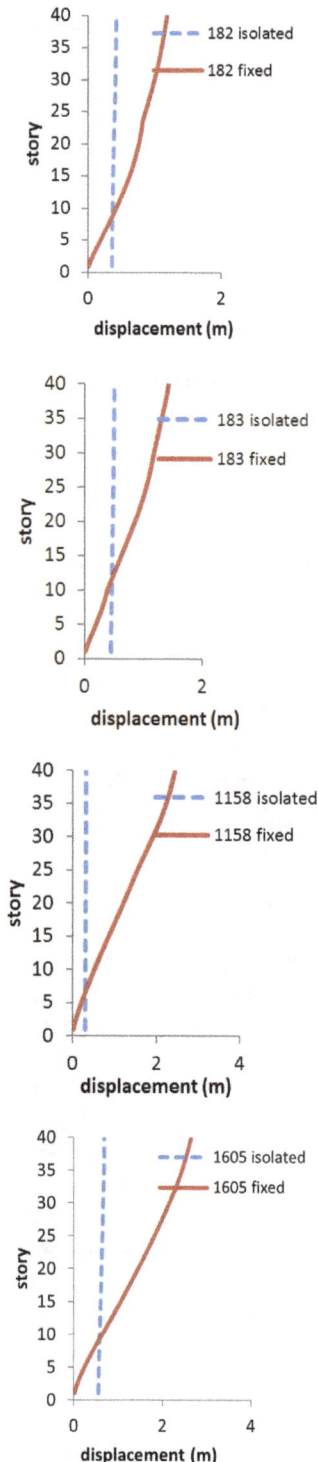

Figure H. *Floor displacements due to response spectrum and time history analysis.*

6. Conclusion

The benefits of implementing base isolation in the 40-story steel office building were clearly shown by the results of this study, including:

- Reduction of floor accelerations, in-story drifts and base reactions by more than 90% percentage.
- Improvement of structural seismic performance levels.

- Lowering the coefficients of friction of the TFP bearings is the most effective way to improve seismic performance (i.e. reduce the superstructure's response values, including floor accelerations and in-story drifts) when implementing base isolation in a tall, flexible building.
- Using TFP bearings with larger radii of curvature (R) leads to a more flexible (smaller lateral stiffness) isolation system and improves seismic performance, although larger bearing sizes are also more expensive.
- Isolator system efficiency in decreasing story displacement, in-story drift, story acceleration and base reactions was directly proportional to earthquake magnitude; and if two EQ have the same magnitude, then efficiency of isolator was reversely proportional to motion acceleration.
- Isolator system efficiency in decreasing base reactions was higher than its efficiency in decreasing in-story drift, story acceleration.

References

[1] T.K. Datta, Control of dynamic response of structures. Proc. of The Indo-US Symposium on Emerging Trends in Vibration and Noise Engineering, 1996, pp.18-20.

[2] R. Randa, T.T. Soong, Parametric study and simplified design of mass dampers. Engineering Structures, Vol. 20, No. 3, 1998, pp. 193 -204.

[3] Kelly J.M. Aseismic Base Isolation: Its History and Prospects. Proc. 1st World Congress on Joints and Bearings,1981, ACI-SP-70, 1: 549-586.

[4] Mostaghel N. and Tanbakuchi J., Response of Sliding Structures to Earthquake Support Motion, Earthquake Engineering and Structural Dynamics, 11, 1983, 729-748

[5] Nicholas R. Marrs (2013). Seismic performance comparison of a fixed-base versus a base isolated office building.

[6] Dao ND, Ryan KL, Verification of a Three-dimensional Triple Pendulum Bearing Element with General Friction Model Using Full Scale Test Data. Submitted toEarthquake Engineering and Structural Dynamics, 2012.

[7] Dao ND. Experimental and Analytical Studies of the Seismic Response of a Full-scale 5-story Steel Moment Frame Building Isolated by Triple Pendulum Bearings.Dissertation,University of Nevada – Reno, 2012

[8] U.S. Geological Survey Earthquake Hazards Program: Seismic Hazard Maps and Data. (2012). http://geohazards.usgs.gov/designmaps/ww/.

[9] American Society of Civil Engineers – Code 7: Minimum Design Loads for Buildings and Other Structures. (2010). 21 June 2011.

[10] Khloud El-Bayoumi, Modeling of triple friction pendulum bearings in sap2000. International Journal of Advances in Engineering & Technology, 2015, Vol. 8 issue 1.

[11] Pacific Earthquake Engineering Research Center (PEER). Web-Based PEER Ground Motion Database (Beta Version) [Interactive web-site database]. Berkeley, California (2010). http://peer.berkeley.edu/peer_ground_motion_database.

15

Seismic Analysis (Non-linear Static Analysis (Pushover) and Nonlinear Dynamic) on Cable - Stayed Bridge

Mohammad Taghipour[1], Hesamoldin Yazdi[2]

[1]Industrial Engineering, Science & Research Branch of Islamic Azad University, Tehran, Iran
[2]Civil engineering, non-profit institution of higher education, Aba - Abyek, Qazvin, Iran

Email address:
mohamad.taghipour@srbiau.ac.ir (M. Taghipour), hesamyazdi18@yahoo.com (H. Yazdi)

Abstract: Pushover analysis application development is greatly increased in recent years and numerous advanced methods to evaluate the seismic pushover are provided. Because these methods have been proposed mainly for building structures and given the fundamental differences between the behavior of bridge structures and buildings using pushover methods on the bridge structure with the uncertainties faced. 1. First, the effect of non-linear pushover results (time history) 2. Choose a target displacement due to the finite element model should be examined in order to understand the prediction of the seismic capacity. Thus a pushover analysis is presented for evaluation of seismic bridge pylons and deck where the effects of displacement and deformation of the plastic joints, structural changes in the modal characteristics of change used plastic forms and effects of higher modes can be seen clearly. The method is able to accurately approximate the dynamic response of the nonlinear analysis. Ultimately this method compared with analysis time history.

Keywords: Seismic Analysis, Non-linear Static Analysis, Non-linear Dynamic Analysis, Cable Stayed Bridge

1. Introduction

Cable-stayed bridge represents key points of the transport networks and, consequently, they are thoughts to remain nearly elastic under the design seismic action, typically dampers to control the response as located in seismic-prone areas. However, several cable-stayed bridges are in Greece, Brazil, US and China also allow some structural damage in the towers in order to reduce response uncertainties under unexpectedly large earthquakes. There are many cable-stayed bridges without seismic devices which are exposed to large earthquakes and elastic design. By this method ductility demand along the towers is acceptable, and it is used to give the elastic response of the deck. Non-Linear Response History Analysis is undoubtedly the most rigorous methodology to deal with inelasticity in dynamic design, this type of analysis with responding to the ambiguities of non-linear static analysis is a complete method. The model, design and analysis in non-linear dynamic are in accordance to the local and international regulations. As this method is complex and time-consuming, non-linear static analysis as occurred at short time, can be a good alternative instead of non-linear dynamic analysis. In recent years, researchers have attempted to use non-linear static analysis instead of using non-linear dynamic analysis and several seismic design guidelines were published. Their main goal is to estimate the nonlinear seismic response by static calculations, and it means pushing the structure up to certain target displacement using load patterns reducing the computational cost drastically.

In design regulations, in Euro code 8, it is assumed that the response of a multi degree-of-freedom structure can be turned into the responses of an single degree-of-freedom model, structure response is dominated by the first mode of vibration and the single by single of these modes are combined with square root of squares, etc. and general mode of responses is achieved and this method is compared with non-linear dynamic mode. One of the results of non-linear static analysis is load-displacement chart or capacity curve achieving base shear to displacement and it is called push-over curve and with high speed of this analysis compared to other analyses, with displacement of floors and other parameters, has received much attention from engineers. Many people have work on bridge in non-linear static analysis in various pushover methods as adaptive pushover, pushover, modal pushover, etc. (Freeman et al., 1975 [1], Fajfar 1988 [2]. These studies had some limitations from that time and many discoveries have been made in this field and most of existing ambiguities in bridge are eliminated. In 1940, 1950, the effect of structure period on changes earthquake force on structure was discovered and interfered in structure calculations and the analysis of structures was based on its elastic response. In 1960, 1970 decades, structure ductility designs were emerged based on the evidence of tests and experiences that designed structures with good details can resist against earth movements. In decades 1980, 1990, it was shown that the damage on structure form structural member with strain and for non-structure members with relative displacement can be consistent. There is no clear relation between structure resistance and relevant damages. It seems that design of structures instead of resistance should be based on ductility and displacement. This concept develops many different seismic design methods as based on structure ductility capacity. This design method is performance-based method. IN the past two decades, many studies have been conducted on displacement-based methods and the initial years studies were mostly on reinforced concrete bridges. In recent years, these studies are generalized to different bridges. During 2005-2008, a great research team called Relous was responsible for development of guidance of design method based on displacement. To have access to study goals and performing a consistent project, 11 Universities of Italy participated in this project. Based on the review of literature, each of Universities promoted a part of this project. For example, Naply University of Italy was responsible to research about steel structures. Based on the results of this extensive project (Calvey Sullivan in 2009 [3]), draft of design code was provided based on displacement and its test version was published.

2. Design Methods

Direct design based on displacement

The direct design method based on displacement with the aim of eliminating the disadvantages of design method based on force was developed. The direct design is based on displacement for main structure of single degree of freedom with the same displacement is the maximum main structure and by features defined for this single degree of freedom, main multi-degree of freedom is designed. In displacement method, the goal is formation of plastic hinges in required locations and lack of formation of these hinges in other good locations. The moments and shears achieved for structure under shear force effect based on mode of first shape of vibration inelastic due to the effect of high vibration modes and conservative view in calculation of structure members should be increased to be adaptable with the real behavior of structure in earthquake. Thus, capacity design method is used to design based on displacement to increase the number of moments in plastic hinges and other locations and the shears are increased with applying some coefficients. The resistance values of distribution of shear force according to the first inelastic mode should be increased by applying extra coefficient of maximum bending capacity and amplification coefficient of the effect of higher modes to achieve from basic resistance to design resistance [4].

$$\emptyset_s S_D = \emptyset^\circ W S_E \tag{1}$$

$\emptyset_s$ is resistance reduction coefficient and for bending moment in similar plastic hinges can be considered and for other forces and moments in other locations is smaller than 1. For bridges in locations except plastic moment, only extra coefficient of bending strength is considered. In the design of beams, the effect of higher modes is not observed,amplificationcoefficient of higher models is not considered but due to the effect of high modes in vertical response and increase of gravity moments, this is better in exact design of capacity. For beams, both extra coefficients of bending capacity and the increasing effect of higher modes are considered. For final moments of columns and shear of columns, applying extra coefficients of bending capacity and increase of the effect of higher modes should be applied. The amplification coefficient of the effect of higher modes for moment is 1.8 and 1.3 for shear force. Generally, the design is determined based on displacement of required resistance in plastic hinges as our good goal in design is fulfilled based on displacement values and then the values of calculated resistances should be investigated by capacity design method and relevant coefficients are applied to be sure plastic hinges are not occurred in unsuitable locations and unsuitable elastic deformations to change plastic shear deformation are not occurred in structure. The general trend of design by displacement method is raised as this trend is applied for all types of structures. Single degree-of-freedom model (SDOF) is considered for frame or different types of structures and a bilinearcurve as the response of an equivalent single degree-of-freedom model to lateral force as displacement. In this chart, an initial stiffness K_i and then stiffness after yield rK_i are introduced. However, seismic design method introduces structure force with elastic features and before yield as initial stiffness and elastic dampingand design trend based on structure displacement is introduced with equivalent stiffness in maximum displacement and equivalent viscousdamping is a combination of elastic damping and hysteretic damping and hysteretic damping depends upon the absorbed energy by structure during elastic and cyclic behavior. As shown in Figure, equivalent viscous damping is equal to demand ductility and type of required structure of the chart [5].

By design displacement as maximum structure displacement and by equivalent viscous damping of the set of elastic displacement response spectrum charts, effective period of structure of equivalent single degree of freedom is achieved. It should be considered that maximum displacement and effective height He is considered. Effective stiffness Ke in maximum displacement is computed by following equation:

$$K_e = 4\pi^2 \left(\frac{M_e}{T_e^2} \right) \tag{2}$$

The effective mass of structure in the first model is inelastic vibration and shear force is obtained as:

$$F = K_e * \Delta d \tag{3}$$

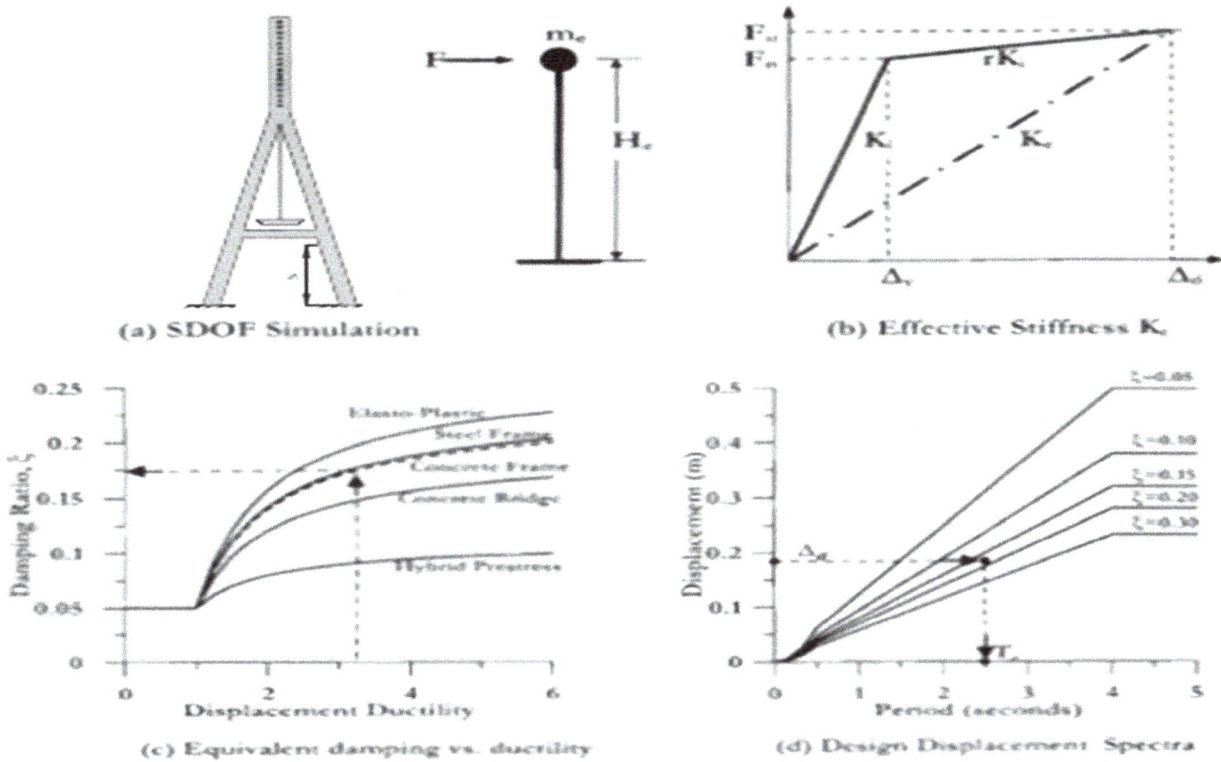

Figure 1. The stages of direct design method based on displacement [5].

3. Non-linear Static Analysis

Pushover static analysis

In this method, lateral load is imposed on structure gradually with a definite pattern (e.g. triangular load) and the structure is allowed to be yielded gradually (continuous yielding of various components). This loading is continued till the displacement of peak point of structure achieves target displacement. Normally, the top of structure is used as the indicator point and by achieving the roof to target displacement, analysis is stopped.

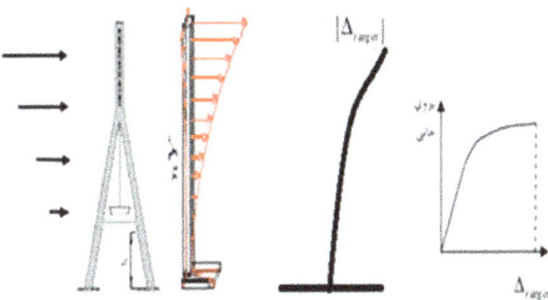

Figure 2. Pushover analysis to reach target displacement.

Total structure displacement = Pushover analysis and determining force chart [6]

Target displacement based on FEMA 356 as total displacement is computed by following equation.

$$\delta_t = C_0 C_1 C_2 C_3 S_a \frac{T_e^2}{4\pi^2} g \qquad (4)$$

4. Non-linear Dynamic Analysis (Time History)

Time dynamic analysis (time history) is used to determine immediate response of structure under accelerogram and includes two different methods of elastic-linear and inelastic (non-linear). This code besides recognizing time dynamic analysis method recommends the followings: 1- The comparison of the results of elastic analysis by standard spectrum or specific design spectrum with what is achieved by elastic time dynamic analysis and the probable difference reasons are justified in a complete technical report. The response values should be modified as it was said. 2- Damping ratio in linear-elastic calculations is 5% and in non-linear calculations based on specialized recommendations and non-linearity of structure components behavior can be considered. This thesis applies New mark method for non-linear analysis of time history and average acceleration in with coefficientsβ=0.25 and α=0.5 are considered. The dynamic analysis is Full Transient. In this analysis, mass matrix coefficients and stiffness matrix are computed. At first, modal analysis is done to achieve structure frequencies and then with damping assumption 5% and using frequency of first and second model of structure and equations (Chopra

1994), mentioned coefficients are computed [6].

$$\begin{cases} a = \dfrac{2\xi w_i w_j}{w_i + w_j} \\ b = \dfrac{2\xi}{w_i + w_j} \end{cases} \qquad (5)$$

5. Modelling

At first, cable stayed structure is divided into various parts as modeled and designed separately. Cable bridge structure includes a deck and a series of Rigid links and top and bottom pylons and beam of deck and beams of connection between pylons and concrete between Rigid link should be modeled and designed. The thickness of cables is 0.05. The cable forms are of two types: 1- Cables kept in 1.3 above Pylon, 2- Cables divided with similar distance in the entire Pylon. In this project, bridge deck design is of great importance. The length of entire deck in this project is 320m and width of deck as 22m and the deck is modeled as symmetrical and the features of materials are as followings.

Table 1. *The materials features for base and deck.*

Concrete tensile strength	$500^{kg}/_{cm2}$
Elasticity module	$335^{t}/_{cm2}$
Poisson coefficient	0.2
Thermal coefficient	0.00001

The features of cables are as followings:

Table 2. *Features of cables in Cable Bridge.*

Nominal diameter	**15.7mm**
Nominal tensile strength	$17.7^{t}/_{cm^2}$
Nominal weight	1.18kg/m
Elasticity module	$1950^{t}/_{cm2}$
Allowed tension	0.6u.t.s

At first, we should plot the above pylon and then bottom pylon and the beam located between bottom pylon and deck on beam, then a series of small beams as Rigid Link are perpendicular to deck beam and with joint on its start and end and some joints are dedicated to upper pylon and cables are connected top to bottom and between these connections, concrete is located and the following Figure shows the general view.

Non-linear analyses are performed on Pylon and the results are achieved.

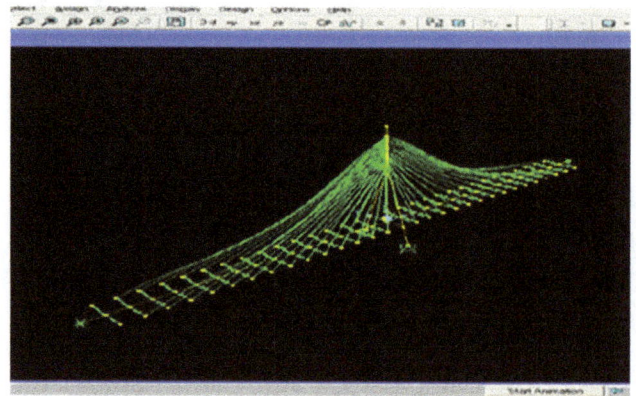

Figure 3. *General view of Cable Bridge in Sap software.*

5.1. A Summary of Pushover Analysis Stages in Software

1. Definition of plastic hinges for the members playing important role in tolerating seismic load.
2. Dedication of defined plastic hinges to relevant elements.
3. Definition of combining gravity loads based on regulation
4. Definition of lateral load patterns based on regulation.
5. Applying gravity loads combination and lateral load patterns in the combination of gravity loads by pushover analysis and evaluation of plastic hinges acceptance criteria in structure elements.

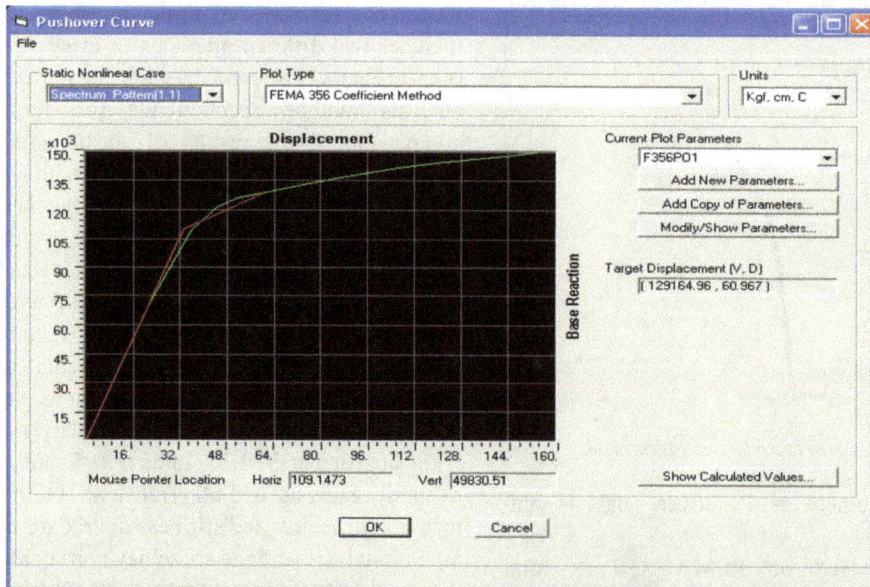

Figure 4. *Capacity spectrum based on existing code (FEMA 356).*

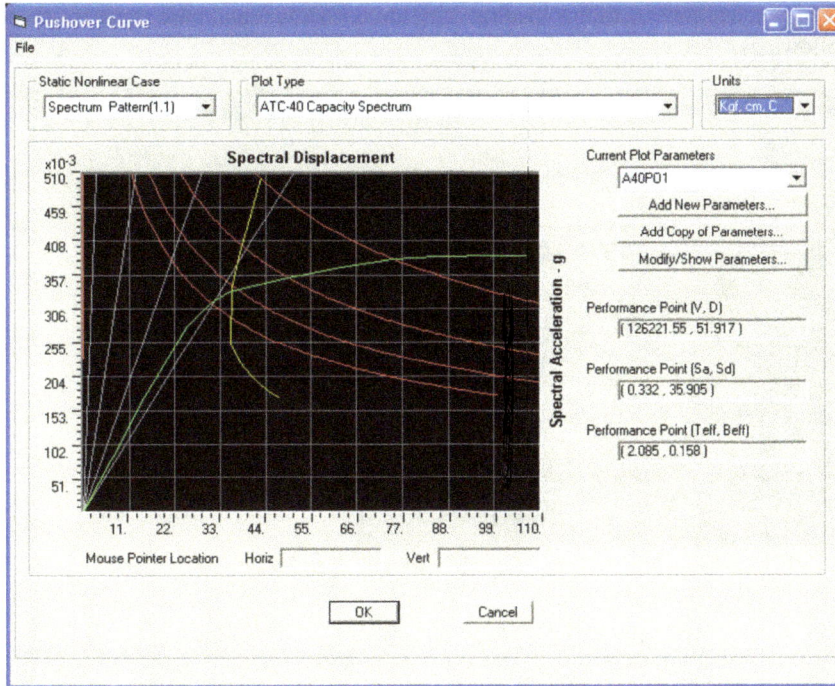

Figure 5. Capacity spectrum based on ATC440.

Based on above stages, the most important result of non-linear static analysis, capacity spectrum is achieved based on FEMA 356 and ATC 440 CODE.

As shown in the above Figure, target displacement for our cable bridge pylon is 60.96 cm of pushover analysis. As shown, base shear is 129.16 Ton. In this analysis, distribution of our forces is consistent with the forces of spectrum analysis. According to ATC 440, it is as followings.

As shown in the above Figure, our performance point is 51.917cm and base shear is 126.22Ton and is consistent with base shear value of FEMA 356.

The analysis of pushover with uniform load is as followings.

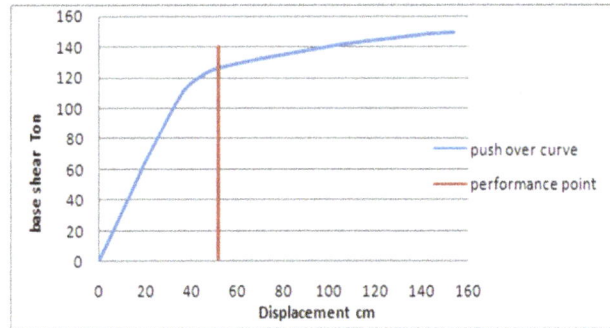

Figure 6. Determining displacement based on spectrum loading.

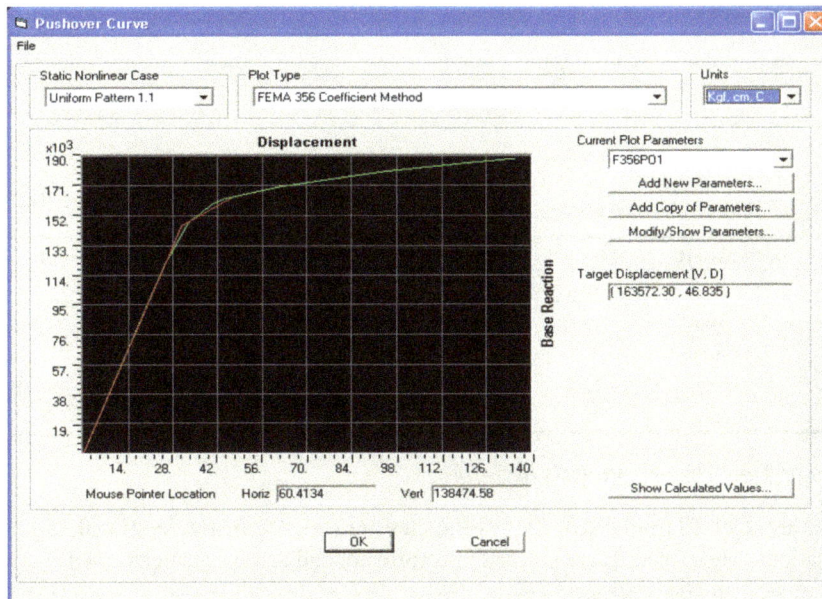

Figure 7. Capacity spectrum based on required loading in accordance to FEMA356.

It is observed that target displacement under uniform loading of pylon is 46.83cm and base shear is computed as 163.57 Ton but based on ATC440 is as followings.

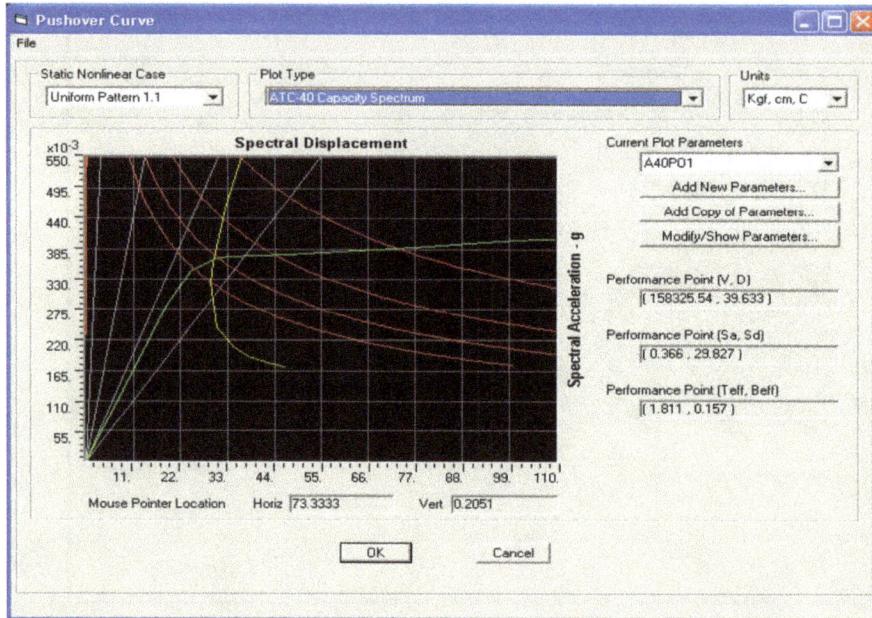

Figure 8. *Capacity spectrum based on required loading in accordance to ATC 440.*

Also, the performance point of uniform loading in pushover analysis is 39.63cm and base shear is 158.32 Ton.

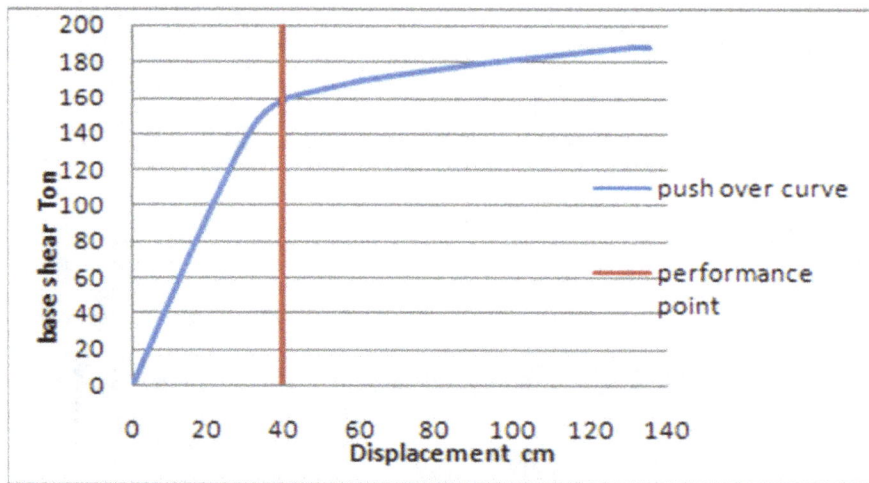

Figure 9. *Determining displacement based on uniform loading.*

Table 3. *The comparison of the results in combination of different loads.*

Case	Load combine	Load pattern	Target displacement (FEMA)		Performance point	
			Displacement cm	Base share ton	Base shear ton	Displacement cm
Pushover	0.9D	Spectrum	61.4	124	52	121
		Uniform	47.4	156	39	152
	1.1(D+L)	Spectrum	60.9	129	51	126
		Uniform	46.8	163	39	158

5.2. A Summary of the Stages of Time History Analysis in Software

At first, some accelerograms are selected from PEER site [7] and they are scaled with acceleration 0.35g and they are averaged and compared with Iran regulations. Then, records are entered into software and analysis is performed.

The applied records in time history analysis are as:

Table 4. *Applied records in time history analysis.*

Record	Hour	Date	Intensity (Mercalli)
Northridge	12.31	17/1/1994	6.7
Kobe	20.46	16/1/1995	6.9

At first, accelerograms are scaled to the design spectrum.

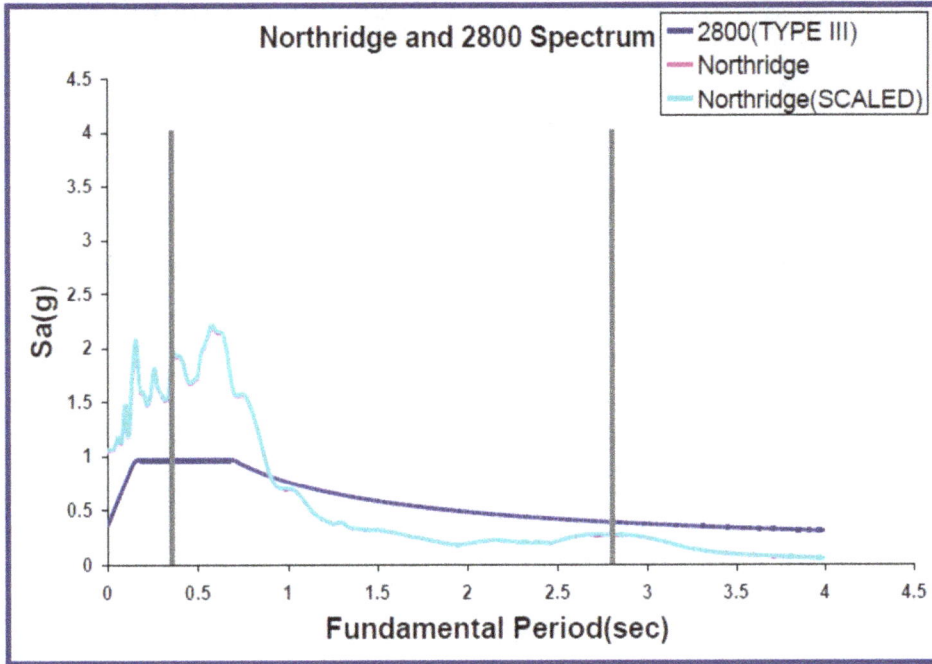

Figure 10. *Scaling accelerograms with design spectrum in Northridge earthquake.*

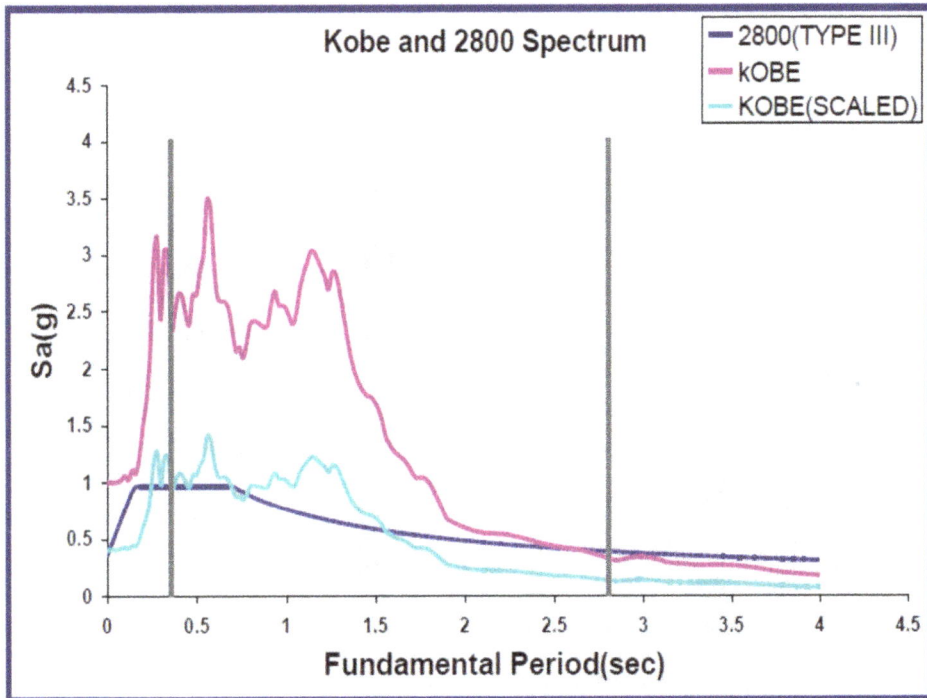

Figure 11. *Scaling accelerograms with design spectrum in Kobe earthquake.*

The results of time history analysis for Northridge and Kobe are shown. The time history chart of displacement of top of pylon of software is also shown.

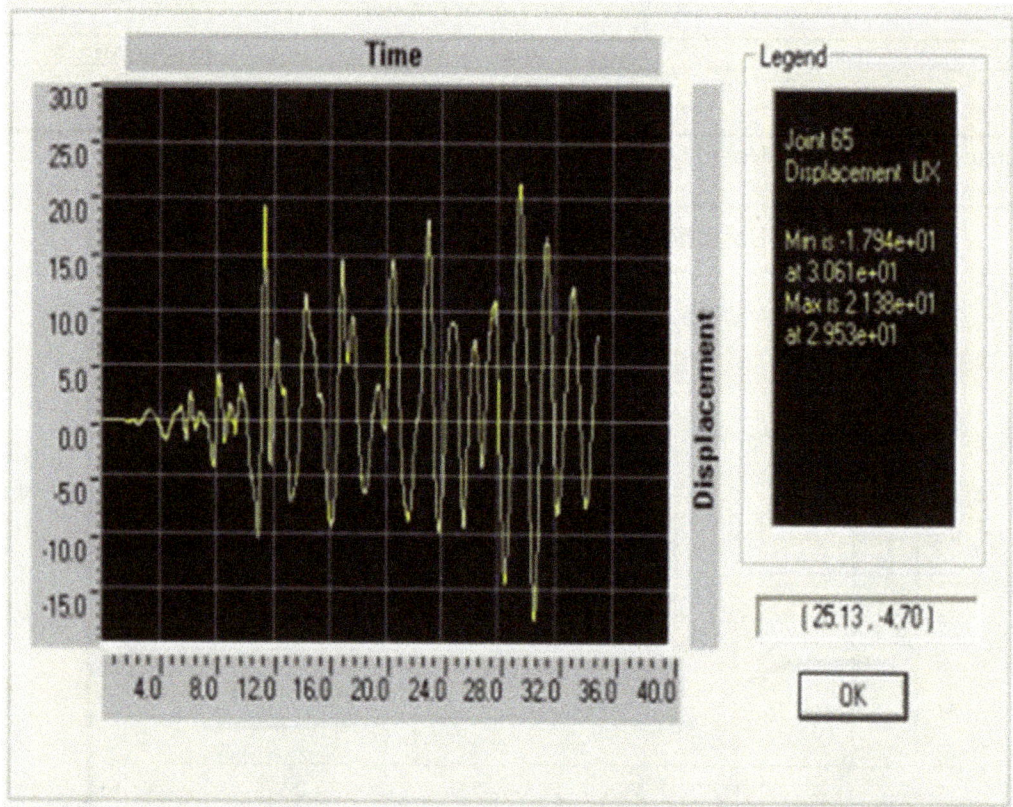

Figure 12. *Displacement of top of pylon in time history analysis in Northridge earthquake.*

The maximum displacement in Northridge earthquake is 31.38but in pushover analysis, target displacement is 47.4.
Max Dis. (31.38 cm) < Target Displacement (47.4 cm)

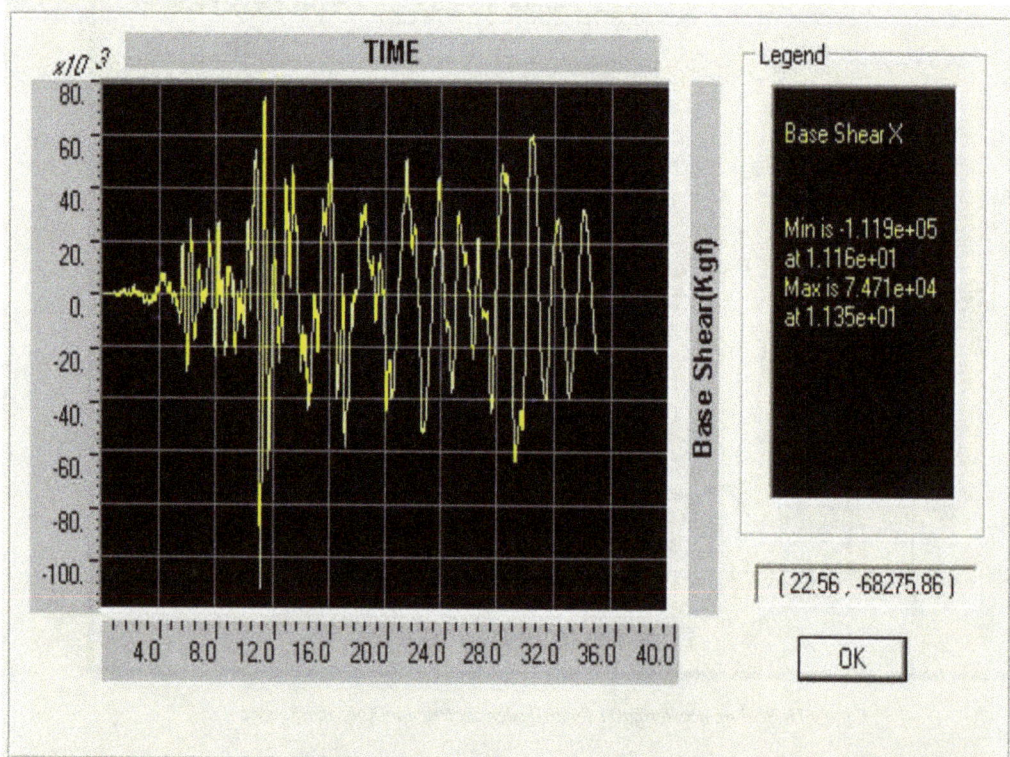

Figure 13. *Base shear of top of Pylon in time history analysis in Northridge earthquake.*

In Northridge earthquake, the maximum base share is 119 Ton but in pushover analysis is 163. By comparison of pushover analysis and time history, the following results are achieved.

Table 5. The comparison of pushover analysis displacement and time history analysis for Northridge record.

Load combination	Load pattern of pushover analysis	Maximum target displacement (cm)	Maximum performance displacement (cm)	Maximum displacement of Northridge record (cm)
0.9D	Spectral	61.47	52.132	17.94
	Uniform	47.408	39.61	
1.1(D+L)	Spectral	60.96	51.917	21.38
	Uniform	46.83	39.63	

Base shear in Northridge record is as followings:

Table 6. The comparison of base shear of pushover analysis and time history analysis for Northridge record.

Load combination	Load pattern of pushover analysis	Maximum base shear (Ton) FEMA	Maximum base shear (Ton) ATC40	Maximum base shear of Northridge
0.9D	Spectral	124.19	121.86	74.71
	Uniform	156.57	152.88	
1.1(D+L)	Spectral	129.16	126.22	111.9
	Uniform	163.57	158.32	

The results in Kobe earthquake are as followings.

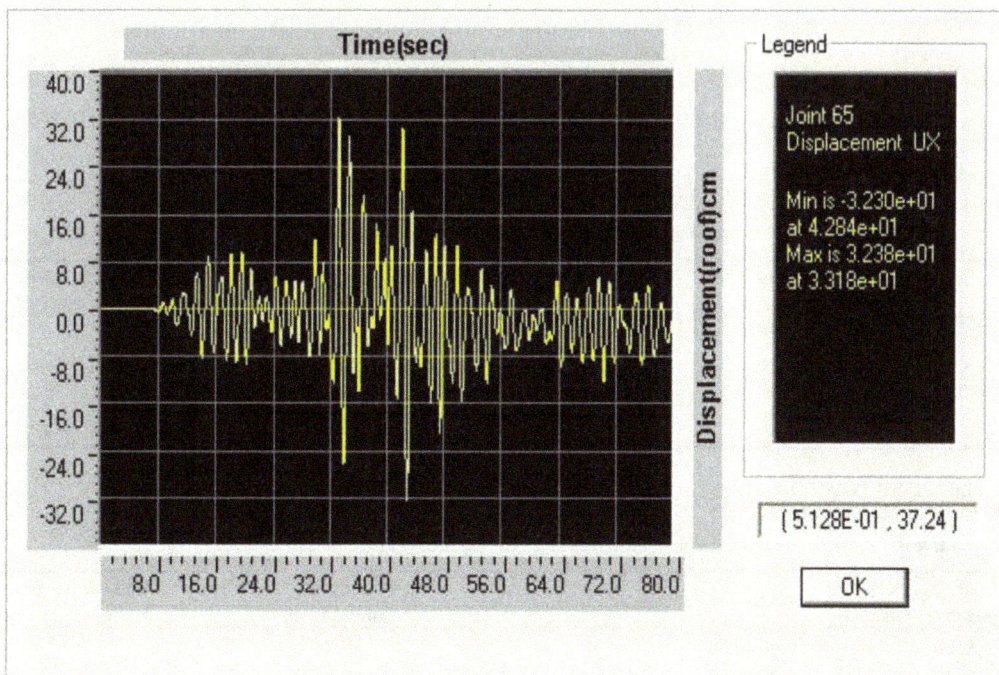

Figure 14. The displacement of top of Pylon in time history analysis in Kobe earthquake.

Maximum displacement is 33.18 and 47.4 in pushover.

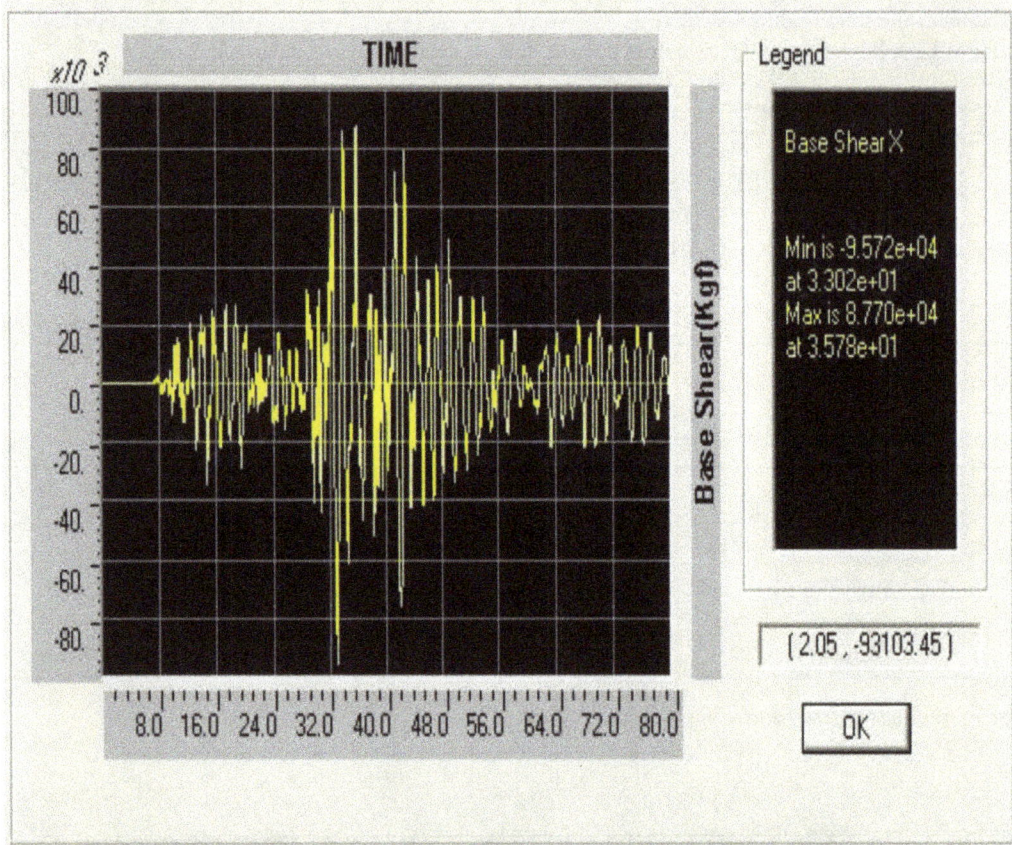

Figure 15. Base share of top of Pylon in time history analysis in Kobe earthquake.

Maximum base shear in Kobe with time history analysis 95.7 and in Pushover is 163.

Table 7. Comparison of displacement of pushover analysis and time history analysis in Kobe record.

Load combination	Load pattern of pushover analysis	Maximum target displacement (cm)	Maximum performance displacement (cm)	Maximum displacement of Kobe record (cm)
0.9D	Spectral	61.47	52.132	32.3
	Uniform	47.408	39.61	
1.1(D+L)	Spectral	60.96	51.917	32.38
	Uniform	46.83	39.63	

But base shear of Kobe record is as followings:

Table 8. Comparison of base share of pushover analysis and time history analysis of Kobe record.

Load combination	Load pattern of pushover analysis	Maximum base shear (Ton) FEMA	Maximum base shear (Ton) ATC40	Maximum base shear of Kobe
0.9D	Spectral	124.19	121.86	95.7
	Uniform	156.57	152.88	
1.1(D+L)	Spectral	129.16	126.22	87.70
	Uniform	163.57	158.32	

6. Conclusion

The comparison of the results of non-linear static analysis and non-linear time history analysis Pylon deformation in non-linear static analysis was formed based on force distribution, based on forces of spectral analysis and uniform distribution, 2-Deformation of higher modes in non-linear static analysis for pylon is not observed, 3-The performance of cable bridge structure in proximity to target displacement for non-linear dynamic analysis based on structure performance level in non-linear static analysis with uniform lateral load. 4- In non-linear static analysis with spectrum load of structure performance in target displacement with time history analysis has no consistency. 5- The required patterns of Pylon deformation distribution in static non-linear analysis are consistent with real deformations during real earthquakes. 6-The disadvantage of non-linear static analysis for evaluation of bridges, namely Cable Bridge is as the effect of higher modes is not considered mostly. 7-Base share of non-linear static analysis in proximity to target displacement doesn't indicate the maximum base share of non-linear dynamic analysis.

References

[1] S. A. Freeman, J. P. Nicoletti, and J. V. Tyrell, 1975, "Evaluations of Existing Buildings for Seismic Risk – A Case Study of Puget Sound Naval Shipyard, Bremerton, Washington," Proc. of the First U.S. Nat. Conf. on Earthq. Engng, Oakland,California, pp. 113-122

[2] P. Fajfar, M. Fischinger, N2 - A method for non-linear seismic analysis of regular buildings. Proc. of the 9th WCEE, August 2-9 Tokyo-Kyoto, Japan, 1988.

[3] T. J. Sullivan. (2012). Formulation of a Direct Displacement-Based Design Procedure for Steel Eccentrically Braced Frame structures.

[4] Priestly M. J. N, Kowalsky M. J, Calvi G. M. (2007). Displacment – Based Design Seismic Design of Structures. IUSS. Press:Pavia

[5] Priestly M. J. N, Kowalsky M. J, Calvi G. M. (2007). Displacment – Based Design Seismic Design Of Structures. IUSS. Press:Pavia

[6] Goel RK Chopra AK (1998). Period formulas for concrete shear wall building. Journal of Structural Engineering ASCE 124(ST4), 426-433.

[7] http://peer.berkeley.edu/

[8] Aprile A, Benedetti A, Grassucci F. Assessment of cracking and collapse for old brick masonry columns. J Struct Eng 2001;127(12):1427–35.

[9] Hernδndez-Montes E, Kwon OS, Aschheim M. An energy-based formulation for first and multiple-mode nonlinear static (pushover) analyses. J Earthq Eng 2004;8(1):69–88.

[10] Fajfar P. A. nonlinear analysis method for performance-based seismic design. Earthq Spectra 2000.

16

An Investigation into Factors Causing Delays in Road Construction Projects in Kenya

Msafiri Atibu Seboru

Department of Education and External Studies, University of Nairobi, Nairobi, Kenya

Email address:

mseboru@yahoo.com (Seboru, A. M.)

Abstract: The majority of road construction projects in Kenya do not get completed within the initially set targets of time. Project delays frustrate the process of development, have an immeasurable cost implication to the society, and also lead to loss of reputation of the parties involved in the projects' execution. The purpose of this study was to investigate the factors causing delays in road construction projects in Kenya. Project delays are a common problem internationally in the construction industry in modern times. Investigating the reasons for delay has become an important contribution to improved construction industry performance. Over seventy percent of projects initiated in Kenya are likely to escalate in time with a magnitude of over fifty percent. The study used purposive sampling technique and survey design. Data was collected using questionnaires which were distributed to consultants and contractors. The data was analyzed using the Relative Importance Index and Spearman's rank correlation. The top five causes of project delays were observed to be payment by client, slow decision making and bureaucracy in client organization, inadequate planning and scheduling, and rain. It is recommended that clients should improve their financial management systems so that they are able to pay contractors in a timely manner. Bureaucracy and red tape should be reduced in client organizations in order to speed up the slow decision making process. Efficient management of the construction process will also lead to a reduction in incidences of claims. Contractors should prepare adequate plans and schedules which can also be used to minimize the effects of rain.

Keywords: Delay Factors, Project Delays, Road Construction Projects

1. Introduction

1.1. Background of the Study

Road construction projects in Kenya are procured through the traditional system where the Consultant Civil Engineer is in charge of design and construction processes on behalf of the Client. According to the traditional system, the design process ought to be completed prior to commencement of construction. The Client commissions the Consultant Civil Engineer who is briefed by the client. The Consultant Civil Engineer then develops the design and prepares contract documents. The tendering process begins by pre-qualifying contractors on the basis of experience, work capacity and past performance. The pre-qualified contractors are then invited to tender. The contract is normally awarded to the lowest bidder. The standard form of contract that is commonly used for civil engineering works in Kenya is FIDIC (International Federation of Consulting Engineers). The Consultant Civil Engineer appoints a Resident Engineer to be permanently based on site to supervise the project. The Consultant Civil Engineer delegates some of his duties and powers under the contract to the Resident Engineer. The Resident Engineer holds monthly site meetings with the Contractor. The Consultant Civil Engineer and the Client usually attend these meetings. At the practical completion stage, an inspection is carried out and the project is handed over to the Client. The Contractor is expected to make good any defects within the Defects Liability Period that normally lasts for one year.

The study of causes of delay in road construction projects in Kenya was important because time is one of three pillars of construction project management: time, cost and quality. A study on project delays was expected to lead to a better understanding of the causes of inefficiency in road construction projects. Once the most important causes of significant delay causing factors are identified, the parties to the projects shall then be able to channel their energies and resources to the specific factors thereby reducing delays to the projects. The study on road construction is important in the Kenyan context because roads contribute to economic growth and poverty reduction.

The study was restricted to the construction phase of the project only, although it is acknowledged that decisions made in other stages prior to the commencement of construction phase do affect the construction process. As argued by Talukhaba (1999), the conditions prevailing during design can be controlled, because unlike construction, design occurs in an enclosed environment. For instance, it may not be easy to control rain in a construction site. Therefore, the greatest challenges in implementation of projects are likely to be experienced during the construction phase. The data collection was restricted to consulting engineering firms and contractors situated in Nairobi. Besides the majority of consulting engineering firms and contractors operating in Kenya are based in Nairobi. Therefore, the study was concerned with the delays observed during the construction process. The delays that occur during the design process are not part of this study. The following were the assumptions of the study as adapted from Talukhaba (1999): Firstly, there was homogeneity in the management of projects under study and therefore project management as a factor had a constant and equal impact on project delays since all the projects were implemented using the traditional procurement system; secondly, there were no differences in the traditional procurement system that could be attributed to the process such that projects implemented using the system had an equal chance of showing similar characteristics in performance; thirdly, the productivity of workers in the projects being studied was the same, hence variation in delays due to differences in worker productivity was negligible; fourthly, the output of the plant and equipment in the projects being studied was the same, hence variation in delays due to differences in machine output was negligible; and lastly, the project contract period the parties agreed to adhere to was adequate for the completion of the project, and therefore the delays observed by the projects in the sample were not as a result of contract time underestimation.

Fan and Kang (2005) assert that roads contribute to economic growth and poverty reduction. Road infrastructure impacts on overall economic growth, agricultural growth, urban growth, urban poverty reduction, and rural poverty reduction. Without infrastructure, efficient markets, adequate health care, a diversified rural economy, and sustainable economic growth will remain elusive. Effective development strategies require good infrastructure as their backbone. Transportation infrastructure is an effective factor of production. Power consumption and health conditions are positively correlated with the availability of road infrastructure. Most of Africa's poor trade performance is the result of weak infrastructure. The availability and quality of road infrastructure also influences food prices. Road investments help the poor through their impact on the rural non-farm economy. An increase in paved roads is positively and significantly related to growth in Gross Domestic Product (GDP) per capita in urban areas.

Fan and Kang (2005) continue to argue that the different phases of a highway project have different impacts. The construction period not only creates tremendous work opportunities, but also improves the skill of local people employed on the project. In the post-construction period, highways promote the development of goods production in poor regions, increase the volume of trade, reduce transportation costs, and improve social services. Highway construction also increases farm incomes.

Nadiri and Mamuneas (1998) argue that an increase in the stock of highway capital has an initial direct productivity effect on business: it reduces the total cost of producing a given level of output in almost all industries. Cost reductions permit products to be sold at lower prices and lower prices can be expected to lead to output growth. Road investments have a significant effect on the production sector's demand for labour, capital, and materials.

1.2. Research Problem

The majority of road construction projects in Kenya do not get completed within the initial set targets of time. Talukhaba (1999) argues that project delay frustrates the process of development, has an immeasurable cost to the society, and also leads to loss of reputation of the parties involved in the concerned projects. Mbatha (1986) and Talukhaba (1988) revealed that time performance of construction projects in Kenya was poor to the extent that over seventy percent of projects initiated in Kenya were likely to escalate in time with a magnitude of over fifty percent.

One of the most widely used measures of project success is time taken to complete the project. Talukhaba (1999) carried out research on factors causing construction project delays in Kenya based on a case study of high rise building projects in Nairobi but did not cover road construction projects. Road construction projects are more mechanised than building projects. Some of the materials used in road construction projects are different from the materials used in building projects. Activities in road projects are more exposed to the weather than activities in building projects. The Civil Engineer is largely the sole player in road projects with minimal involvement of other construction professionals as opposed to building projects where the various professionals play significant roles. Road projects are essentially public projects whereas building projects could either be public or private. It is therefore the purpose of this study to investigate the factors causing delays in road construction projects in Kenya.

1.3. Objectives of the Study

The objectives of this study were:
1. To document the range of identified causes of delay in completing road construction projects in Kenya;
2. To document the most important causes of delay in road construction projects in Kenya; and
3. To document identified differences in perception of contractors and consultants regarding causes of delay in delivering projects by the intended completion date.

1.4. Research Hypothesis

Ho: Road construction project delays are not caused by exogenous and endogenous factors.

Ha: Road construction project delays are caused by exogenous and endogenous factors.

Exogenous factors are external factors to the project such as politician's interference, inflation and interest rates. Endogenous factors are internal factors in the project such as contractor's cash flow, design change by engineer, and inadequate planning/scheduling.

1.5. Significance of the Study

This work is important because time is one of three pillars of construction project management: time, cost and quality. A study on project delays will lead to a better understanding of the causes of inefficiency in road construction projects. Once the most significant delay causing factors are identified, the parties to the projects shall then be able to channel their energies and resources to the specific factors thereby reducing delays to the projects.

Walker (1994) carried out an investigation in Australia on construction time performance and concluded that through improving its productivity, the construction industry can have an important role in promoting national competitiveness, and therefore in defending living standards and achieving a satisfactory rate of growth. The benefits from such improvement would include increased attractiveness of Australia as a location for investment in new plants or projects. Measures that prevent or slow steps toward improving building and construction industry productivity are, in effect, an attack on the employment prospects and future welfare of Australian workers. Such measures would also be an attack on the potential performance of Australian industry and the economy generally.

The above view can also be applicable to Kenya and reinforces the argument for attention to construction time performance. The study on road construction is important in the Kenyan context because roads contribute to economic growth and poverty reduction.

2. Literature Review

Delays in construction projects are still very common in most parts of the world even with the introduction of modern management techniques. Studies conducted on the causes of construction project delays in 22 different countries of the world have been examined.

Talukhaba (1999) carried out an investigation into factors causing construction project delays in Kenya and found out that the major causes of delay were: Clients payment; Architect's instructions; Client's instructions; Rock; and Underground water. Assaf, Al-Khalil and Al-Hazmi (1995) studied the causes of delay in large building construction projects in Saudi Arabia and revealed that the most important causes of delay were: Approval of shop drawings; Delays in payments to contractors and the resulting cash-flow problems

during construction; Design changes; Conflicts in work schedules of subcontractors; and Slow decision making and executive bureaucracy in the Owners' organisations. Mansfield, Ugwu and Doran (1994) studied the causes of delay and cost overruns in construction projects in Nigeria and the results showed that the most important factors were: Financing and payment for completed works; Poor contract management; Materials shortages; and Improper planning.

Al-Tabtabai (2002) conducted a study on causes of delays in construction projects in Kuwait and found out that the major causes of delay were: Slow financial and payment procedures; Slow decision-making process; Limited authority among supervision staff; Risk allocation mainly on the contractor; and Lack of design drawings coordination. Memon, Rahman and Azis (2012) conducted a study on time and cost performance in construction projects in Malaysia and revealed that only 21% of public sector projects and 33% of private sector projects were completed within time. The results of the study showed that the most important delay factors were: Design and Documentation Issues; Financial Resource Management; Project Management and Contract Administration; Contractors Site Management; and Information and Communication Technology.

Owolabi et al. (2014) studied the causes and effects of delay on project construction delivery time in Nigeria. They stated that seven out of ten projects in Nigeria suffered delays in their execution. The results of the study indicated that the following were the five major causes of delay: Lack of funds to finance the project to completion; Changes in drawings; Lack of effective communication among the parties involved; Lack of adequate information from consultants; and Slow decision making. In Ghana, Frimpong, Oluwoye and Crawford (2003) carried out a research on Causes of delay and cost overruns in construction of groundwater projects in developing countries. The researchers indicated that 75% of the projects in Ghana exceeded the original project schedule. The study revealed that the most important causes of delay were: Monthly payment difficulties; Poor contract management; Material procurement; Inflation; and Contractor's financial difficulties. In Morocco, Challal and Tkiouat (2012) researched on the causes of deadline slippage in construction projects and found out the five major causes of delay were: Errors in initial budget assessment; Architecture and engineering volatility program (multiple modification requests); Site hazards; Failure of an actor; and Insufficiency or lack of prior study and feasibility.

Alinaitwe, Apolot and Tindiwensi (2013) carried out a study on causes of delays and cost overruns in Uganda's public sector construction projects and the results showed the major causes as: Change of work scope and/or changes in material specifications; High inflation, insurance and interest rates; Poor monitoring and control, due to incompetent and/or unreliable supervisors; Delayed payment to contractors, subcontractors and/or suppliers; and Fuel shortages. Memon (2014) conducted a study on contractor perspective on time overrun factors in Malaysian construction projects and the major factors causing delays were: Frequent design changes;

Change in the scope of the project; Financial difficulties of owner; Delays in decisions making; and Unforeseen ground conditions. In India, Desai and Bhatt (2013) studied the critical causes of delay in residential construction projects and found out that the most important delay factors were: Original contract duration was too short; Legal disputes between various parties; Ineffective delay penalties; Delay in progress payments by owner; and Delay to furnish and deliver the site to the contractor by the owner.

Sweis, Sweis, Hammad and Shboul (2008) studied delays in construction projects in Jordan and the major causes of delay were: Financial difficulties faced by the contractor; Too many change orders from owner; Poor planning and scheduling of the project by the contractor; Presence of unskilled labour; and Shortage of technical professionals in the contractor's organization. In India, Ravisankar, Anandakumar and Krishnamoorthy (2014) conducted a study on the quantification of delay factors in the construction industry. The researchers indicated that time overrun vary between 50% and 80% for projects completed worldwide. The study revealed that the most important causes of delay were: Shortage of unskilled and skilled labour; Design changes by owner or his agent during construction; Fluctuation of prices; High waiting time for availability of work teams; and Rework due to errors. Shanmugapriya and Subramanian (2013) investigated significant factors influencing time and cost overruns in Indian construction projects. The researchers indicated that 60% of projects in India suffered time overruns. The study found out that the following were the most significant factors causing time overruns: Material market rate; Contract modification; Rework of bad quality performance; Unclear specification; and Dependence on freshers to bear the whole responsibility.

Kholif, Hosny and Sanad (2013) analyzed time and cost overruns in educational building projects in Egypt and found out that the major causes of time overruns were: Political insecurity (instability); Financial difficulties of contractor; Escalation of material prices (inflation); High cost of skilled labour; and Difficulties in getting work permits from government. Kagiri and Wainaina (2008) studied time and cost overruns in power projects in Kenya and revealed that the major causes of time overruns were: Delayed payment to contractor; Employer cash flow problems; Delays in disbursement of funds by financiers; Bureaucracy of government agencies; and Delay of access to site. In Sri Lanka, Dolage and Rathnamali (2013) carried out a study causes of time overrun in construction phase of building projects and found out that the following were the major causes of time overrun: Delay in progress payment by clients; Inaccurate planning and scheduling of projects by contractors; Rainy weather; Non availability of experienced technical staff of contractor; and Excessive work in hand of the contractors. Sweis (2013) investigated factors affecting time overruns in public construction projects in Jordan and revealed that the major causes of delay were: Too many change orders from owner; Poor planning and scheduling of the project by the contractor; Ambiguities and mistakes in specifications and

drawings; Slow decision making from owner; and Poor qualification of consultants, engineers and staff assigned to the project.

In South Africa, Baloyi and Bekker (2011) researched on causes of construction cost and time overruns and revealed that the following were the most important causes of time overruns: Incomplete drawings; Design changes; Clients' slow decision-making; Late issue of instructions; and Shortage of skilled labour. Alaghbari, Kadir and Salim (2007) studied the significant factors causing delay of building construction projects in Malaysia and found out that the major causes of delay were: Owners' financial difficulties and economic problems; Contractors' financial problems; Late supervision and slowness in making decisions; Consultants' slowness in giving instructions; and Lack of materials on market. Mohammed and Isah (2012) carried out a study on the causes of delay in Nigerian construction industry and the results showed that the major causes of delay were: Improper planning; Lack of effective communication; Design errors; Shortage of supply like steel, concrete; and Slow decision-making. Fugar and Agyakwah (2010) researched on delays in building construction projects in Ghana and found out that the most important causes of delay were: Delay in honouring payment certificates; Underestimation of cost of projects; Underestimation of complexity of projects; Difficulty in accessing bank credit; and Poor supervision. Kikwasi (2012) studied the causes and effects of delays and disruptions in construction projects in Tanzania and the results showed that the following were the major causes of delay: Design changes; Delays in payment to contractors; Information delays; Funding problems; and Poor project management.

Ibironke, Oladinrin, Adeniyi and Eboreime (2013) analysed the non-excusable delay factors influencing contractors' performance in Nigeria and revealed that the major delay factors were: Insufficient amount of equipment; Inaccurate time estimates; Monthly payment difficulties; Change orders; and Inaccurate cost estimates. Wong and Vimonsatit (2012) studied the factors affecting construction time in Australia and the results showed that the following were the major factors affecting construction time: Skills shortage, Financial difficulties; Shortage of labour; Unrealistic deadlines for project completion; and Unforeseen ground conditions. Hoai, Lee and Lee (2008) researched on delay and cost overruns in large construction projects in Vietnam and revealed that the most important causes of delay were: Poor site management and supervision; Poor project management assistance; Financial difficulties of owner; Financial difficulties of contractor; and Design changes.

Ayudhya (2011) evaluated the common delay causes of construction projects in Singapore and found out that the major causes of delay were: Delay in progress payment by owner; Adverse weather conditions; Main contractor financial problems; Evaluation of completed works; and Acts of God. Faridi and El-Sayegh (2006) studied the significant factors causing delay in the construction industry in the United Arab Emirates and revealed that the following were the most important factors causing delay: Preparation and approval of drawings; Inadequate early planning of the project; Slowness of the owner's decision-making process; Shortage of manpower; and Poor supervision and poor site management. Mahamid (2013) researched on the causes of time overrun causes in road construction projects in Palestine and found out that the major causes of time overrun were: Segmentation of the West Bank and limited movement between areas; Political situation; Progress payments delay by owner; Lack of efficient equipment; and Difficulties in financing project by contractor. In Malaysia, Abdulla, Rahman and Azis (2010) studied the causes of delay in Construction Projects and the results showed that the major causes of delay were: Cash flow and financial difficulties faced by Contractors; Contractor's poor site management; Ineffective planning and scheduling by Contractors; Inadequate Contractor experience; and Shortage of site workers.

El-Razek, Bassioni and Mobarak (2008) conducted a study on causes of delay in building construction projects in Egypt and found out that the most important causes of delay were: Financing by contractor during construction; Delays in contractor's payment by owner; Design changes by owner or his agent during construction; Partial payments during construction; and Non-utilization of professional construction/contractual management. Sambasivan and Soon (2007) researched on the causes and effects of delays in Malaysian construction industry and the results showed the following major causes of delay: Improper planning; Site management; Inadequate contractor experience; Finance and payments of completed work; and Subcontractors. Kamanga and Steyn (2013) studied the causes of delay in road construction projects in Malawi and revealed that the following were the major causes of delay: Shortage of fuel; Insufficient contractor cash flow / difficulties in financing projects; Shortage of foreign currency (importation of materials and equipment); Slow payment procedures adopted by client in making progress payments; and Insufficient equipment. Patil, Gupta, Desai and Sajane (2013) researched on the causes of delay in Indian transportation infrastructure projects and the results showed that the following were the most important causes of delay: Delay due to Land Acquisition; Environmental issues related with project; Financial closure; Change orders by the client; and Poor site management and supervision by Contractor.

Akogbe, Feng and Zhou (2013) studied delay factors for development construction projects in Benin and found out that the major causes of delay were: Contractor's financial capability; Owner's financial difficulties; Poor subcontractor performance; Materials procurement; and Changes in drawings. Mustapha (2013) researched on the factors of delays in project delivery in Ghana and found out that the major delay factors were: Delay in honouring payment certificates; Delay by sub-contractors; Fluctuation of prices; Difficulty in accessing bank credit; and Client initiated variations. Asiamah and Asiamah (2013) conducted a study on causes of delays in construction of public buildings in Ghana and revealed that the following were the most

important causes of delay: Method of construction; Long bureaucratic process of honouring certificates; Variation orders; Cash flow problems; and Lackadaisical attitude to decision making. Andi, Lalitan and Loanata (2010) carried out a study on factors causing delays in structural and finishing works in Indonesia and the results showed that the following were the major causes of delay: Slow contractor's payment; Design changes during construction; Force Majeure; Bad weather; and Slow delivery of material. According to Kivaa (2000), one cause of poor time performance in construction projects in Kenya is the inadequacy of initial contract periods. These have been found to be inconsistently and erroneously calculated. The initial contract period is estimated using the estimator's personal intuition, which is based on his skill and past experience. The method does not consider, objectively and accurately, all the factors that influence the construction time of a project.

3. Research Methodology

The study dealt with road projects in Kenya but whose consultants and contractors were based in Nairobi. The population of road projects and the identity of the consultants and contractors involved in the projects were obtained from the records at the headquarters of Ministry of Roads and Public Works in Nairobi. A list of consulting engineering firms was obtained from the secretariat of the Association of Consulting Engineers of Kenya situated in Nairobi. The total number of Consulting Engineering firms that were approached was 15 whereas the number of Contractors was 16. These were the firms that dealt with design and construction of roads and were based in Nairobi. Survey research design was used and questionnaires were delivered to participants in person to obtain primary data. Participants filled in the questionnaires in their own time without any assistance from the researcher. The questionnaires were collected from the participants after a period of about three weeks. This approach removed any undue pressure from the respondents and gave them the freedom to fill in the questionnaires as truthfully as possible.

The Likert rating scale was used in the questionnaire. An opportunity was left to the respondents to include and rate variables that might have been missed out. The variables in the questionnaire were adapted from the studies cited in the literature review. In total, 141 variables were identified. The 141 variables were grouped into 25 broad categories to facilitate objective analysis. The grouping considered variable relationships with each other as well as similarity in characteristics. For example, variables such as clay, rock, underground water and fossils were grouped together under the broad category of subsoil conditions. The variables identified were thus grouped under design problems, subsoil conditions, financial/economic problems, subcontractors problems, contractual disputes, permits/licences, weather conditions, labour availability, and site accidents. Other variables included politics, materials availability, proximity to required resources, equipment availability, shop drawings, site layout, industrial relations disputes, and sample of materials / work approvals. The rest of the variables included material testing, natural hazards, underground services, working environment, supervision / management staff availability, advanced technology, suppliers, and client problems.

The questionnaire was designed in such a way that the stratification of the data was easy for analysis. The questions involved recording the contribution of each variable to delay in the project on a rating scale of: 1 – Very Low; 2 – Low; 3 – Slightly Low; 4 – Average; 5 – Slightly High; 6 – High; and 7 – Very High. The questions concentrated on past phenomena on the project. The interest was to show how the past events had affected the projects. The question that was asked in the questionnaire was "During the period of construction, what was the contribution of the given factors to delay in the project?".

According to Talukhaba (1999) researchers on construction projects have in many cases worked with relatively small sample sizes. For example, Nkado (1992) (as cited in Talukhaba, 1999) investigated information system for the building industry with a sample of 29 cases. Ogunlana, Promkuntong and Jearkjirm (1996) (as cited in Talukhaba, 1999) investigated the causes of delay in projects in Thailand basing their research on a sample of 12 projects. Uher (1996) (as cited in Talukhaba, 1999) investigated the cost estimating practices in Australia construction industry using a sample of 10 projects. The sample of 31 consultants and contractors used in this study is therefore well above what has been used in other studies in the construction industry elsewhere. On the other hand, the sample size determines the type of statistics to be applied. Therefore, by using a suitable statistical tool, some of the problems that could be associated with the sample size are minimised. The 31 cases in this study produced 61.3% positive response. Kothari (1990) indicates that the percentage of responses for survey-type research is as low as 20 to 30%. Two methods were used to analyze the data: Relative Importance Index (RII) and Spearman's rank correlation. The data that was collected was presented in a tabular form. Analysis was then carried out on the data and the results were presented in a tabular form as well.

Kometa, Olomalaiye and Harris (1994) (as cited in Sambasivan & Soon, 2007) used the relative importance index method to determine the relative importance of various causes of delays. The same method was adopted in this study. The seven-point scale ranged from 1 (very low) to 7 (very high) was adopted and transformed to Relative Importance Index (RII) for each factor as follows:

$$RII = \Sigma W/(A*N)$$

Where 'W' is the weighting given to each factor by the respondents (ranging from 1 to 7); 'A' is the highest weight (7 in this case); and 'N' is the total number of respondents. The RII value had a range from 0 to 1 (0 not inclusive). The higher the value of RII, the more important was the cause of delays.

Assaf and Al-Hejji (2006) (as cited in Kamanga & Steyn, 2013) indicate that the Spearman's rank correlation is a relationship measure among different parties or factors and the strength and direction of the relationship. This study uses Spearman's rank correlation to show the level of agreement between two parties.

$$rs = 1 - 6\Sigma d^2/(n^3 - n)$$

Where 'rs' is the Spearman's rank correlation coefficient; 'd' is the difference in ranking between any two parties; and 'n' is the number of factors. The correlation coefficient varies between +1 and −1, where +1 implies a perfect positive relationship (agreement), while −1 results from a perfect negative relationship (disagreement). Sample estimates of correlation close to unity in magnitude imply good correlation, while values near zero indicate little or no correlation.

4. Findings and Discussions

The combined views for all delay factors for consultants and contractors are shown in Table 1. The overall top five causes of delay indentified by both consultants and contractors were: Payment by client (RII = 0.759); Slow decision-making and bureaucracy in client organization (RII = 0.699); Claims (RII = 0.609); Inadequate planning / scheduling (RII = 0.602); and Rain (RII = 0.579). The ten most important views of the consultants are shown in Table 2. The top five causes of delay identified by consultants were: Payment by client (RII = 0.738); Slow decision-making and bureaucracy in client organization (RII = 0.595); Inadequate planning / scheduling (RII = 0.595); Different site conditions (RII = 0.583); and Proximity to borrow pit (0.571). The ten most important views of the contractors are shown in Table 3. The top five causes of delay identified by contractors were: Slow decision-making and bureaucracy in client organization (RII = 0.755); Payment by client (RII = 0.673); Engineer's certificates (RII = 0.633); Claims (RII = 0.633); and Rain (RII = 0.612). The Spearman's rank correlation coefficient between the consultants and contractors was calculated and found to be 0.64 which showed that there was a positive correlation between the views of consultants and contractors.

Payment by client is overall the most important cause of delay. According to consultants, payment by client is also the most important delay factor. However, according to contractors the most important cause of delay is slow decision-making and bureaucracy in client organization. Overall, the second most important delay factor is slow decision-making and bureaucracy in client organization. This delay factor also appears second in the view of consultants. However, in the view of contractors, the second most important cause of delay is payment by client. Overall, the third most important delay factor is claims. According to consultants the third most important delay factor is inadequate planning / scheduling. Contractors view Engineer's certificates as the third most important cause of delay. Inadequate planning / scheduling is overall the fourth most important cause of delay. According to consultants, the fourth most important delay factor is different site conditions. For contractors, the fourth most important delay factor is claims. Overall, the fifth most important factor is rain. This delay factor also appears fifth in the view of contractors. However, in the view of consultants, the fifth most important cause of delay is proximity to borrow pit. The findings of this research should be of importance to both local and international engineering and construction firms entering the road construction market in Kenya.

Table 1. Relative Importance Index and Ranking of Delay Factors for both Consultants and Contractors

Delay Factors	W1	W2	W3	W4	W5	W6	W7	W8	W9	W10	W11	W12	W13	W14	W15	W16	W17	W18	W19	Total Weight ΣW	Highest Weight (A)	Total No. of Resp (N)	$RII = \Sigma W/(A*N)$	Rank
1. Design Problems																								
a) Conflicting design information	4	3	1	1	5	6	4	5	4	2	6	1	1	4	1	4	5	4	6	73	7	19	0.549	10
b) Timeliness of revised drawings issue	3	3	1	1	5	4	4	5	1	2	1	1	2	1	1	5	6	4	5	55	7	19	0.414	41
c) Missing information	2	2	1	1	5	4	3	2	4	2	6	1	1	5	1	6	7	2	2	63	7	19	0.474	26
d) Dimensional inaccuracies	4	2	1	1	3	4	3	2	1	2	5	1	1	1	1	4	4	2	2	49	7	19	0.368	71
e) Design change by Client	4	2	1	1	1	4	5	4	5	2	6	1	3	5	1	2	7	1	7	68	7	19	0.511	13
f) Design change by Engineer	2	2	1	1	4	5	4	4	2	3	4	3	4	3	1	3	7	4	7	68	7	19	0.511	13
g) Different site conditions	6	5	1	1	6	6	6	6	2	3	6	1	3	5	1	3	4	5	1	77	7	19	0.579	5
2. Subsoil Conditions																								
a) Clay	4	4	1	3	6	1	6	4	1	2	6	1	1	1	3	4	2	2	2	60	7	19	0.451	28
b) Rock	1	6	1	1	1	1	2	4	1	2	4	1	2	1	2	5	2	5	6	52	7	19	0.391	55
c) Underground water	3	5	1	1	1	2	2	3	2	4	5	1	2	1	1	3	2	4	6	54	7	19	0.406	47
d) Fossils	1	1	1	1	1	1	2	1	1	2	2	1	1	1	1	1	1	2	1	25	7	19	0.188	141
3. Financial/Economic Problems																								
a) Payment by Client	6	7	5	7	2	5	1	7	7	6	6	3	2	1	4	7	7	7	5	101	7	19	0.759	1
b) Engineer's Certificates	4	2	1	1	4	2	1	2	1	2	2	1	2	3	4	7	5	7	3	56	7	19	0.421	34
c) Contractor's cash flow	5	7	6	1	4	1	1	2	3	7	2	4	2	1	6	6	2	4	4	70	7	19	0.526	12
d) Interest Rates	2	6	1	1	5	5	1	2	6	5	2	1	1	1	1	6	2	5	3	58	7	19	0.436	31
e) Inflation	2	3	1	1	5	5	1	2	5	5	4	3	1	4	2	3	2	5	5	63	7	19	0.474	26
f) Insolvencies and bankruptcies	2	1	1	1	1	1	1	1	1	1	4	7	1	1	1	1	1	4	1	39	7	19	0.293	109
4. Subcontractors Problems																								
a) Late nomination of subcontractors	1	2	1	1	4	3	1	2	3	1	4	1	1	1	1	1	2	4	2	40	7	19	0.301	105
b) Poor workmanship	1	3	1	1	4	1	2	4	2	1	4	1	1	1	1	1	2	4	2	41	7	19	0.308	99
c) Labour problems	1	5	1	2	3	1	7	4	2	1	1	1	1	1	1	1	2	7	2	44	7	19	0.331	88
d) Material problems	1	3	1	1	3	1	6	4	5	1	1	1	1	1	1	1	5	5	5	47	7	19	0.353	77
e) Late payment	1	6	1	1	4	1	1	7	7	1	5	1	1	1	1	1	7	7	6	65	7	19	0.489	21
f) Relationship with main contractor	1	4	1	1	4	1	1	6	1	1	4	1	1	1	1	1	4	4	1	43	7	19	0.323	93
g) Conflicts in work schedules of subcontractors	1	3	1	1	2	1	1	5	1	1	2	1	1	1	1	1	2	6	1	35	7	19	0.263	129
5. Contractual Issues																								
a) Poor workmanship	2	6	1	1	2	7	1	4	2	2	6	1	1	1	1	2	2	4	2	54	7	19	0.406	47
b) Variations	4	6	2	1	6	4	2	4	4	5	4	3	3	3	1	3	6	6	4	75	7	19	0.564	7
c) Claims	2	7	1	1	6	5	3	4	7	5	4	1	2	3	4	6	6	6	4	81	7	19	0.609	3
d) Inadequate planning/scheduling	2	7	1	1	5	6	6	5	2	4	7	4	1	1	3	6	3	5	4	80	7	19	0.602	4
6. Permits/Licenses																								
a) Central Government	1	1	1	1	1	1	7	6	1	1	3	1	1	1	1	6	7	6	2	52	7	19	0.391	55
b) Local Authority	3	5	1	1	3	1	7	7	2	1	4	1	1	1	1	6	7	7	2	65	7	19	0.489	21
c) Kenya Power & Lighting	4	3	1	1	5	4	1	6	3	3	2	1	1	1	1	7	2	7	1	56	7	19	0.421	34
d) Telkom Kenya	4	5	1	1	5	4	1	6	1	1	2	1	1	1	1	1	5	7	1	51	7	19	0.383	63
e) NEMA	4	1	1	1	1	1	1	2	1	1	7	1	1	1	1	6	7	5	1	51	7	19	0.383	63
7. Weather Conditions																								

Delay Factors	W1	W2	W3	W4	W5	W6	W7	W8	W9	W10	W11	W12	W13	W14	W15	W16	W17	W18	W19	Total Weight ΣW	Highest Weight (A)	Total No. of Resp (N)	RII = ΣW/(A*N)	Rank
a) Rain	7	6	3	7	1	3	3	1	3	3	4	2	4	3	5	4	3	4	7	77	7	19	0.579	5
b) Hot temperatures	4	2	1	1	1	1	3	1	1	1	4	1	2	1	4	1	3	1	4	41	7	19	0.308	99
c) Cold temperatures	4	1	1	1	1	1	3	1	1	1	4	1	1	1	1	1	3	1	4	36	7	19	0.271	126
8. Labour Availability																								
a) Skilled	2	2	4	1	2	2	5	6	1	2	2	3	1	1	4	5	3	5	1	54	7	19	0.406	47
b) Unskilled	1	2	1	1	2	2	6	4	1	2	2	1	1	1	4	2	1	2	1	39	7	19	0.293	109
9. Site Accidents																								
a) Fatal	1	2	1	1	1	1	5	2	1	1	4	1	1	1	1	2	1	2	2	35	7	19	0.263	129
b) Serious injuries	3	2	1	1	1	1	4	2	1	1	4	1	1	1	1	2	2	2	2	37	7	19	0.278	123
c) Minor injuries	4	2	1	1	1	1	1	2	2	1	4	1	1	1	4	3	3	2	2	41	7	19	0.308	99
10. Politics																								
a) Politician's interference	6	5	1	1	6	3	4	7	2	1	5	1	1	1	1	5	6	4	2	67	7	19	0.504	16
b) Political wrangling between workers	1	2	1	1	1	1	6	2	1	1	4	1	1	1	1	2	2	2	2	37	7	19	0.278	123
c) Political decrees	1	2	1	1	3	1	7	5	1	1	4	1	1	1	1	1	6	3	2	47	7	19	0.353	77
d) Civil strife/riots	1	2	1	1	1	1	7	2	1	1	2	1	1	1	1	1	2	3	2	34	7	19	0.256	134
e) Protest action-groups	6	2	1	1	1	1	5	2	1	1	2	1	1	1	1	2	1	4	1	37	7	19	0.278	123
11. Materials Availability																								
a) Cement	5	1	1	1	1	1	6	1	1	2	4	5	2	1	1	6	2	3	3	51	7	19	0.383	63
b) Steel reinforcement	4	2	1	1	1	1	6	1	1	2	4	2	2	1	4	6	2	4	3	52	7	19	0.391	55
c) Timber	2	1	1	1	1	1	6	1	1	2	5	2	2	1	4	6	5	6	2	55	7	19	0.414	41
d) Sand	2	1	1	1	1	1	4	3	2	2	4	1	1	1	4	3	5	5	2	48	7	19	0.361	75
e) Aggregate	2	1	3	1	1	1	7	3	1	2	4	4	1	1	5	4	5	6	3	59	7	19	0.444	30
f) Bitumen	6	6	1	1	3	3	7	1	1	2	4	4	1	1	4	6	3	6	3	67	7	19	0.504	16
g) Murram	7	6	1	4	3	1	5	5	3	2	4	2	1	1	4	6	4	2	3	68	7	19	0.511	13
h) Water	4	2	1	1	1	1	6	4	1	2	4	1	1	1	4	5	7	3	1	54	7	19	0.406	47
i) Graded Crushed Stone	1	3	1	1	1	3	7	3	1	2	4	1	1	1	5	4	5	4	3	55	7	19	0.414	41
j) Hand packed stone	1	3	1	1	5	1	7	2	1	2	4	1	1	1	1	4	5	3	3	51	7	19	0.383	63
k) Lime	1	5	1	1	1	1	7	1	1	2	4	1	1	1	1	5	2	3	4	47	7	19	0.353	77
l) Structural steel	1	5	1	1	1	1	7	1	1	2	4	1	1	1	1	6	5	4	3	51	7	19	0.383	63
m) Explosives	1	3	1	1	1	1	7	1	1	2	4	1	1	1	1	5	2	5	3	46	7	19	0.346	80
n) Theft of materials	4	6	1	1	1	1	1	3	2	2	4	2	1	1	4	3	5	6	1	53	7	19	0.398	54
o) Wastage of materials	2	2	1	1	1	1	3	3	1	2	5	1	1	1	4	3	2	6	1	46	7	19	0.346	80
p) Double handling of materials	3	3	1	1	1	1	7	3	2	2	4	1	1	1	3	3	5	5	1	52	7	19	0.391	55
q) Disposal of construction waste	6	3	1	1	5	5	6	3	1	2	4	1	2	1	1	2	2	3	3	56	7	19	0.421	34
12. Proximity to Required Resources																								
a) Borrow pit	7	3	1	4	6	5	7	5	1	4	4	1	2	1	1	5	6	4	3	74	7	19	0.556	8
b) Quarry	6	4	1	1	6	5	7	5	1	4	4	2	2	1	1	5	7	4	4	74	7	19	0.556	8
c) Asphalt batching plant	6	3	1	1	6	3	1	4	1	4	4	1	1	1	1	4	6	2	4	58	7	19	0.436	31
13. Equipment Availability																								
a) Crane	1	2	1	1	6	1	7	3	1	3	4	1	1	1	4	3	1	4	3	52	7	19	0.391	55
b) Concrete mixer	1	4	1	1	2	1	1	1	1	3	4	1	1	1	4	3	1	2	1	38	7	19	0.286	117
c) Excavator	1	4	1	1	1	1	1	1	1	3	4	3	1	4	1	6	1	2	2	43	7	19	0.323	93
d) Concrete vibrator	1	4	1	1	1	1	1	1	1	3	4	1	1	1	4	1	1	2	1	35	7	19	0.263	129
e) Crushing plant	4	5	1	1	1	1	1	2	1	3	4	3	1	3	1	6	1	4	3	50	7	19	0.376	68
f) Asphalt Paver	1	5	1	1	1	1	1	2	1	3	4	1	1	5	1	5	1	2	1	42	7	19	0.316	96
g) Roller	5	6	4	1	1	1	1	1	1	3	4	3	1	3	1	5	1	2	1	49	7	19	0.368	71
h) Grader	4	7	4	1	1	1	1	1	1	3	4	2	1	3	1	6	1	2	1	49	7	19	0.368	71
i) Bulldozer	3	5	1	1	1	1	1	1	1	3	4	2	1	4	1	6	1	2	1	44	7	19	0.331	88
j) Wheel loader	3	5	1	1	1	1	1	1	1	3	4	2	1	4	1	6	1	2	1	44	7	19	0.331	88
k) Dump Truck	7	6	5	1	1	2	1	1	1	3	4	3	1	1	1	6	1	2	1	52	7	19	0.391	55
l) Dumper	1	3	1	1	1	1	1	1	1	3	4	3	1	3	1	6	1	2	1	40	7	19	0.301	105
m) Bowser	5	6	1	1	1	1	1	1	1	3	4	2	1	2	1	4	1	2	1	44	7	19	0.331	88
n) Lorry	4	5	1	1	1	1	1	1	1	3	4	4	1	1	1	5	1	2	2	44	7	19	0.331	88
o) Pick-up	4	3	1	1	1	1	1	1	1	3	4	1	1	2	1	5	1	2	1	39	7	19	0.293	109
p) Low loader	5	3	1	1	1	1	1	2	1	3	4	1	1	3	1	3	1	4	1	42	7	19	0.316	96

Delay Factors	W1	W2	W3	W4	W5	W6	W7	W8	W9	W10	W11	W12	W13	W14	W15	W16	W17	W18	W19	Total Weight ΣW	Highest Weight (A)	Total No. of Resp (N)	RII = ΣW/(A*N)	Rank
q) Conveyor	1	1	1	1	1	1	1	2	1	3	4	1	1	2	1	3	1	2	1	33	7	19	0.248	136
r) Cement Silo	1	1	1	1	1	1	1	2	1	3	4	1	1	1	1	1	1	2	1	30	7	19	0.226	138
s) Compressor	4	4	1	1	1	2	1	1	1	3	4	1	1	2	1	3	1	2	1	39	7	19	0.293	109
t) Pump	4	3	1	1	1	2	1	1	1	3	4	1	1	2	1	3	1	2	1	38	7	19	0.286	117
u) Plate Compactor	1	3	1	1	1	1	1	2	1	3	4	1	1	2	1	3	1	2	1	35	7	19	0.263	129
v) Chips Spreader	4	3	1	1	1	1	1	2	1	3	4	3	1	5	1	4	1	3	1	45	7	19	0.338	84
w) Theft of machines/machine parts	3	6	1	1	1	1	1	2	2	3	4	2	1	4	1	2	1	5	1	46	7	19	0.346	80
14. Shop Drawings																								
a) Preparation	1	2	1	1	6	5	5	6	1	1	6	1	2	1	4	2	6	4	3	64	7	19	0.481	24
b) Approval	1	2	1	1	5	5	5	5	2	1	5	1	3	1	4	2	6	6	4	65	7	19	0.489	21
15. Site Layout																								
a) Access to site entry/exit points	1	2	1	1	3	4	1	2	3	1	4	1	2	1	4	2	2	4	2	45	7	19	0.338	84
b) Congestion at site entry/exit points	1	2	1	1	3	4	1	2	1	1	4	1	2	1	4	3	2	5	2	45	7	19	0.338	84
c) Storage space	1	2	1	1	2	4	1	2	1	1	3	1	1	1	4	3	2	5	3	42	7	19	0.316	96
16. Industrial Relations Disputes																								
a) Wages	5	4	3	4	1	2	6	4	2	1	4	3	1	4	4	3	2	5	2	64	7	19	0.481	24
b) Safety	3	4	1	1	1	2	6	5	2	1	4	1	1	1	4	3	6	4	2	56	7	19	0.421	34
c) Welfare	3	4	1	1	1	2	6	6	2	1	4	1	1	1	4	3	2	5	2	54	7	19	0.406	47
d) Site communication	4	3	1	1	1	2	6	5	2	1	4	1	1	6	4	2	4	4	2	58	7	19	0.436	31
e) Labour camps	3	4	1	1	1	1	6	4	2	1	4	1	1	4	4	2	2	4	2	52	7	19	0.391	55
f) Discipline of workers	2	3	1	1	2	1	6	5	2	1	4	3	1	2	6	2	3	5	1	55	7	19	0.414	41
g) Absenteeism	3	3	1	1	1	1	6	4	3	1	4	3	1	1	6	2	2	6	1	54	7	19	0.406	47
17. Sample of Materials/Work Approvals																								
a) Client approval	1	2	1	1	1	1	6	7	3	1	4	1	2	3	4	2	2	5	4	55	7	19	0.414	41
b) Engineer approval	1	2	1	1	3	6	2	4	2	1	2	1	2	3	4	5	3	5	4	54	7	19	0.406	47
c) Re-work	3	2	1	4	5	3	4	3	3	4	4	2	2	4	1	2	3	4	2	60	7	19	0.451	28
18. Material Testing																								
a) Cement	1	2	1	1	1	1	2	5	1	1	2	2	3	1	1	2	2	5	2	38	7	19	0.286	117
b) Steel reinforcement	1	2	1	1	1	1	2	5	1	1	2	2	3	1	1	2	3	5	2	39	7	19	0.293	109
c) Timber	1	2	1	1	1	1	2	5	1	1	4	1	3	1	1	2	2	5	2	41	7	19	0.308	99
d) Sand	2	2	1	1	1	1	2	5	1	1	2	1	3	2	1	1	2	5	2	38	7	19	0.286	117
e) Aggregate	2	2	1	1	1	1	2	5	1	1	2	2	3	2	1	2	2	5	2	40	7	19	0.301	105
f) Bitumen	6	2	1	1	2	1	2	5	1	1	2	1	4	2	1	2	2	5	2	45	7	19	0.338	84
g) Murram	7	2	1	1	4	1	2	5	1	1	2	2	2	3	1	3	2	5	2	49	7	19	0.368	71
h) Water	1	2	1	1	1	1	2	5	1	1	2	1	1	1	1	1	2	5	2	34	7	19	0.256	134
i) Graded Crushed Stone	1	2	1	1	1	1	2	5	1	1	2	3	2	4	1	2	2	5	2	41	7	19	0.308	99
j) Hand packed stone	1	2	1	1	4	1	2	5	1	1	2	1	2	1	1	2	2	5	2	39	7	19	0.293	109
k) Lime	1	2	1	1	1	1	2	5	1	1	2	1	1	1	1	2	2	5	2	35	7	19	0.263	129
l) Structural Steel	1	2	1	1	1	1	2	5	1	1	2	1	3	1	1	2	3	5	2	38	7	19	0.286	117
m) Asphalt	5	2	1	1	3	1	2	5	1	1	4	1	4	3	1	2	2	5	2	50	7	19	0.376	68
n) Soil	1	2	1	1	3	1	2	5	1	1	2	1	2	1	1	3	2	5	2	39	7	19	0.293	109
o) Rock	1	2	1	1	2	1	2	5	1	1	2	1	2	1	1	3	2	5	2	38	7	19	0.286	117
19. Natural Hazards																								
a) Fire	1	1	1	1	1	1	1	2	1	1	2	1	1	1	1	2	2	4	1	28	7	19	0.211	139
b) Floods	2	3	3	7	1	1	6	2	3	1	4	1	1	1	1	2	5	5	3	56	7	19	0.421	34
c) Landslides	2	2	1	4	1	1	1	2	1	1	3	1	1	1	1	1	3	4	2	36	7	19	0.271	126
d) Wind	1	1	1	1	1	1	1	2	1	1	2	1	1	1	1	1	2	4	1	27	7	19	0.203	140
20. Underground Services																								
a) Electricity cables	3	2	1	1	6	6	1	7	2	7	4	1	2	3	6	1	2	6	1	66	7	19	0.496	20
b) Telephone cables	3	2	1	1	6	6	1	7	2	7	4	1	1	2	6	1	5	6	1	67	7	19	0.504	16
c) Water pipes	5	3	1	1	6	6	1	7	3	7	4	1	2	3	6	2	2	6	1	71	7	19	0.534	11
d) Petroleum fuel pipes	1	1	1	1	6	1	1	2	1	2	4	1	1	1	1	1	2	6	1	39	7	19	0.293	109
e) Stormwater pipes	1	3	1	1	2	1	1	7	1	7	4	1	1	4	1	1	2	6	1	50	7	19	0.376	68
f) Data cables	1	1	1	1	1	1	1	7	1	7	4	1	1	1	6	1	1	6	1	48	7	19	0.361	75

Delay Factors	W1	W2	W3	W4	W5	W6	W7	W8	W9	W10	W11	W12	W13	W14	W15	W16	W17	W18	W19	Total Weight ΣW	Highest Weight (A)	Total No. of Resp (N)	RII = ΣW/(A*N)	Rank
21. Working Environment																								
a) Requirement for restrictive working hours	2	2	1	1	2	3	4	2	2	1	4	1	1	1	1	2	1	5	1	41	7	19	0.308	99
b) Ambient noise conditions	2	2	1	1	2	3	1	2	1	1	4	1	1	1	1	1	2	4	1	36	7	19	0.271	126
c) Ambient light conditions	2	2	1	1	2	1	1	2	1	1	4	1	1	1	1	1	2	3	1	33	7	19	0.248	136
22. Supervision/Management Staff Availability																								
a) Resident Engineer	6	5	1	1	1	1	1	5	1	2	4	1	1	4	4	2	6	4	1	55	7	19	0.414	41
b) Inspector of Works	4	3	1	1	1	1	1	5	1	2	4	1	1	1	4	1	5	4	1	46	7	19	0.346	80
c) Site Agent	2	7	4	1	1	1	1	4	1	2	4	1	1	3	4	3	7	4	1	56	7	19	0.421	34
d) General Foreman	3	3	4	1	1	1	1	4	1	2	6	2	1	3	4	3	5	4	1	56	7	19	0.421	34
23. Advanced Technology																								
a) Computers	2	2	1	1	1	2	1	2	1	1	4	1	1	4	4	2	3	2	1	40	7	19	0.301	105
b) Automated machines	2	2	1	1	1	2	1	2	1	1	4	1	1	4	4	2	6	2	1	43	7	19	0.323	93
24. Suppliers																								
a) Late nomination of suppliers	4	2	1	1	3	2	7	6	1	3	4	1	1	1	1	2	2	4	2	52	7	19	0.391	55
b) Late deliveries of materials by suppliers	7	2	1	1	3	2	7	4	2	3	5	3	1	6	1	5	2	5	2	67	7	19	0.504	16
25. Client Problems Slow decision making and bureaucracy in Client Organization	7	7	1	6	2	1	1	7	6	5	6	1	3	6	6	7	7	5	3	93	7	19	0.699	2

Table 2. Relative Importance Index and Ranking of Delay Factors for Consultants

Delay Factors	W1	W2	W3	W4	W5	W6	W7	W8	W9	W10	W11	W12	Total Weight ΣW	Highest Weight (A)	Total No. of Resp(N)	RII = ΣW/(A*N)	Rank
1. Design Problems																	
Different site conditions	6	5	1	1	6	6	6	6	2	3	6	1	49	7	12	0.583	4
2. Financial/Economic Problems																	
Payment by Client	6	7	5	7	2	5	1	7	7	6	6	3	62	7	12	0.738	1
3. Contractual Issues																	
a) Variations	4	6	2	1	6	4	2	4	4	5	4	3	45	7	12	0.536	8
b) Claims	2	7	1	1	6	5	3	4	7	5	4	1	46	7	12	0.548	6
c) Inadequate planning/scheduling	2	7	1	1	5	6	6	5	2	4	7	4	50	7	12	0.595	2
4. Weather Conditions																	
Rain	7	6	3	7	1	3	3	1	3	3	4	2	43	7	12	0.512	10
5. Materials Availability																	

Murram	7	6	1	4	3	1	5	5	3	2	4	2	43	7	12	0.512	10
6. Proximity to Required Resources																
a) Borrow pit	7	3	1	4	6	5	7	5	1	4	4	1	48	7	12	0.571	5
b) Quarry	6	4	1	1	6	5	7	5	1	4	4	2	46	7	12	0.548	6
7. Underground Services																
Water pipes	5	3	1	1	6	6	1	7	3	7	4	1	45	7	12	0.536	8
8. Client Problems																
Slow decision making and bureaucracy in Client Organization	7	7	1	6	2	1	1	7	6	5	6	1	50	7	12	0.595	2

Table 3. Relative Importance Index and Ranking of Delay Factors for Contractors

Delay Factors	W13	W14	W15	W16	W17	W18	W19	Total Weight ΣW	Highest Weight (A)	Total No. of Resp (N)	RII = ΣW/(A*N)	Rank
1. Design Problems												
a) Design change by Client	3	5	1	2	7	1	7	26	7	7	0.531	7
b) Design change by Engineer	4	3	1	3	7	4	7	29	7	7	0.592	6
2. Financial/Economic Problems												
a) Payment by Client	2	1	4	7	7	7	5	33	7	7	0.673	2
b) Engineer's Certificates	2	3	4	7	5	7	3	31	7	7	0.633	3
3. Contractual Issues												
a) Variations	3	3	1	3	6	6	4	26	7	7	0.531	7
b) Claims	2	3	4	6	6	6	4	31	7	7	0.633	3
4. Weather Conditions												
Rain	4	3	5	4	3	4	7	30	7	7	0.612	5
5. Materials Availability												
Timber	2	1	4	6	5	6	2	26	7	7	0.531	7
6. Shop Drawings												
Approval	3	1	4	2	6	6	4	26	7	7	0.531	7
7. Sample of Materials/Work Approvals												
Engineer approval	2	3	4	5	3	5	4	26	7	7	0.531	7
8. Client Problems												
Slow decision making and bureaucracy in Client Organization	3	6	6	7	7	5	3	37	7	7	0.755	1

5. Conclusions and Recommendations

This study was aimed at finding the causes of delay in road construction projects in Kenya. There were 141 causes of delay that were extracted from the literature on the subject. The 141 causes of delay were divided into 25 broad categories. The overall top five causes of delay indentified by both consultants and contractors were: Payment by client; Slow decision-making and bureaucracy in client organization; Claims; Inadequate planning / scheduling; and Rain. The Spearman's rank correlation coefficient between the consultants and contractors was found to be 0.64 which showed that there was a positive correlation between the views of consultants and contractors. The range of causes of delay in completing road construction projects in Kenya was identified. The most important causes of delay in road construction projects in Kenya were also identified. The differences in perception of contractors and consultants regarding causes of delay in delivering projects by the intended completion date were identified as well. Therefore, the objectives of this study were substantially accomplished.

The Null Hypothesis which states that road construction project delays are not caused by exogenous and endogenous factors is rejected. Hence the Alternative Hypothesis which states that road construction project delays are caused by exogenous and endogenous factors is accepted.

Clients should improve their financial management systems so that they could be able to pay contractors in a timely manner. Kamanga and Steyn (2013) suggest that while it is common practice for contracts to include a performance guarantee clause, there should also be a payment guarantee clause so that if a duly issued payment certificate is not paid by the client within the stipulated period, the contractor may demand his payment from the guarantor. Bureaucracy and red tape should be reduced in client organisations in order to speed up the slow decision making process. Claims should be settled quickly so that they do not become a source of delays. Contractors should prepare adequate plans and schedules during execution of road projects. During the rainy season, contractors should plan to execute activities that are not normally affected by the rain in order to mitigate delays.

The following are recommended areas for further research: Investigation of factors that cause delays in road projects during the design stage; Investigation of the impact of BOOT (Build Own Operate Transfer) method of procurement on construction time performance; Impact of usage of old equipment on road construction projects on construction time performance; and Impact of usage of concrete surfacing in lieu of asphalt on roads on construction time performance.

References

[1] Abdullah, M. R., Rahman, I. A., & Azis, A. A. A. (2010). Causes of Delay in MARA Management Procurement Construction Projects. *Journal of Surveying, Construction & Property, 1*(1), 123 – 138.

[2] Akogbe, R. T. M., Feng, X., & Zhou, J. (2013). Importance and Ranking Evaluation of Delay Factors for Development Construction Projects in Benin. *KSCE Journal of Civil Engineering, 17*(6), 1 – 10.

[3] Alaghbari, W., Kadir, M. R., & Salim, A. (2007). The significant factors causing delay of building construction projects in Malaysia. *Engineering Construction and Architectural Management Journal, 14*(2), 192 – 206.

[4] Alinaitwe, H., Apolot R., & Tindiwensi, D. (2013). Investigation into the Causes of Delays and Cost Overruns in Uganda's Public Sector Construction Projects. *Journal of Construction a. in Developing Countries, 18*(2), 33–47.

[5] Al-Tabtabai, H. M. (2002). Causes for Delays in Construction Projects in Kuwait. *Engineering i. Journal of the University of Qatar, 15*, 19 – 37.

[6] Andi, A., Lalitan, D. & Loanata, V. R., (2010). Owner and Contractor Perceptions Toward Factors Causing Delays in Structural and Finishing Works. *Civil Engineering Dimension Journal, 12*(1), 8 – 17Asiamah, A. D. A., & Asiamah, O. K. A. (2013). Management of Government Funded Construction Projects in Ghana: Stakeholders' Perspective of Causes of Delays in Construction of Public Buildings. *Journal of*

International Institute for Science, Technology and Education, 3(12) 149 – 156.

[7] Assaf, S. A., & Al-Hejji, S. (2006). Causes of Delay in Large Construction Projects. *International Journal of Project Management, 24*(4), 349 – 357. doi: 10.1016/j.ijproman.2005.11.010

[8] Assaf, S. A., Al-Khalil, M., & Al-Hazmi, M. (1995). Causes of Delay in Large Building Construction Projects. *Journal of Management in Engineering, 11*(2), 45 – 50 Retrieved from

[9] Ayudhya, B. I. N. (2011). Evaluation of Common Delay Causes of Construction Projects in Singapore. *Journal of Civil Engineering and Architecture, 5*(11), 1027 1034

[10] Baloyi, L., & Bekker, M. (2011). Causes of construction cost and time overruns: The 2010 FIFA World Cup stadia in South Africa. *Acta Structilia, 18*(10), 51 – 67.

[11] Challal, A., & Tkiouat, M. (2012). Identification of the Causes of Deadline Slippage in Construction Projects: State of the Art and Application. *Journal of Service Science and Management, 5*, 151 – 159

[12] Desai, M., & Bhatt, R. (2013). Critical Causes of Delay in Residential Construction Projects: Case Study of Central Gujarat Region of India. *International Journal of Engineering Trends and Technology, 4*(4), 762 – 768.

[13] Dolage, D. A. R., & Rathnamali, D. L. G. (2013). Causes of Time Overrun in Construction Phase of Building Projects: A Case Study on Department of Engineering Services of Sabaragamuwa Provincial Council. *Engineer Journal of the Institution of Engineers of Sri Lanka. 46*(03), 9 – 18. doi: 10.4038/engineer.v46i3.6780

[14] El-Razek, M. E. A., Bassioni, H. A., & Mobarak, A. M. (2008). Causes of Delay in Building Construction Projects in Egypt. *Journal of Construction Engineering and Management, 134*(11), 831 – 841.

[15] Fan, S., & Kang, C. C. (2005). Road Development, Economic Growth, and Poverty Reduction in China, International Food Policy Research Institute Retrieved from http://www.ifpri.org/sites/default/files/publications/rr138.pdf

[16] Faridi, A. S., & El-Sayegh, S. M. (2006). Significant Factors Causing Delay in the United Arab Emirates Construction Industry. *Construction Management and Economics Journal, 24*(11), 1167 – 1176.

[17] Frimpong, Y., Oluwoye, J., & Crawford, L. (2003). Causes of delay and cost overruns in construction of groundwater projects in developing countries; Ghana as a case study. *International Journal of Project Management, 21*, 321 326

[18] Fugar, F. D. K., & Agyakwah, B. A. B. (2010). Delays in Building Construction Projects in Ghana, *Australasian Journal of Construction Economics and Building, 10*(1), 103 - 116.

[19] Hoai, L. L., Lee, Y. D., & Lee, J. Y. (2008). Delay and Cost Overruns in Vietnam Large Construction Projects: A Comparison with Other Selected Countries. *KSCE Journal of Civil Engineering, 12*(6), 367 – 377.

[20] Ibironke, O. T., Oladinrin, T. O., Adeniyi, O., & Eboreime, I. V. (2013). Analysis of Non- Excusable Delay Factors Influencing Contractors' Performance in Lagos State, Nigeria. *Journal of Construction in Developing Countries, 18*(1), 53 – 72.

[21] Kagiri, D., & Wainaina, G. (2008). Time And Cost Overruns in Power Projects in Kenya: A Case Study of Kenya Electricity Generating Company Limited. Paper presented at the 4th International Operations Research Society of Eastern Africa (ORSEA) Conference.

[22] Kamanga, M. J., & Steyn, W. (2013). Causes of Delay in Road Construction Projects in Malawi. *Journal of The South African Institution of Civil Engineering, 55*(3), 79 – 85.

[23] Kholif, W., Hosny, H., & Sanad, A. (2013). Analysis of Time and Cost Overruns in Educational Building Projects in Egypt. *International Journal of Engineering and Technical Research, 1*(10).

[24] Kikwasi, G. J. (2012). Causes and Effects of Delays and Disruptions in Construction Projects in Tanzania. *Australasian Journal of Construction Economics and uilding,   Conference Series, 1*(2),52 – 59.

[25] Kivaa, P. T. (2000). Developing a Model for Estimating Construction Period: A Survey of Building Projects in Nairobi (Masters Thesis). University of Nairobi

[26] Kometa, S. T., Olomolaiye, P. O., & Harris, F. C. (1994). Attributes of UK Construction Clients Influencing Project Consultants' Performance. *Construction Management and Economics Journal, 12*(5), 433 -443.

[27] Kothari, C. R. (1990). *Research Methodology: Methods and Techniques,* 2nd Edition, New Delhi, New Age International (P) Ltd

[28] Mahamid, I. (2013). Frequency of Time Overrun Causes in Road Construction in Palestine: Contractors' View. *Organization, Technology and Management inConstruction Journal, 5*(1), 720 – 729

[29] Mansfield, N. R., Ugwu, O. O., & Doran, T. (1994). Causes of Delay and Cost Overruns in Nigerian Construction Projects. *International Journal of Project Management, 12*(4), 254 – 260.

[30] Mbatha, C. M. (1986). Building Contract Performance: A Case Study of Government Projects in Kenya (Masters Thesis). University of Nairobi

[31] Memon, A. H. (2014). Contractor Perspective on Time Overrun Factors in Malaysian Construction Projects. *International Journal of Science, Environment and Technology, 3*(3), 1184 – 1192

[32] Memon, A. H. Rahman, I. A., & Azis, A. A. A., (2012). Time and Cost Performance in Construction Projects in Southern and Central Regions of Peninsular Malaysia. *International Journal of Advances in Applied Sciences, 1*(1), 45 – 52.

[33] Mohammed, K. A. & Isah, A. D., (2012). Causes of Delay in Nigerian Construction Industry. *Interdisciplinary Journal of Contemporary Research in Business, 4*(2) 785 – 794

[34] Mustapha, Z. (2013). Accelerated Factors of Delays on Project Delivery in Ghana: A Case Study of Cape Coast Metropolis. *Journal of Applied Sciences & Environmental Sustainability, 1*(1), 54 – 61.

[35] Nadiri, M. I., & Mamuneas, T. P. (1998). Contribution of Highway Capital to Output and Productivity Growth in the U.S. Economy and Industries, United States Department of Transportation, Federal Highway Administration.

[36] Nkado, R. N. (1992). Construction Time Information System for the Building Industry. *Construction Management and Economics Journal, 10*(6), 489 – 509.

[37] Ogunlana, S. O., Promkuntong, K., & Jearkjirm, V., (1996). Construction Delays in a fast-growing Economy: Comparing Thailand with other Economies.*International Journal of Project Management, 14*(1), 37– 45

[38] Owolabi, J. D., Amusan, L. M., Oloke, C. O., Olusanya, O., Tunji, O. P., Owalabi, Omuh, I. (2014). Causes and effects of Delay on Project Construction Delivery Time. *International Journal of Education and Research, 2*(4), 197 – 208

[39] Patil, S. K., Gupta, A. K., Desai, D. B., & Sajane, A. S. (2013). Causes of Delay in Indian Transportation Infrastructure Projects. *International Journal of Research in Engineering and Technology, 02*(11), 71 80

[40] Ravisankar, K. L., Anandakumar, S., & Krishnamoorthy, V. (2014). Study on the Quantification of Delay Factors in Construction Industry. *International Journal of Emerging Technology and Advanced Engineering, 4*(1), 105–113

[41] Sambasivan, M., & Soon, Y. W. (2007). Causes and Effects of Delays in Malaysian Construction Industry.*International Journal of Project Management, 25*, 517 –526.

[42] Shanmugapriya, S., & Subramanian, K. (2013). Investigation of Significant Factors Influencing Time and Cost Overruns in Indian Construction Projects. *International Journal of Emerging Technology and Advanced Engineering, 3*(10), 734 – 740.

[43] Sweis, G. J. (2013). Factors Affecting Time Overruns in Public Construction Projects: The Case of Jordan. *International Journal of Business and Management, 8*(23), 120 – 129.

[44] Sweis, G., Sweis, R., Hammad, A. A., & Shboul, A. (2008). Delays in Construction Projects: The Case of Jordan. *International Journal of Project Management, 26*, 665 674.

[45] Talukhaba, A. A. (1988). *Time and Cost Performance of Construction Projects* (Masters Thesis). University of Nairobi

[46] Talukhaba, A. A. (1999). An Investigation into Factors Causing Construction Project Delays in   Kenya: A Case Study of High Rise Building Projects in Nairobi (PhD Thesis). University of Nairobi

[47] Uher, T. E. (1996). Cost Estimating Practices in Australian Construction. Engineering, Construction and Architectural Management Journal, 3(1), 83 – 95.

[48] Walker, D. H. T., (1994), An Investigation into Factors that Determine Building Construction Time Performance (PhD Thesis). Royal Melbourne Institute of Technolog Australia.

[49] Wong, K., & Vimonsatit, V. (2012). A Study of the Factors ffecting Construction Time in Western Australia. Scientific Research and Essays, 7(40), 3390 – 3398.

17

Effect of Design Ductility on the Progressive Collapse Potential of RC Frame Structures Designed to Eurocode 8

Mark Adom-Asamoah[1, *], Nobel Obeng Ankamah[2]

[1]Department of Civil Engineering, Kwame Nkrumah University of Science and Technology, Kumasi, Ghana
[2]Department of Civil Engineering, Sunyani Polytechnic, Sunyani, Ghana

Email address:

markadomasamoah@gmail.com (M. Adom-Asamoah), obeng86@hotmail.com (N. O. Ankamah)

Abstract: Progressive collapse is the cause of most structural failures around the world. The US General Service Administration (GSA) has presented guidelines for the assessment of the vulnerability of building structures to progressive collapse. It has been established in literature that the philosophy of ductility and redundancy used in seismic design is beneficial in resisting progressive collapse but not accounted for in these guidelines. The GSA methodology is particularly suited to seismic codes which allows for a constant member rotation but may be unsuitable to other codes that makes provision for ductility level. In this study, an investigation into the progressive collapse potential of RC framed structures designed to the seismic design code, EC 8, with varying design ground accelerations and ductility classes under different column loss scenarios was done. Based on the EC 8, a criteria for maximum plastic rotations and dynamic multiplies for progressive collapse analysis was proposed. These proposed criteria, together with the GSA criteria, were used to investigate the designed structures. The EC 8 criteria proved that buildings designed for higher ductilities yield at lower loads but undergo greater deformations and absorbs more energy to resist collapse. On the other hand, buildings designed for lower ductilities have higher yield loads but undergo lower deformations before collapse. Higher PGAs result in higher yield strengths but does not necessarily deformation capacity. This effect of ductility was not seen with the GSA criteria since a constant rotation capacity was recommended for all the buildings regardless of design ductility. It was also found that the removals of a corner column possess the greatest threat to progressive collapse on a building.

Keywords: Design Ductility, Progressive Collapse, RC Frames, Eurocode 8

1. Introduction

Progressive collapse is a phenomenon that involves the damage of a structural element resulting in the collapse of a disproportionately large part of the structure or the entire structure. Being a result of common conditions such as accidental impact, construction defects, structural overloads and failure of foundations, it is the cause of most structural collapse around the world (Wardhana and Hadipriono, 2003). The United States General Service Administration (GSA) and the Department of Defence (DoD) have presented practical guidelines (GSA, 2003 and DoD, 2005) for progressive collapse analysis of building. The basic technique of progressive collapse analysis of buildings known as alternative path method is adopted by both guidelines. This method involves the instantaneous removal of a load bearing member and its consequence on the ability of the modified structure to attain equilibrium. Currently, four methods of analysis namely linear static, linear dynamic, nonlinear static, and nonlinear dynamic are recommended for the alternate path method (Qazi et al., 2015, Patel and Parikh, 2013). Even though the nonlinear dynamic analysis procedure is believed to be the most accurate, the complexity of analysis and the extensive computing time involved do not lend itself suitable for design office use. Therefore, the nonlinear static (pushdown) analysis is still very relevant in investigating the collapse behaviour of a structure (Patel, 2014). As the lateral pushover analysis is widely used to evaluate structural properties such as yield stress, lateral stiffness, maximum lateral load resistance, and ultimate lateral displacement, it is expected that similar useful information may be obtained by the pushdown analysis for progressive collapse (Mohamed, 2015).

Numerical studies by Tsi and Lin (2008) and laboratory experiments by Tsitos (2008) have shown that seismically designed RC and steel frame buildings have high resilience to progressive collapse. It is also known that a seismically designed structure relies on its ductility and redundancy properties to limit the damaged to the initially affected zones. As per the EC 8 provisions, seismic forces are dissipated by varying levels of damping energy or ductile behaviour. A structure or structural element is considered ductile if it undergoes large deformations beyond the yield point without breaking. It is worthy of note that in seismic engineering, ductility is expressed in terms of maximum available ductility possessed by a structure and the ductility demand of the seismic action. EC 8 has rules for construction to achieve the ductile behaviour modelled for predefined critical areas (Elghazouli, 2009). These in turn ensure that a preferred plastic behaviour is achieved instead of a brittle mode of failure (concrete shear, concrete crushing and reinforcement pull-out).

There are three levels of energy absorption, known as ductility classes, according to EC 8. These ductility classes, coupled with levels of seismic design load ensure different seismic resistance levels by trading between available designs seismic forces and ductility demand (Fig. 1).

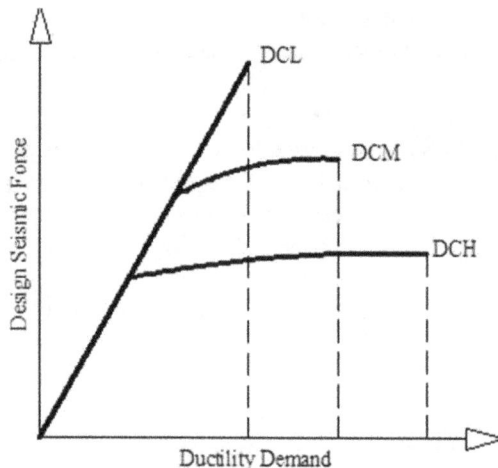

Figure 1. Variation of design seismic force with ductility demand.

These classes, defined by a behaviour factor q, are as Ductility Class Low (DCL), Ductility Class Medium (DCM) and Ductility Class High (DCH). DCL does not require delayed ductility, but resistance to seismic loading is achieved through the strength capacity of the structures. The design case is inherently elastic ($q \leq 1.5$). DCM allows high level of ductility and there are responsive design demands ($1.5 < q \leq 4$). DCH allows even higher levels of ductility with responsive strict and complicated design demands ($4.0 < q \leq 6.0$).

The interpretation of the code requirements is that these three ductility classes are equivalent regarding the performance of the structure under the design seismic action (Kappos and Penelis, 2010). It is expected that for a structure under the same design seismic force, whilst the design

seismic lateral force increases from DCH to DCL, the damping energy (ductility) reduce from DCH to DCL (see Fig.1). This assures that the three classes are equivalent in terms of energy absorbed, which is measured as the area under the respective force-displacement curves. In order to achieve this energy equivalence, EC8 allows different rotation capacities for different ductility classes before failure.

The GSA guidelines recommend that a nonlinear progressive collapse analyses be performed with a maximum rotation of 0.035 radians for RC frames. Therefore, in the progressive collapse analysis of structures designed to EC 8, there is a disagreement between the EC8 and the GSA provisions. Different maximum plastic hinge rotations are required for the accurate prediction of the progressive collapse potential of structures based on the provisions of EC 8. In this study, the required maximum rotation capacities corresponding to the various EC8 ductility classes and their effect on the progressive collapse potential of EC8 designed buildings were investigated.

Again, according to the GSA the nonlinear static procedure requires a dynamic multiplier (DM) to account for dynamic (inertial) effects. The GSA recommends a constant DM of 2.0, which is applied directly to the progressive collapse load combination. Marchis (2013) suggested that the values for DMs are affected by the ductility capacity of the structures. An investigation into the variation of DMs based the ductility class of the structure was also done in this study. The calculated DMs for the various structures were used in the analysis.

2. Description of the Structure

Six 11-storey reinforced concrete spatial frame models were designed to EC 8 with ground accelerations of 0.10g, 0.15g and 0.25g and varying ductility classes and hence behaviour factors. The values of the corresponding behaviour factors are shown in Table 1.

Table 1. Details of structural configurations.

Structure Reference	Design PGA (g)	Ductility Class	Behaviour (q) Factor
L0.10	0.10	Low	1.50
H0.15	0.15	High	5.85
M0.15	0.15	Medium	3.90
L0.15	0.15	Low	1.50
H0.25	0.25	High	5.85
M0.25	0.25	Medium	3.90

Each building had four bays in the x-direction and three bays in the y-direction as shown in Fig. 2. The structures were regular in plan and elevation. They had uniform column spacing of 5.0m giving an overall dimension of 20.0m in the x-direction and 5.0m in the y-direction giving an overall dimension of 15m. Storey height was 3.0m for all floors giving a total building height of 33.0m for all. All beam-column connections were modelled as fully rigid whilst the foundations were modelled as fixed. Table 2 shows the detailed description of the member section dimensions used in the models. All dimensions are in millimetres.

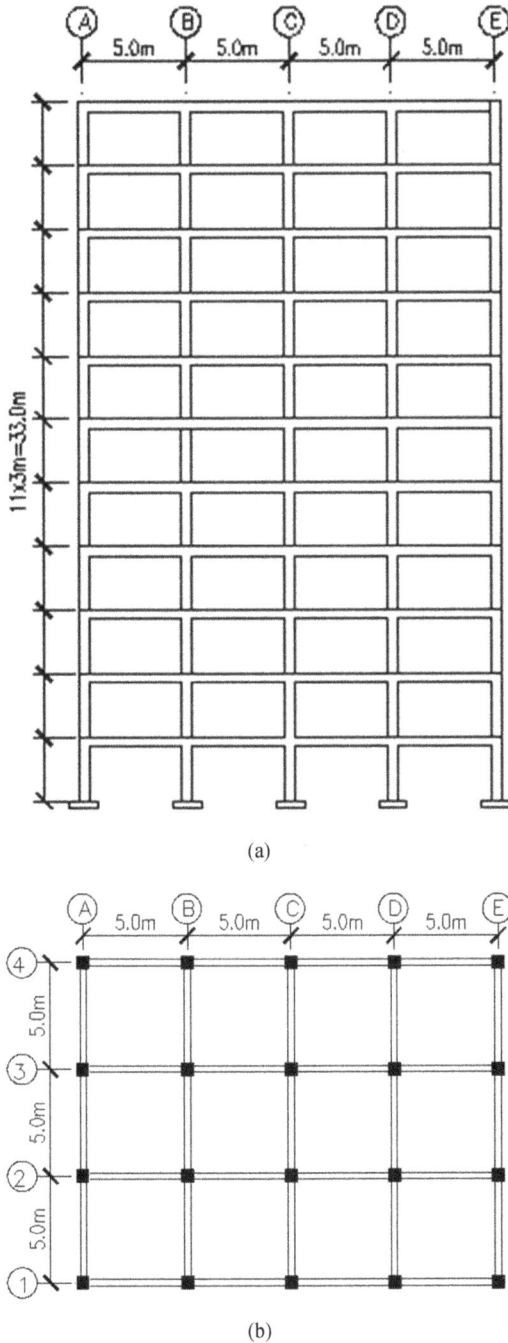

(a)

(b)

Figure 2. *Plan and elevations of building.*

Table 2. *Member cross-section dimensions*

Dimension	Columns (mm)			Beams (mm)
	Internal	External	Corner	
Height	800	600	550	600
Width	800	600	550	300

3. Analysis Parameters

3.1. Rotation Capacity

For progressive collapse analyses according to the GSA,

the moment-hinge properties shown in Fig. 3 were used. The collapse point is represented as Point C on the curve represents and was assigned a rotation of 0.035rad, as recommended by the GSA for RC frames. The slope from point B to C was taken as 10% of the elastic slope to accounts for strain hardening; the seismic code ASCE 41(2006) indicates that the slope should be taken as a small percentage between 0% and 10%. Point D corresponds to the residual strength 0.2 of the ultimate strength. Since the GSA (2003) does not specify a value for point E as the failure limit, a value of 0.07 radians was considered as an average value (0.04 rad - 0.10 rad) given by the DoD (2009).

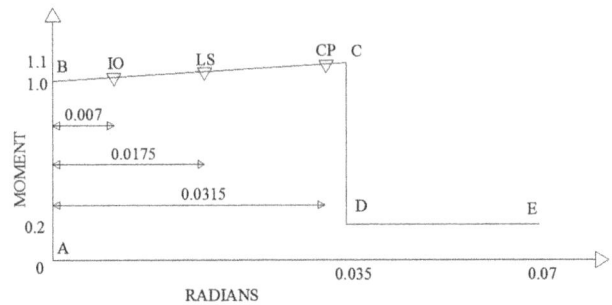

Figure 3. *GSA Moment-rotation behaviour of hinges*

FEMA 356 (2000) performance levels were used to monitor the performance of the structures at different stages as load is applied. These include Immediate Occupancy (IO), Life Safety (LS), and Collapse Prevention (CP). For this study, these three points as 0.2Δ, 0.5Δ and 0.9Δ respectively based on provisions of FEMA 356 (2000) were defined. The Symbol "Δ" is the length of plastic hinge plateau.

Rotation capacities reflecting design ductility, according to EC 8, is illustrated in Fig. 4.

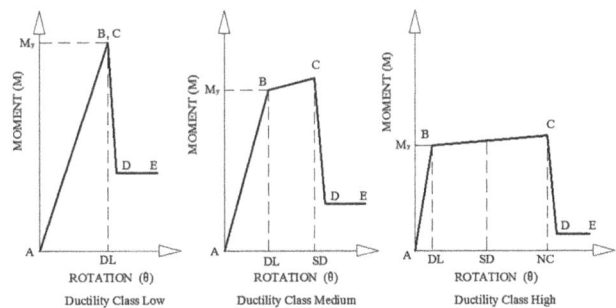

Figure 4. *EC8 Moment-rotation behaviour of hinges*

Thus, for high ductility class structures, members are detailed for the Near Collapse (NC) performance level with a rotation capacity θ_u, expressed as

Medium ductility structures are detailed for the Significant Damage (SD) performance level, which allows for 75% of the ultimate rotation capacity θ_u. Ductility is not accounted for in Low ductility structures. Therefore, a member is considered failed when it deforms beyond the Damage Limitation (DL) performance level i.e. chord rotation equal to the yield rotation θ_y, expressed as

This EC 8 criterion was also considered for progressive

collapse analysis. The rotation capacities were calculated by the program based on the seismic design results. The calculated average values for beams directly above the removed columns were as 0.007, 0.026, 0.019, 0.009, 0.026 and 0.020 for L0.10, H0.15, M0.15, L0.15, H0.25 and M0.25 respectively.

3.2. Columns Removal Scenarios

The GSA guidelines specify four column loss scenarios in the assessment of the progressive collapse of a building as shown in Fig. 5. These are:

- Case 1: An exterior column near the middle of the short side of the building
- Case 2: An exterior column near the middle of the long side of the building
- Case 3: A column located at the corner of the building
- Case 4: A column interior to the perimeter

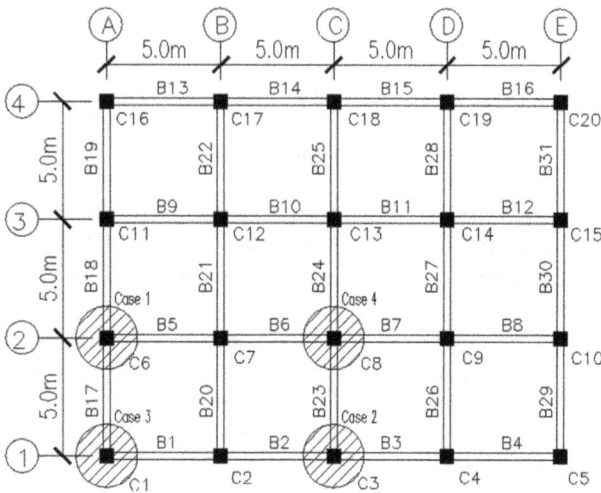

Figure 5. *Typical plan of reinforced concrete structure.*

3.3. Dynamic Multiplier (DM)

To determine the dynamic multipliers (DMs) for nonlinear static analysis, the nonlinear dynamic alternate path analysis was performed for all column loss cases under a load of $(G_k+0.25Q_k)$. The maximum displacements at the points of column removal were noted. The nonlinear static

"pushdown" analysis was also performed by setting the displacements achieved in the nonlinear dynamic analysis as the target displacements. The load factor at which the nonlinear static analysis produced the same displacement as the nonlinear dynamic analysis is noted as the dynamic multipliers (DM). The DMs determined for all the models in all four column-loss cases are presented in Table 3.

It is observed that the DMs are typically less than the 2.0 provided by the GSA guidelines. Thus, the use of a load factor of 2.0 would overestimate the collapse vulnerability of the structures. In addition, the DMs vary with the different design parameters. They decrease with increasing design ductility class and decreasing peak ground acceleration (PGA

Table 3. *DMs for all the models in all four column loss cases.*

Model	Case 1	Case 2	Case 3	Case 4
H0.15	1.47	1.53	1.01	1.50
L0.10	1.55	1.59	1.36	1.6
H0.15	1.47	1.53	1.01	1.5
M0.15	1.51	1.56	1.06	1.52
L0.15	1.55	1.60	1.37	1.63
H0.25	1.54	1.58	1.13	1.58
M0.25	1.55	1.59	1.25	1.6

4. Analysis Methods

4.1. Nonlinear Dynamic Analysis (ND)

In performing the ND analysis, a uniformly distributed gravity load of $(G_k+0.25Q_k)$ was applied to the structure. The damping ratio used in the dynamic analysis was 5%. Before the column removal, a nonlinear static analysis of the model was undertaken subjected to applied gravity load of $(G_k+0.25Q_k)$. With the structure in static equilibrium, the target column was then removed instantaneously. In order to simulate the instantaneous removal of a column, the column was replaced with equivalent reaction obtained from a nonlinear static analysis of the building under the load of $(G_k+0.25Q_k)$ applied to the whole structure (Fig. 6). The time for removal was set to 1.0ms. The response of the structure was observed until the structure became relatively stable.

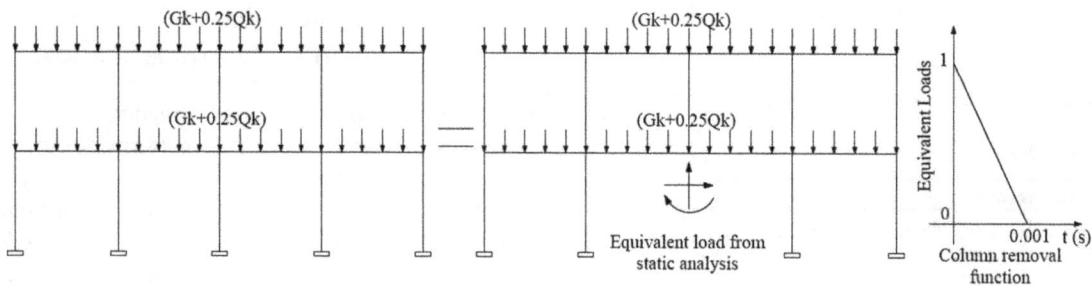

Figure 6. *Definition of Nonlinear Dynamic load case.*

Plastic hinge were assigned to both ends of each member. Studies done by Choi and Kim (2011) and Yi et. al (2008) have shown that the progressive collapse of RC framed

structures is controlled by the flexural failure of beams. Therefore, only flexural failure hinges were used here. The maximum plastic hinge rotations and displacements at the points of column removal during the analyses were noted and compared against the acceptable criteria of the GSA and EC 8. Vertical displacements at the points of column removal were monitored until the structure achieved relative stability.

4.2. Nonlinear Static Analysis (NS)

The NS procedure involves a stepwise increase of vertical loads, until a maximum amplified load of $DM(G_k+0.25Q_k)$ is reached or the structure collapses (Fig. 7). This method has the advantage of accounting for nonlinear effect without sophisticated hysteresis material model and time-consuming time-history analysis. Though it is unable to consider the dynamic effect due to the sudden loss of columns, it is useful in determining the elastic and failure limits of the structure (Taewan et al, 2009).

In this study, the displacement controlled pushdown analysis was carried out by increasing the applied load to increase the vertical displacement at the location of the removed column until collapse. The stepwise load increase was only applied on the bay with a lost column since load amplification due to inertia would directly affect this bay. A constant unamplified load of $(G_k+0.25Q_k)$ was applied in the other bays to ensure a more accurate prediction of the dynamic effect (Taewan et al, 2009). For easier analysis and comparison of results for different models and column loss locations, the base shear was replaced by a load factor. The load factor at any step was determined as the ratio of the base reaction at that step to the base reaction at a load of $1.0(G_k+0.25Q_k)$.

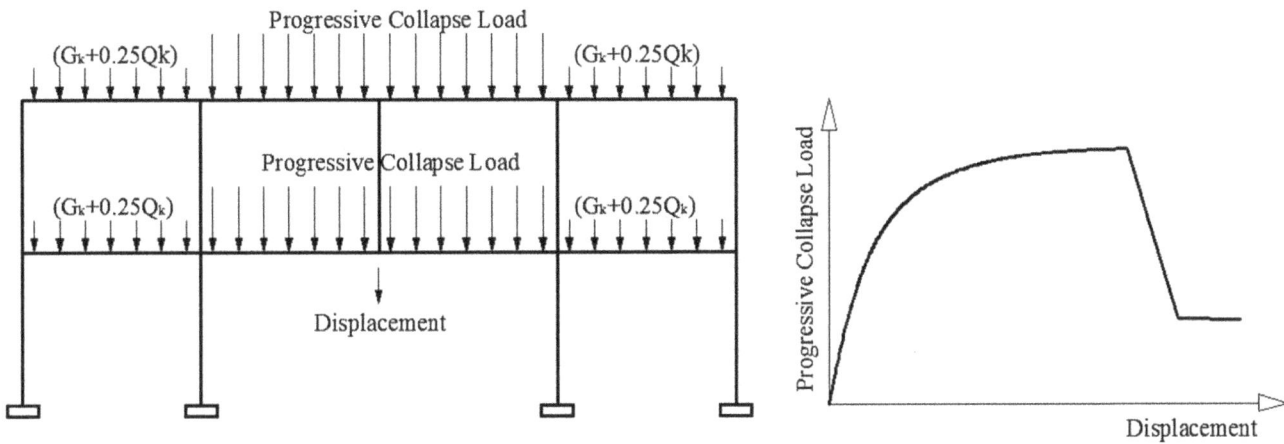

Figure 7. *Definition of Nonlinear Static load case.*

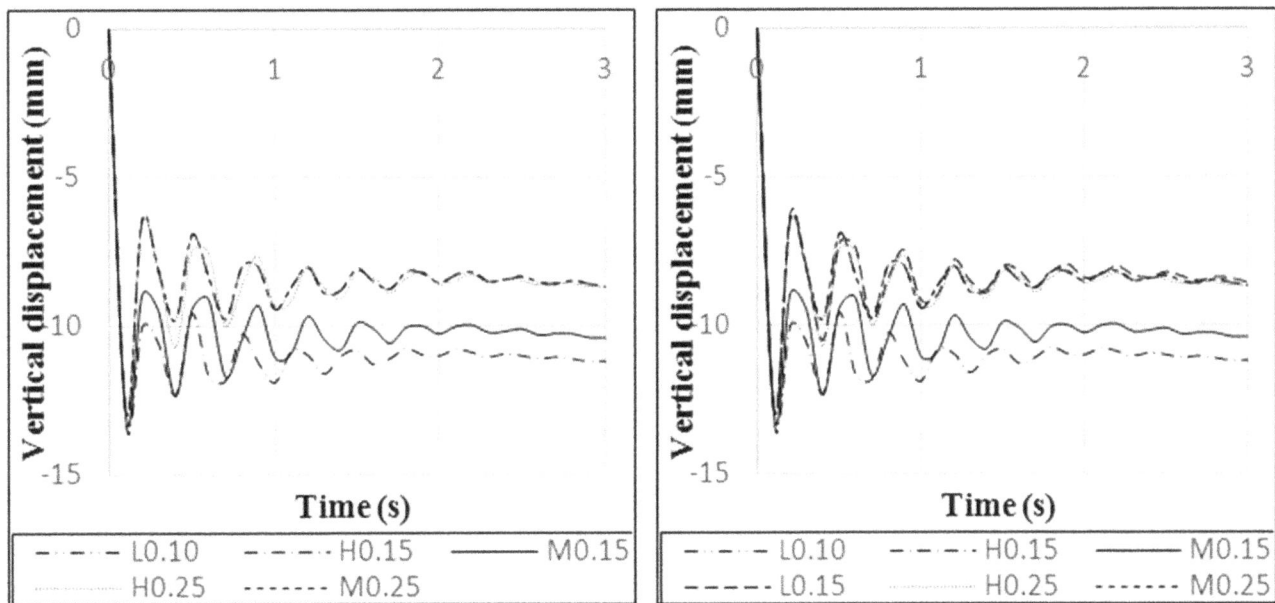

(a) GSA   (b) EC8

Figure 8. *Vertical displacement history for Case 1*

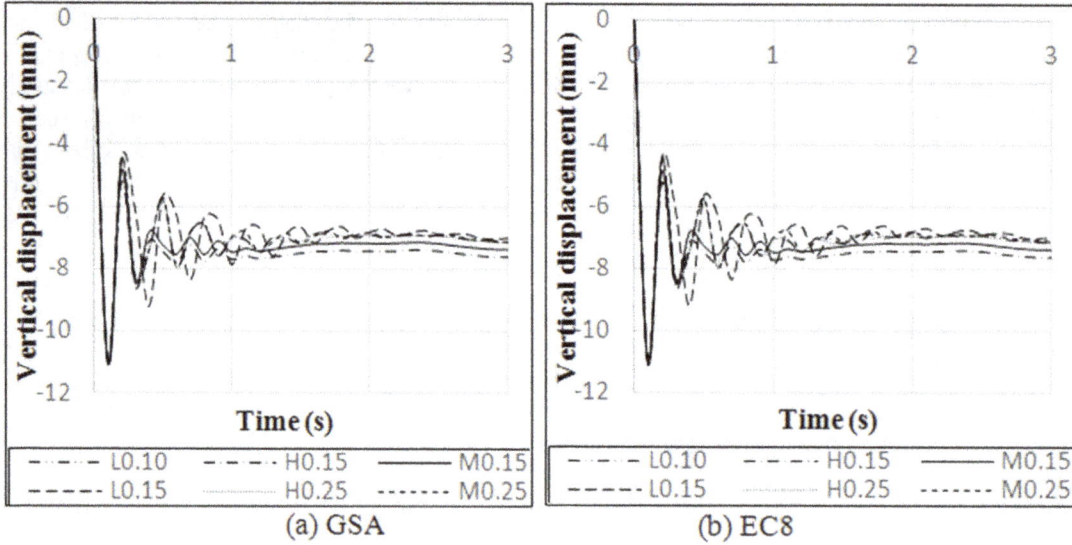

Figure 9. Vertical displacement history for Case 2.

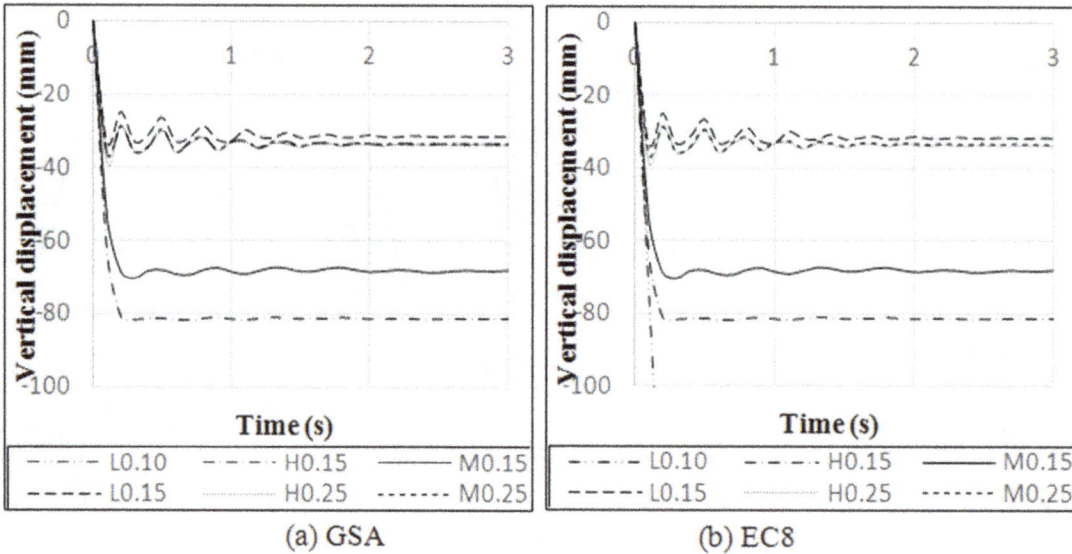

Figure 10. Vertical displacement history for Case 3.

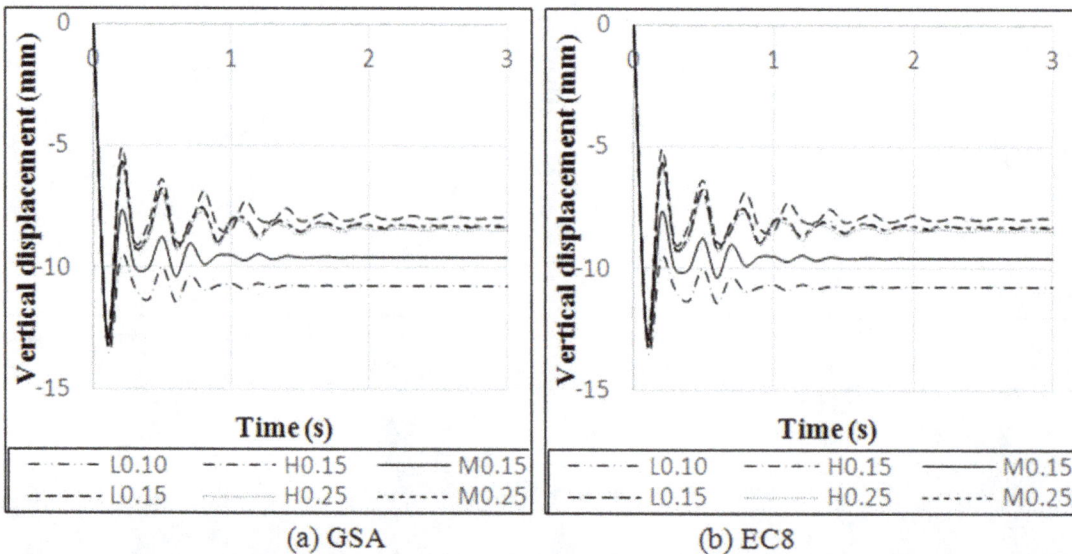

Figure 11. Vertical displacement history for Case 4.

5. Discussion of Results

5.1. Nonlinear Dynamic Analysis

Fig. 8-11 shows a comparison of the vertical displacements at the points of column removal with time for the GSA and EC8 methods. The maximum and final vertical displacements at the points of column removal are presented in Table 4.

The observed patterns of displacements were the same for both the GSA and the EC 8 methods except for the removal of a corner column (Case 3) from model L0.10. In all damage cases, vertical displacements peaked at approximately 0.11 seconds. Vibrations phased out gradually in different times. It was also observed that models designed with higher ductilities stabilized much earlier than models with lower ductilities. This is due to the early formation of plastic hinges in the beams, detuning the structures by elongating the period and consequently damping. The gradual stabilization of the vibration suggests low damping.

From the GSA analysis, plastic hinges developed in all models. However, all hinges rotations were all less than 0.035 radians. Collapse was therefore not experienced in any

buildings. From the EC8 method, displacement at the point of column removal was infinite in the instance of the loss of a corner column (Case 3) from model L0.10. This indicates that the frame had failed under the dynamic load. The maximum displacement recorded for this instance by the GSA method was 36.31 mm whereas the maximum allowed was 175 mm (corresponding to a rotation capacity of 0.035 radians), hence the structure did not collapse. In the EC8 method, maximum allowed displacement was 35 mm (corresponding to a rotation capacity of 0.007 radians); hence, the structure collapsed before the displacement of 36.31 mm could be achieved. Collapse was not observed in any other building.

The generally higher displacements observed in Case 3 could be attributed to the fact that this case had the highest redistribution of axial loads. In this case, only two columns adjoin the removed column, therefore having percentage-redistributed loads of 49.9% and 50.0% as shown in Table 4. It is worthy of note that Cases 1, 2 and 4 all have a minimum of three columns immediately surrounding the removed column.

Table 4. Displacements from Nonlinear Dynamic Analysis (mm) (= Failed).*

Model	Case 1		Case 2		Case 3		Case 4	
	EC8	GSA	EC8	GSA	EC8	GSA	EC8	GSA
H0.15	13.52	13.52	11.24	11.24	68.72	68.72	13.44	13.44
L0.10	13.12	13.12	11.13	11.13	36.31*	36.31	13.17	13.17
H0.15	13.52	13.52	11.24	11.24	81.5	81.5	13.44	13.44
M0.15	13.21	13.21	11.22	11.22	70.13	70.13	13.42	13.42
L0.15	-	13.07	11.24	11.24	33.2	33.2	12.93	12.93
H0.25	12.99	12.99	11.20	11.20	38.63	38.63	12.96	12.96
M0.25	12.98	12.98	11.19	11.19	36.18	36.18	12.95	12.95

It must be noted that the observation from the proposed EC 8 is contrary to that made from the GSA analysis. This is because, GSA permits plastic rotation up to 0.035 radians regardless of the ductility class and design parameters of the structure. This demonstrates that the GSA method overestimates the progressive collapse resistance of structures designed for low ductility. In this damage case, the other models were found resilient to progressive collapse since the plastic hinge rotations fell below the corresponding performance threshold.

5.2. Nonlinear Static Analysis

The vertical displacement and corresponding vertical loads for all the buildings are plotted in Figures 12 - 15 for the various column loss scenarios. The collapse resistance of a building was as its ability to carry the ultimate load of $DM(G_k+0.25Q_k)$ before collapse. The maximum load factors for the models under the various column loss scenarios for both GSA and EC8 methods are presented in Table 6.

When a corner column was removed (Case 3), only model L0.15 was able to resist collapse as per the GSA method, with a load factors of 2.07. All the other models failed at load

factors of 1.30, 1.16, 1.34, 1.32 and 1.42 for models L0.10, H0.15, M0.15, H0.25 and M0.25 respectively. Progressive collapse is therefore expected in these models for Case 3. For the EC8 method, only model L0.10 failed before the ultimate load was reached. Progressive collapse is therefore expected here. All other buildings were found resilient.

When a column was removed from a location interior to the perimeter (Case 4), model H0.15 failed under the loading for the GSA method, however the EC 8 method did not find any building venerable to progressive collapse in this case. This is because, even though the EC 8 criteria had lower plastic hinge rotation capacities, it also had lower DMs.

From the GSA load-displacement curves (Figures 12-15) and Table 5, it is observed that the resistance to progressive collapse of the structures increased with PGAs and lower ductility classes. This is because structures with higher PGAs and lower ductility classes had the highest amount of reinforcement from the seismic design and the GSA analysis is only based on strength and not ductility. Thus, the increase and decrease in ductility capacity with higher and lower design ductility classes was not accounted for.

From the EC 8 curves, it is observed that the structures resisted progressive collapse by two main mechanisms.

These are strength and ductility. Low ductility class models had an advantage of high strength and therefore had higher yield loads and ultimately higher collapse loads. High ductility models yielded at lower loads, but were able to undergo relatively higher deformations, enabling them to absorb more energy and prevent collapse. This indicates that assessing structures design for different ductility with the same plastic rotations will result in overestimating the resistance of the low ductility buildings or underestimating the resistance of the high ductility buildings. In addition, the conclusion that designing at low ductility increases the structures resistance to collapse from other studies (Ioani et al (2007), Ioani and Cucu (2010)).

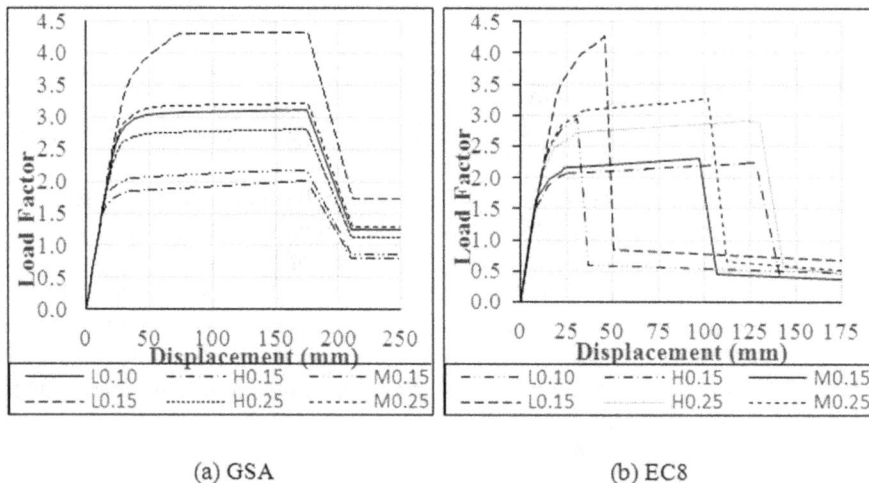

(a) GSA   (b) EC8

Figure 12. Load-displacement relations for Case 1.

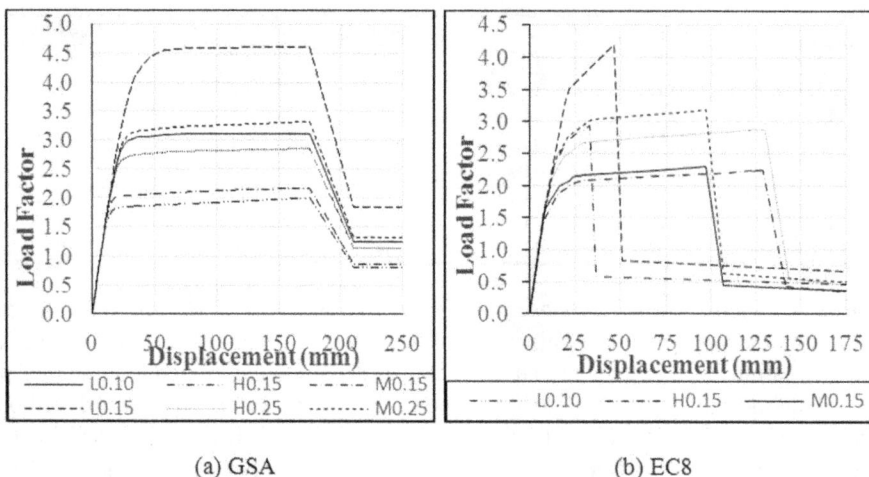

(a) GSA   (b) EC8

Figure 13. Load-displacement relations for Case 2.

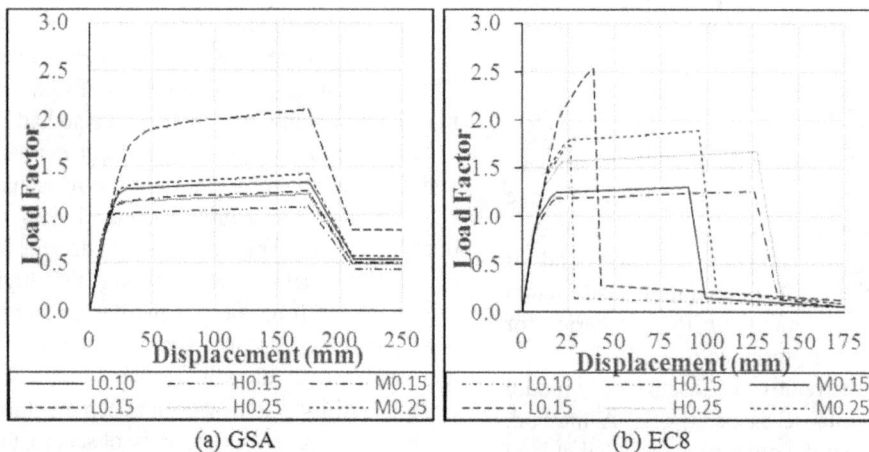

(a) GSA   (b) EC8

Figure 14. Load-displacement relations for Case 3.

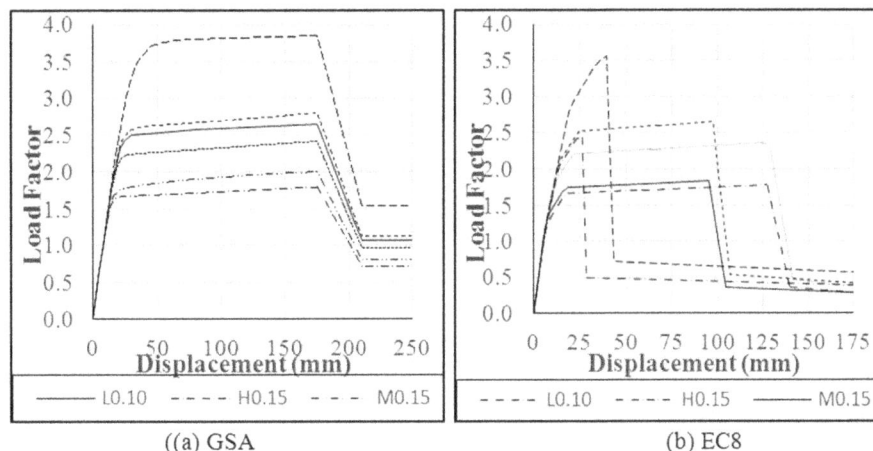

((a) GSA   (b) EC8

Figure 15. *Load-displacement relations for Case 4*

Table 5. *Maximum load factors (* = Failed)*

Model	Case 1		Case 2		Case 3		Case 4	
	EC 8	GSA	EC 8	GSA	EC 8	GSA	EC 8	GSA
L0.10	2.96	3.08	2.93	3.12	1.35	1.30*	2.44	2.57
H0.15	2.24	2.13	2.23	2.17	1.10*	1.16*	1.77	2.15
M0.15	2.32	2.22	2.30	2.22	1.15	1.34*	1.84	2.04
L0.15	4.28	4.35	4.19	4.59	2.21	2.07	3.56	3.81
H0.25	2.91	3.12	2.87	3.22	1.46	1.32*	2.35	2.61
M0.25	3.29	3.41	3.18	3.42	1.64	1.42*	2.65	2.80

6. Conclusion

The various analyses carried out in this study came up with some major conclusions as elaborated below:

- The maximum plastic rotations required for nonlinear analysis vary according to the seismic design parameters. EC 8 allows rotations up to the Damage Limitation (DL), Significant Damage (SD) and Near Collapse (NC) performance levels for buildings designed for Low, Medium and High ductility classes respectively. The maximum allowed rotations also increased with design ductility and PGA. Therefore, the constant rotation capacity of 0.035 radians provided by GSA (2003) for all structures is inaccurate. This demonstrates the effect of seismic design on progressive collapse and disproves the assertion made by many researchers that buildings designed to lower ductility classes better resist progressive collapse.

- For the EC 8 nonlinear static method, different DMs were determined for all instances of collapse. It was shown that the required DMs for the nonlinear static method were lower than that for linear static method and also less than 2.0. They ranged from 1.22 to 1.63. This indicates that the use of DMs of 2.0 overestimated the demand on the structures in progressive collapse analysis.

- The location of lost column also influenced the progressive collapse susceptibility of the building. The vulnerability of all the buildings was similar in Case 1, 2 and 4. All the buildings were found most vulnerable

when a corner column was removed (Case 3). It is therefore recommended that based on this limited study and other research works, the progressive collapse analysis on regular medium storey RC frame buildings could be limited to the case 3 column removal scenario.

References

[1]   ASCE-41, (2006): "Seismic Rehabilitation of Existing Buildings", American Society of Civil Engineers, Reston, VA, USA.

[2]   Choi H, Kim J, (2011): "Progressive collapse-resisting capacity of RC beam-column sub-assemblage", Magazine of Concrete Research, Vol. 63, No.4.

[3]   DOD (2005): "Design of building to resist progressive collapse". Unified Facilities Criteria (UFC) 4-023-03, Department of Defence, USA.

[4]   Elghazouli A, (2009): "Seismic Design of buildings to Eurocode 8", Spon Press, London, United Kingdom.

[5]   EN 1998 (2004): Eurocode 8: "Design of structures for earthquake resistance", European Committee for Standardization.

[6]   FEMA 356 (2000): "Prestandard and Commentary for the Seismic Rehabilitation of Buildings", Federal Emergency Management Agency, Washington, USA.

[7]   GSA (2003): "GSA Progressive Collapse Analysis and Design Guidelines for New Federal Office Buildings and Major Modernizations Projects", General Services Administration, USA.

[8]   Ioani A. M, Cucu H. L, (2010): "Resistance to progressive collapse of RC structures: principles, methods and designed models", Computational Civil Engineering 2010, Iasi, Romania.

[9]   Ioani A. M, Cucu H. L, Mircea C (2008): "Seismic Design vs. Progressive Collapse: A Reinforced Concrete Framed Structure Case Study", Innovation in Structural Engineering and Construction.

[10] Kappos A, Penelis G. G, (2010): "Earthquake resistant concrete structures", CRC Press, pp607.

[11] Marchis A. G, Moldovan T. S, Ioani A. M, (2013): "The influence of the seismic design on the progressive collapse resistance of mid-rise RC framed structures", Acta Technica Napocensis: Civil Engineering & Architecture Vol. 56, No. 2.

[12] Mohamed O. A. (2015): "Calculation of load increase factors for assessment of progressive collapse potential in framed steel structures", Case studies in Structural Engineering, Vol 3.

[13] Patel B. R (2014): "Progressive collapse analysis of RC Buildings using Nonlinear Static and Nonlinear Dynamic Method", International journal of Emerging Technology and Advanced Engineering, Vol 4(9).

[14] Patel PV, Parikh RD (2013), "Various procedures for progressive collapseanalysis of steel framed buildings", The IUP Journal of Structural Engineering, 6, p17-31.

[15] Qazi AU, Majid A, Hameed A, Ilyas M (2015): "Nonlinear progressive collapse analysis of RC frame structure", Pak. J. Engg. And Applied Science, 16, p121-132.

[16] Taewan K. Jinkoo K, Junhee P. (2009): "Investigation of Progressive Collapse-Resisting capability of Steel Moment Frames Using Push-Down Analysis", Journal of Performance of Constructed Facilities, Vol. 23, No. 5.

[17] Tsitos A, Mosqueda G, Filiatrault A, Reinhorn A. M., (2008): "Experimental investigation of progressive collapse of steel frames under multi-hazard extreme loading", The 14th World Conference on Earthquake Engineering, October 12-17, 2008, Beijing, China.

[18] Wardhana K, Hadipriono F. C, (2003): "Study of Recent Building Failures in the United States", Journal of Performance of Constructed Facilities, Volume 17, No. 3.

[19] Yi W. J, He Q. F, Xiao Y, Kunnath S. K, (2008): "Experimental study on Progressive Collapse-resistant behaviour of reinforced concrete frame structures", ACI Structural Journal, Vol.105, No.4, pp.433-438.

18

A comparison between two field methods of evaluation of liquefaction potential in the Bandar Abbas City

Mohammad Naderi Pour[*]**, Adel Asakereh**

Department of Civil Engineering, University of Hormozgan, Bandar Abbas, Iran

Email address:
m.naderi2020@yahoo.com (M. N. Pour), asakereh@yahoo.com (A. Asakereh)

Abstract: The geotechnical characteristics of the soil layers are one of the main factors influencing liquefaction potential of the ground. In common usage, *liquefaction* refers to the loss of strength in saturated, cohesionless soils due to the build-up of pore water pressures during dynamic loading. The following five screening criteria, are recommended for completing a liquefaction evaluation: Geologic age and origin, Fines content and plasticity index, Saturation, Depth below ground surface and Soil penetration resistance. The liquefaction resistance of soils can be evaluated using laboratory tests such as cyclic simple shear, cyclic triaxial, cyclic torsional shear, and field methods such as Standard Penetration Test (SPT), Cone Penetration Test (CPT), and Shear Wave Velocity (Vs). The present study is aimed at comparing the results of two field methods used to evaluate liquefaction resistance of soil, i.e. SPT and CPT. It is concluded that the liquefaction evaluation methods based on the SPT data show more conservative results compared with those based on the CPT data.

Keywords: Liquefaction Potential, Standard Penetration Test (SPT), Pore Water Pressure, Dynamic Loading

1. Introduction

Liquefaction is the phenomena when there is loss of strength in saturated and cohesion-less soils because of increased pore water pressures and hence reduced effective stresses due to dynamic loading. It is a phenomenon in which the strength and stiffness of a soil is reduced by earthquake shaking or other rapid loading.

A more precise definition as given by Sladen et al (1985) states that "Liquefaction is a phenomena wherein a mass of soil loses a large percentage of its shear resistance, when

subjected to monotonic, cyclic, or shocking loading, and flows in a manner resembling a liquid until the shear stresses acting on the mass are as low as the reduced shear resistance".

After initial liquefaction if large deformations are prevented because of increased undrained shear strength then it is termed," limited liquefaction" (Finn 1990).

When dense saturated sands are subject to static loading they have the tendency to progressively soften in undrained cyclic shear achieving limiting strains which is known as cyclic mobility (Castro 1975; Castro and Poulos 1979). Cyclic mobility should not be confused with liquefaction. Both can be distinguished from the very fact that a liquefied soil

displays no appreciable increase in shear resistance regardless of the magnitude of deformation (Seed 1979).

Soils undergoing cyclic mobility first soften subjected to cyclic loading but later when monotonically loaded without drainage stiffen because tendency to increase in volume reduce the pore pressures. During cyclic mobility, the driving static shear stress is less than the residual shear resistance and deformations get accumulated only during cyclic loading. However, in layman"s language, a soil failure resulting from cyclic mobility is referred to as liquefaction.

Using the *SPT* data for evaluating liquefaction potential of the soil layers is nearly as long as the phenomenon was first recognized during 1964 Niigata earthquake.

Seed and Idriss (1971) developed the first experimental method based on the *SPT* data to evaluate the liquefaction potential of the ground during strong earthquakes.

Nevertheless, there are some deficiencies and shortcomings with the *SPT*, the most important of which can be summarized as follows:

- The repeatability of the test cannot be guaranteed.
- The soil profilecannot be detected continuously
- The pore pressure cannot be measured during the test.
- The sensitivity of the device to changing soil profile is

sometimes poor.

- The influence of pore pressure fluctuations due to blow effects of the system on the test results cannot be considered.
- The theoretical interpretations about the test results cannot be implemented.

Although the effect of these factors on the accuracy and reliability of the test results are not the same, some of them may considerably influence the measured data. In contrast to *SPT, CPT* is also another in situ testing device and technique that can be used for the same purpose, without having the above mentioned problems. However the complexity of the system and the more energy and time consuming of operation relative to the *SPT,* have caused it less popular and common in practice.

There are some initial requirements for each site to be under consideration in this study.

The results of the *SPT* and *CPT* studies must have been available and the points at which these tests are carried out cannot be far from each other.

Considering these fact, some different sites in the southern parts of Iran have been selected. These sites were located on the Hormozgan province near the coastal region of the Persian Gulf. The ground in these areas is usually consisted of deposits belonging to Testiary and Quaternary geological periods. The soil layers in these sites are between sandy silts to silty sands and can be classified as fine granular soils *(PI ≤ 5%).* The water table in these sites are between 1.5-3.0 m depths and the densification of the top layers can be categorized between medium to loose.

The seismicity of the regions is relatively high compared with other areas of the country.

2. Effective Factors in Liquefaction

Liquefaction is most commonly observed in shallow, loose, saturated cohesionless soils subject to strong ground motions in earthquakes. Unsaturated soils are not subject to Liquefaction because volume compression does not generate excess pore water pressure.

Since liquefaction phenomena arises because of the tendency of soil grains to rearrange when sheared, any factor that prevents the movement of soil grains will increase the liquefaction resistance of a soil deposit.

Stress history is also crucial in determining the liquefaction resistance of a soil. Over consolidated soils (i.e. the soils that have been subjected to greater static pressures in the past) are more resistant to particle rearrangement and hence liquefaction as the soil grains tends to be in a more stable arrangement.

Liquefaction resistance of a soil deposit increases with depth as overburden pressure increases. That is why soil deposits deeper than about 15m are rarely found to have liquefied (Krinitzky et al.1993).

Characteristics of the soil grains like distribution of shapes, sizes, shape, composition, etc. influence the susceptibility of a soil to liquefy (Seed 1979). While sands or silts are most

Commonly observed to liquefy, gravelly soils have also been known to have liquefied.

Rounded soil particles of uniform size are mostly susceptible to liquefaction (Poulus et al.1985). Well graded soils, due to their stable inter-locking configuration, are less prone to liquefaction.

Clays with appreciable plasticity are resistant to relative movement of particles during shear cyclic shear loading and hence are usually not prone to pore water pressure generation and liquefaction.

Ishihara (1993) gave the theory that non-plastic soil fines with dry surface texture do not create adhesion and hence do not provide appreciable resistance to particle rearrangement and liquefaction. Koester (1994) stated that sandy soils with appreciable fines content may be inherently collapsible, perhaps because of greater compressibility of the fines between the sand grains.

3. Recommended Screening Criteria for Liquefaction Potential

The following five screening criteria are recommended for completing a liquefaction evaluation:

- *Geologic age and origin.* If a soil layer is a fluvial, lacustrine or aeolian deposit of Holocene age, a greater potential for liquefaction exists than for till, residual deposits, or older deposits.
- *Fines content and plasticity index.* Liquefaction potential in a soil layer increases with decreasing fines content and plasticity of the soil. Cohesionless soils having less than 15 percent (by weight) of particles smaller than 0.005 mm, a liquid limit less than 35 percent, and an in situ water content greater than 0.9 times the liquid limit may be susceptible to liquefaction (Seed and Idriss, 1982).
- *Saturation.* Although low water content soils have been reported to liquefy, at least 80 to 85 percent saturation is generally deemed to be a necessary condition for soil liquefaction. The highest anticipated temporal phreatic surface elevations should be considered when evaluating saturation.
- *Depth below ground surface.* If a soil layer is within 50 feet of the ground surface, it is more likely to liquefy than deeper layers.
- *Soil Penetration Resistance.* Seed et al, 1985, state that soil layers with a normalized SPT blowcount [(N1)60] less than 22 have been known to liquefy. Marcuson et al, 1990, suggest an SPT value of [(N1)60] less than 30 as the threshold to use for suspecting liquefaction potential. Liquefaction has also been shown to occur if the normalized CPT cone resistance (qc) is less than 157 tsf (15 MPa) (Shibata and Taparaska, 1988).

If three or more of the above criteria indicate that liquefaction is not likely, the potential for liquefaction can be dismissed. Otherwise, a more rigorous analysis of the liquefaction potential at a facility is required.

4. Liquefaction Analysis

If potential exists for liquefaction at a facility, additional subsurface investigation may be necessary. Once all testing is complete, a factor of safety against liquefaction is then calculated for each critical layer that may liquefy. A liquefaction analysis should, at a minimum, address the following:

- Developing a detailed understanding of site conditions, the soil stratigraphy, material properties and their variability, and the areal extent of potential critical layers. Developing simplified cross sections amenable to analysis. SPT and CPT procedures are widely used in practice to characterize the soil (field data are easier to obtain on loose cohesionless soils than trying to obtain and test undisturbed samples). The data needs to be corrected as necessary, for example, using the normalized SPT blowcount [(N1)60] or the normalized CPT. The total vertical stress (so) and effective vertical stress (so') in each stratum also need to be evaluated. This should take into account the changes in overburden stress across the lateral extent of each critical layer, and the temporal high phreatic and piezometric surfaces,
- Calculation of the force required to liquefy the critical zones, based on the characteristics of the critical zone(s) (e.g., fines content, normalized standardized blowcount, overburden stresses, level of saturation),
- Calculation of the design earthquake's effect on each potentially liquefiable layer should be performed using the site-specific in situ soil data and an understanding of the earthquake magnitude potential for the facility, and
- Computing the factor of safety against liquefaction for each liquefaction susceptible critical layer.

5. The Liquefaction Evaluation Method Used in the Study

Although there are different methods for evaluating liquefaction potential of the sand layers using *SPT* and *CPT* data, in order to avoid scattering the results, one of them which proven to be the most appropriate one, and has been used in many cases by different researchers, has been selected and used as below:

5.1. Robertson and Wride Method

This method is in fact based on the method, originally suggested by Seed and Idriss (1971). In this method the values of tip resistance of the CPT and also the number of SPT blows, are corrected in terms of the fine content according to one of the two following ways:

$$(N_1)_{60CS} = K_s (N_1)_{60} \qquad (1)$$

In which

$K_s = 0.025FC + 0.875$ for $5\% \leq FC \leq 35\%$, $PI \leq 5\%$ &
$K_s = 1$ for $FC \leq 5\%$, $PI \leq 5\%$
where FC is the fines content measured from laboratory

gradation tests on retrieved soil samples and *PI* is Plasticity Index of the soil. $(N_1)_{60}$ is *SPT* blow counts corrected for overburden stress.

The tip resistance of the *CPT* can be corrected by these equations:

$$(q_{c1N})_{cs} = K_c q_{c1N} \qquad (2)$$

In which
If $I_c \leq 1.64$, $K_c = 1.0$
If $I_c \geq 1.64$, $K_c = -0.403I_c^4 + 5.581I_c^3 - 21.63I_c^2 + 33.75I_c - 17.88$
I_c is the soil behavior type index obtained by using an Iterative Method and q_{c1N} is the cone penetration resistance corrected for overburden stress.

In the second way, which has been developed in 1997, the following equations can be used to correct the *SPT* numbers and also the *CPT* tip resistance, respectively.

5.2. Seed and Idriss Method

The following equations, developed by I.M. Idriss with assistance from H.B. Seed are recommended for correcting standard penetration resistance determined for silty sands to an equivalent clean sand penetration resistance:

$$(N_1)_{60CS} = \alpha + \beta(N_1)_{60} \qquad (3)$$

where α and β are coefficients determined from the following equations:

$\alpha = 0$ for $FC \leq 5\%$
$\alpha = \text{Exp}.[1.76 - (190/FC)^2]$ for $5\% \leq FC \leq 35\%$ &
$\alpha = 5.0$ for $FC \geq 35\%$
$\beta = 1.0$ for $FC \leq 5\%$
$\beta = [0.99 - (FC^{1.5}/1000)]$ for $5\% \leq FC \leq 35\%$ &
$\beta = 1.2$ for $FC \geq 35\%$
And for *CPT*:

$$(q_{c1N})_{CS} = q_{c1N} + \Delta(q_{c1N}) \qquad (4)$$

in which
$\Delta(q_{c1N}) = K_{CPT}(q_{c1N})_{CS}$
$\Delta(q_{c1N}) = [K_{CPT}/(1 - K_{CPT})](q_{c1N})$
Where
$K_{CPT} = 0$ for $AFC \leq 5\%$
$K_{CPT} = 0.0267(AFC - 5)$ for $5\% \leq AFC \leq 35\%$ &
$K_{CPT} = 0.8$ for $AFC \geq 35\%$
Where the *AFC* is Apparent Fine Content, to be determined as follows:
If $I_c < 1.26$ apparent fines content (FC%)= 0
If $1.26 \leq I_c \leq 3.5$ apparent fines content (FC%)= $1.75I_c^{3.25} - 3.7$
If $I_c > 3.5$ apparent fines content (FC%)= 100
This method has been used in the present study.

6. Comparison between Analysis Results

The comparison between the results of analysis has been made in terms of calculated safety factors, based on SPT data and CPT data belong to each site under consideration.

A linear regression has been used to correlate the analysis results and the correlation factors have been considered as the degree of relationship between these two methods. The safety factors against liquefaction using the Robertson and Wride method [7] for all sites have been calculated and shown in Figure (1).

As can be seen the results are very scattered. In ten points the absolute differences between their safety factors are more than 1.0 (ABS> 1.0). If they are ignored, the correlation factor will increase significantly, but this factor is still very small. The above points only cover 20% of all information points, see Figure (1).

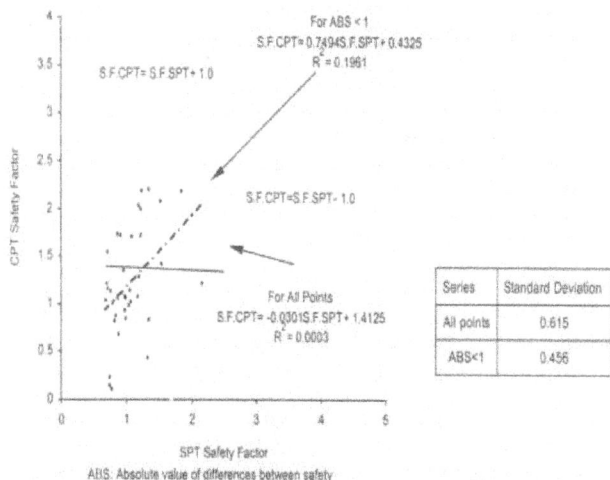

Figure 1. Comparison between safety factors against liquefaction using the method suggested by Robertson & Wride [8].

According to the general results of this study, as far as the fine non-cohesive soils are concerned, in spite of highly scattered results, an overall conclusion can be derived, in the way that the liquefaction potential evaluation of the ground by SPT data would be more conservative (Pessimistic) than that obtained by CPT data (Optimistic), see Figure (1).

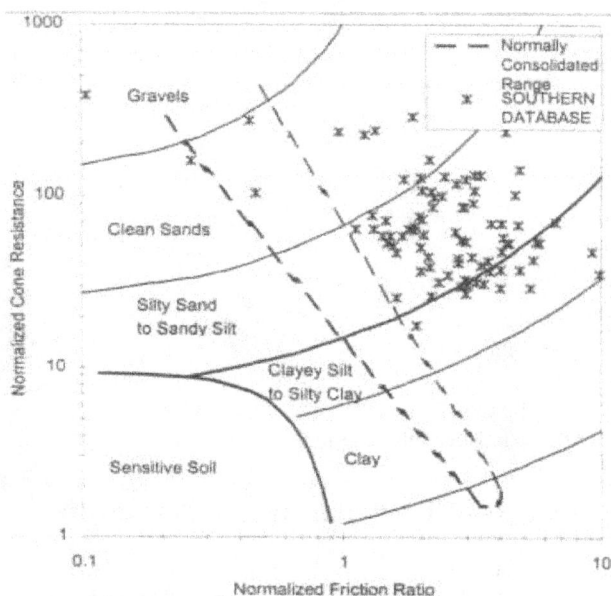

Figure 2. Soil Classification Based on CPT results.

As it was observed in this study, all sites selected were in the sandy silt to silty sand ranges, thus the results can be valid only for these fine granular soils. This classification can be also confirmed by CPT data belonging to the sites, see Figure (2).

Different researchers have focused on liquefaction potentials of susceptible soils in a comparative study by using both *SPT* and *CPT* of the ground layers. Among them Youd and Gilstrap [15] carried out extensive investigations to correlate between liquefaction safety factors based on CPT and SPT data of several sites. They used Robertson- Wride [7] method and obtained important results in their studies. The information points used, mainly belonged to the sites of clean sand to silty sands.

As shown in Figure (3), for $AFC > 50\%$, the suggested graphs by Robertson and Wride give the predicted AFC values less than its real value in term of I_c. This is clear in Youd and Gilsrap studies as well. It has to be noted that the suggested AFC- I_c relation by Robertson-Wride is an average curve, which has been, fitted to an extensive range of many informations points.

In the comparison made between liquefaction safety factors estimated based on the *CPT* and *SPT* data by Youd and Gilstrap, show also a large scattering ($R^2 = 0.5864$), nevertheless, ignoring the points of having ABS > 0.4 and concentrating to the 77% of the remaining points, the correlation factor would be of high value ($R^2 = 0.914$).

The main cause of this difference between the results of Gilstrap and Youd and the results of the current study may be attributed to the quite fine nature of the selected sites in this piece of research.

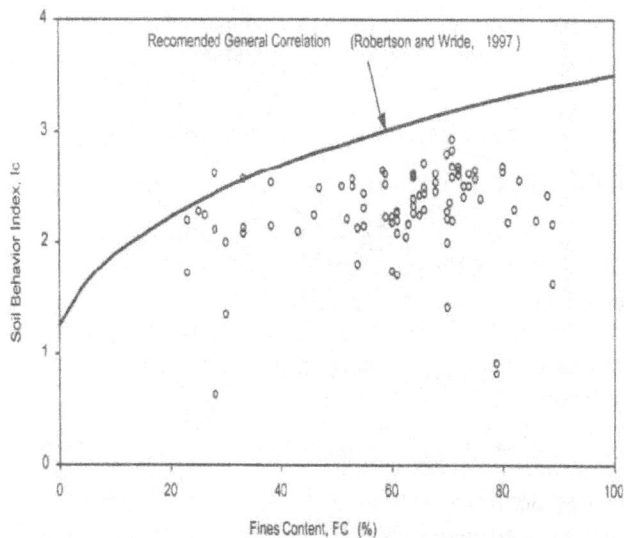

Figure 3. Correlation between fine content, F_c, of the selected sites and soil behavior index, I_c from the closest CPT sounding.

7. Discussion and Conclusion

Soil liquefaction is a major concern for structures constructed with or on sandy soils. Due to the difficulty and the cost of obtaining high quality undisturbed samples, simplified methods based on in-situ tests such as the standard penetration test (SPT) and the cone penetration test (CPT) are preferred by geotechnical engineers for evaluation of liquefaction potential of soils.

In order to compare the liquefaction potentials based on the SPT data with those based on the CPT data, some sites in the southern parts of Iran have been selected and studied. The geotechnical characteristics of these sites have been measured both from SPT and CPT methods, and for the same seismicity condition, the liquefaction potential were estimated using the SPT and CPT based evaluation methods. At the end some correlations were derived between the obtained results and their validities were discussed and justified. Although the correlation factor was found to be very small and the results were highly scattered, it could be concluded that the liquefaction evaluation methods based on the SPT data show more conservative results compared with those based on the CPT data.

References

[1] Castro, G. (1975) "Liquefaction and cyclic mobility of saturated sands". Journal of the Geotechnical Engineering Division, ASCE, 101 (GT6), 551-569.

[2] Finn, W. L., Ledbetter, R. H., and Wu, G. *"Liquefaction in silty soils: design and analysis, Ground failures under seismic conditions"*, Geotechnical Special Publication No 44, ASCE, Reston, 51–79, 1994

[3] Ishihara, K. (1993), *"Liquefaction and Flow Failure during earthquakes (Rankine Lecture)"*. Geotechnique, 43 (3): 351-415, 1993

[4] Koester, J.P. (1994). *"The Influence Of Fine Type And Content On Cyclic Strength"* Ground Failures Under Seismic Conditions, Geotechnical Special Publication No. 44, ASCE, pp. 17-33

[5] Krinitzky et al.(1993)

[6] Poulos, S.J., Castro, G., and France, W. (1985). *"Liquefaction evaluation procedure"*, J. Geotechnical Engineering Div., ASCE, Vol. 111, No.6, pp. 772-792.

[7] Robertson. P.K. and Wride, C.E. (1997*). " Cyclic Liquefaction and its Evaluation Based on SPT and, CPT "*, Final Contribution to the Proceedings of the 1996 NCEER Workshop on Evaluation of Liquefaction Resistance (T.L Youd, Chair).

[8] Robertson, P.K. and Wride, C.E. (1998). *"Evaluating Cyclic Liquefaction Potential Using the Cone* Penetration Test", Canadian Geotechnical Journal, 35(3), 442-459.

[9] Robertson, P.K.(1994*), "suggested terminology for liquefaction"*, An Internal CANLEX Report

[10] Sladen, J. A., D"Hollander, R. D., and Krahn, J. (1985),"*The liquefaction of sands, a collapse surface approach"*, Can. Geotech. J., 22, 564– 578.

[11] Seed, H.B. and Idriss, I.M. (1971). *"Simplified Procedure for Evaluating Soil Liquefaction Potential"*, J. of the Soil Mechanics and Foundations Division, ASCE, 97(SM9), 1249-1273.

[12] Seed, H. B. (1979). *"Soil Liquefaction and Cyclic Mobility Evaluation for Level Ground During Earthquake"*, Journal of Geotechnical Engineering Division, ASCE, Vol 105, No. GT2, pp 201-225.

[13] Selig, E.T., and Chang C.S.(1981), *"soil failure modes in undrained cyclic loading"* J. Geotech. Engg. Div.,ASCE, Vol.107, No.GT5, May, pp 539-551

[14] Youd, T.L. and Idriss, I.M., eds (1997). *"Proceedings of the NCEER on Evaluation of Liquefaction Resistances of Soils Tech. Report NCEER-1997-0022"*, Multidisciplinary Center for Earthquake Engineering Research, Buffalo, New York.

[15] Youd, T.L. and Gilstrap, S.D. (1999). "Liquefaction and Deformation of Silty and Fine Grained Soils", *Procceedings 2nd International Conference on Earthquake Geotechnical Engineering*, Lisbon Portugal, 3,1013-1020.

19

Exploring the Potential of Alternative Pozzolona Cement for the Northern Savannah Ecological Zone in Ghana

Abdul-Manan Dauda

Tamale Polytechnic, Department of Building Technology, Tamale, Ghana

Email address:
ddabdul-manan@tamalepoly.edu.gh

Abstract: This project evaluates the performance of pozzolana cement elements produced from alternative raw materials with a view using them in low-cost housing. It also seeks the reduction of waste from agricultural sources and the cost of sandcrete blocks by using locally available materials. The need to find alternative materials to replace existing conventional ones has necessitated research into substitutes to cement with a view to investigating their usefulness to wholly or partly substitute ordinary Portland cement in the production of sandcrete blocks. This project investigates the possible use of Corn Cob Ash (CCA) as a partial replacement of cement in sandcrete block production. 140 no. 450mmx150mm×225mm solid sandcrete blocks of mix ratio 1:8 were cast, cured and crushed at 7, 14, 21, and 28 days. The corn cob ash was replaced at 0 to 40 percent levels at 5% intervals. The maximum compressive strength of 2.10 N/mm2 was recorded at 30% replacement on the 28th day. After 12 months of exposure under northern savannah climatic conditions, the compressive strength remained stable or even increased with the weathering exposure. The maximum value of 2.10N/mm2 for the 30% replacement level is found suitable and recommended for building construction having attained a 28-day compressive strength of more than 2.0N/mm2 as required by the National Building Code for non load bearing walls.

Keywords: Pozzolana-Cement, Construction Materials, Low-Cost Housing, Sandcrete, Corn Cob Ash, Northern Savannah, Ghana

1. Introduction

The importance of sandcrete blocks as part of local building materials cannot be over emphasized in building and construction industry in Ghana. Sandcrete blocks have been widely used for building construction in the country in general and are increasingly replacing mud and stabilized earth as a walling material in the savannah ecological zone. The rapid adoption of these blocks has encouraged investigations into ways to make it more efficient and affordable. It is also an observable fact that river sand which is an important ingredient in making sandcrete blocks is widely available in the study area as a result of the many rivers and water bodies that traverse it.

Maize is the most important cereal crop produced in Ghana and it is also the most widely consumed staple food in Ghana with increasing production since 1965 (FAO, 2008; Morris *et al.*, 1999). In Ghana, maize is produced predominantly by smallholder resource poor farmers under rain-fed conditions (SARI, 1996). With these quantities comes the accompanying waste from corn husk and cob.

The cost of cement is a major issue for many builders in the study area. Therefore exploring the possible use of a commonly occurring waste material as a cement substitute will constitute an equally major relief for builders.

2. Theoretical Framework

Sandcrete blocks are composite material made up of cement, sand and water, moulded into different sizes. It is widely used in Ghana as a walling unit. The quality of blocks produced however, differs from due to the different methods employed in the production and the properties of the constituent materials.

Cement is a material with cohesive and adhesive properties when mixed with water, which makes it capable of bonding material fragments into a compact whole (Neville, 1996). Cements are classified as calcium silicate and calcium aluminate cement. Calcium silicate cement is further classified into Portland and Slag, while calcium aluminate is classified into High alumina and Pozzolona cement (Jackson and Dhir, 1991). Corn husk and cob has recently been recognized as pozzolona. A pozzolona is a siliceous/ aluminous material which in itself has little or no cementious value, but which will in finely divided form and in the presence of moisture, chemically reacts with calcium hydroxide liberated during the hydration of Portland cement to produce stable, insoluble cementious compound which contributes to its strength and impermeability (Sima, 1974).

Corn husk and cob ash is one of the promising pozzolanic materials that can be blended with Portland cement for the production of durable concrete/ blocks and at the same time it is a value added product. Addition of corn husk/cob ash to Portland cement does not only improve the early strength of concrete/block, but also forms a calcium silicate hydrate (CSH) gel around the cement particles which is highly dense and less porous, and may increase the strength of concrete/block against cracking (Saraswathy and Ha- Won, 2007).

Many countries including Ghana have the problem of shortage of conventional cementing materials. Recently there are considerable efforts worldwide of utilizing indigenous and waste, materials in concrete. One of such materials is the corn husk/cob which under controlled burning, and if sufficiently ground, the ash that is produced can be used as a cement replacement material in concrete (Anwar et al, 2000). Corn husks/cobs are used as building materials; lightweight concrete briquettes have been made from partly burnt husks/cobs. Insulating blocks have also been made with cement and husk ash that resists very high temperatures (Grith, 1974).

Corn cob is the hard thick cylindrical central core of maize (on which are borne the grains or kernels of an ear of corn). Raheem (2010) described Corn cob as the agricultural waste product obtained from maize or corn; which is the most important cereal crop in sub-Saharan Africa. There had been various research efforts on the use of corn cob ash (CCA) and other pozzolana as a replacement for cement in concrete. Olutoge et al(2010); presented a comparative study on fly ash and ground granulated blast furnace slag (GGBS) high performance concrete, Ogunfolami (1995); considered mixing of the CCA with Ordinary Portland cement at the point of need (i.e. on site). Adesanya and Raheem (2010); studied the workability and compressive strength characteristics of Corn cob ash (CCA) blended cement concrete. Adesanya and Raheem (2009); also assessed the development of Corn cob ash (CCA) Blended Cement.

3. Materials and Methods

3.1. Study Area

The data were collected between January and September, 2013 in the five regions (Northern, Upper East, Upper West

and the northern Brong Ahafo and Volta regions as shown in Figure 1 below) of Northern Ghana. This area also referred to as the northern savannah ecological zone in Ghana makes up about 48% of the country's total land area. Rainfall distribution is uni-modal giving a single growing season of 180 to 200 days with an annual mean of 1,100 mm. The dry season starts in November and ends in March/April with maximum temperatures of about 42°C occurring towards the end of the dry season. The data collection was carried out in six districts across the zone, two districts in each of the three regions and one district each from northern Volta and Brong Ahafo.

DISTRICTS OF SADA IN NATIONAL CONTEXT

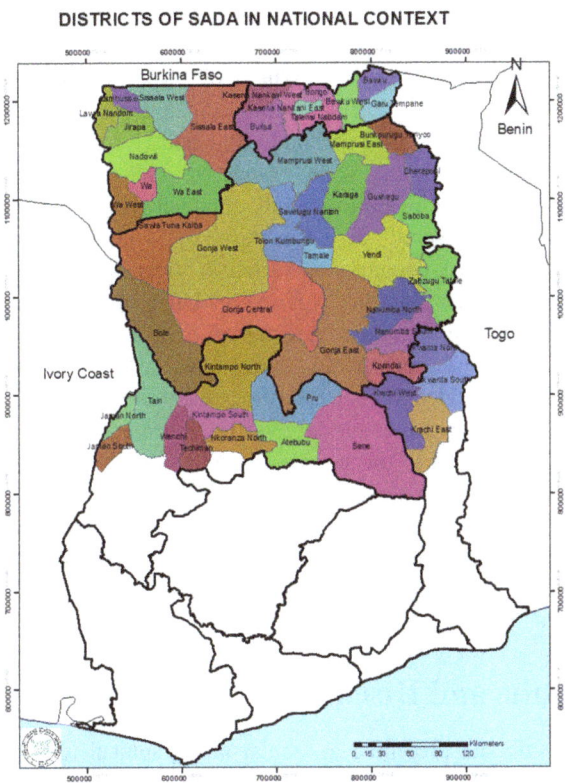

Source: SADA (Savannah Accelerated Development Authority)

Figure 1. A map of Ghana showing the study area.

3.2. Materials

3.2.1. Sand

The Sand used was clean, sharp river sand that was free from clay obtained from Dalung near Tamale in the Northern Region. The sand had a specific gravity of 2.60.

3.2.2. Cement

The cement used was ordinary Portland cement from the Savannah cement company, located in Buipe in the Northern Region with properties conforming to BS 12(1971). The properties of the cement are shown in Table 1.

3.2.3. Water

The water used was potable, fresh, colourless, odourless and tasteless water that is free from organic matter of any type.

3.2.4. Corn Husk/Cob Ash (CCA)

Corn husk and Cob used for this work was obtained from different farms located in the study area. The samples were allowed to dry under the sun for three days. The dried husk and cobs were burned slowly to ash in a gas kiln gradually for 3 days at 1100°C to obtain a finely divided ash, which is then sieved and kept ready for analysis (Rice Husk Ash, 2009). Chemical analysis of CCA was carried out in the laboratory of Chemistry Department at the Kwame Nkrumah University of Science and Technology. The properties are shown in Table 1.

3.3. Manufacture of Sandcrete Blocks

In this study, the solid blocks were manufactured with the use of the fabricating machine. One mix proportion of 1:8 was used in the production of 450mm x 225mm x 225mm sandcrete block. One hundred and forty of 450mmx150mmx225mm solid sandcrete blocks were produced. The quantities of materials obtained from the mix design were measured in each case by volume. The percentage of CCA content was varied in steps of 5% to a maximum of 40%. For the experiment, hand mixing was employed, and the materials were turned over a number of times until an even colour and consistency was attained. Water was then added as required through a water hose, and the materials were further turned over to secure adhesion. It was then rammed into the machine mould, compacted and smoothened off with a steel face tool.

After removal from the machine moulds, the blocks were left on pallets under cover and cured by watering through a fine watering hose. Testing for crushing strength was then carried out at 7, 14, 21 and 28 days.

4. Results and Discussion

The results are presented in tabular and graphical forms. Table 2 shows the results of the compressive strength test at various percentages of CCA contents. Figure 1 shows the plot of compressive strength against percentage CCA content at ages 7, 14, 21 and 28 days. Figure 2: shows the relationship between CCA content and compressive strength.

4.1. Effect of Corn Husk/Cob Ash on Compressive Strength

The compressive strength values show that the inclusion of CCA in the cement matrix does not appreciably enhance the compressive strength of the sandcrete blocks. At 5% replacement of CCA the compressive strength is 2.16N/mm2, which is 15% less than the control value. Figure 2 revealed that there is decrease in compressive strength as the proportion of coconut husk ash increases in the mix. The reduction in compressive strength for the replacement is due to low percentage of Calcium Oxide and Silicate Oxide in coconut husk ash. These two silicates are the main constituent of cement and are mainly responsible for the strength development. So as the cement content is gradually replaced by the coconut husk ash, the quantity of cement for

hydration is reduced thereby the strength of the sandcrete blocks is reduced.

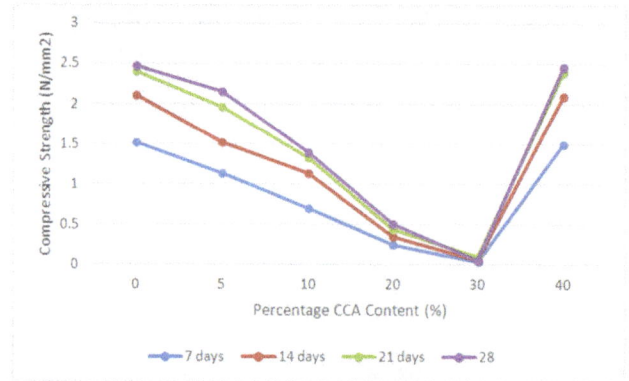

Figure 2. *Diagram showing the relationship between CCA content and compressive strength.*

The Nigeria National Building Code (16) recommended that average strength of 6 sandcrete block shall be 2.00N/mm2 and lowest strength for individual block shall be 1.75N/mm2 sandcrete block (18). For the 5% replacement of CCA in the cement matrix, the least value for the 28th day compressive strength is 2.03N/mm2, higher than recommendation by the code.

4.2. Effect of CCA on Dry Density

There is inconsistency in the dry density of the mix for the ages, but the maximum dry density for all the ages occurred at 15% CHA replacement. The maximum value of 1945.90 Kg/m3 was recorded at 7 day test.

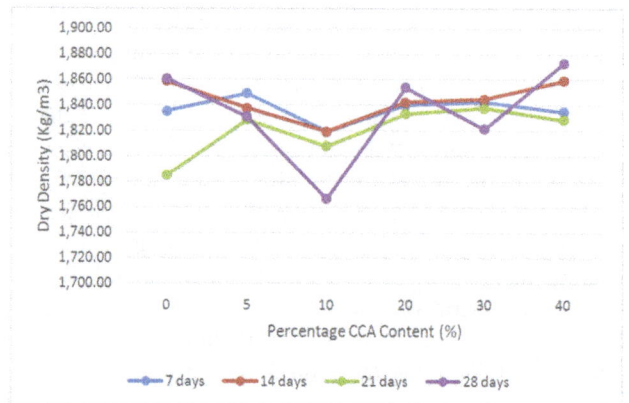

Figure 3. *Diagram showing the relationship between dry density and CCA content.*

4.3. Weathering Conditions

To determine the effect of weathering on the properties on the sandcrete blocks, the prepared samples that attained the desired compressive strength were exposed to the outdoor climatic conditions at various locations of the study area. The samples were exposed 28 days after manufacture in racks inclined 45° to ensure maximum exposure during the rainy and dry seasons.

The exposed specimens recorded stable or slight increases in compressive strength with no visible signs of deterioration.

5. Conclusion

The main conclusions derived from this investigation are as follows:

- Agriculture wastes such as corn cob/ husk ash shows good pozzolanic property in the production of sandcrete blocks.
- The maximum compressive strength of 2.16 N/mm2 was obtained for the sandcrete block specimens at a percentage CCA content of 30%.
- Corn husk/cob ash addition should not exceed 30% of the weight of cement for best results.
- The maximum compressive strength achieved at 5% is more than recommendation of 2.00N/mm2 recommended by National Building Code, for non load bearing walls.

Recommendation

1. Subsequent studies should be done on 0- 40% replacement of cement with corn cob ash and in steps of 5%.
2. Block/Concretes with the presence of ash content should be allowed to cure for 90days, by which pozzolanic activity of ash would have been concluded.
3. The use of locally available materials in infrastructure development will be met with the use of corn cob ash as a construction material and ultimately help meet our millennium development goals (MDG), thereby also enhancing the economic power of the rural dwellers if they are encouraged to plant maize from which these corn cobs could be gotten. The global green environment initiative will also be greatly influenced by the reduction in solid waste disposal.
4. The volume replacement attempted to get high strength concrete should be enhanced with super-plasticizers and a further reduction in the water-cement ratio so that concrete of very high strength can be achieved.

The following recommendations are drawn.

1. Use of other admixtures be incorporated with corn husk/cob ash in order to retard the hydration of water be studied
2. Other raw materials containing slightly higher calcium oxide and alumina could be used to improve the used of RHA as cement replacement.
1. Subsequent studies should be done on 0-40% replacement of cement with corn cob ash and in steps of 5%.
2. Concretes with the presence of ash content should be allowed to cure for 90days, by which pozzolanic activity of ash would have been concluded.
3. The use of locally available materials in infrastructure development will be met with the use of corn cob ash

as a construction material and ultimately help meet our millennium development goals (MDG), thereby also enhancing the economic power of the rural dwellers if they are encouraged to plant maize from which these corn cobs could be gotten. The global green environment initiative will also be greatly influenced by the reduction in solid waste disposal.

4. The volume replacement attempted to get high strength concrete should be enhanced with super-plasticizers and a further reduction in the water-cement ratio so that concrete of very high strength can be achieved.

Acknowledgement

The Authors wish to thank Mr. Abdul-Gafar Hamidu-Billa of the Department of Urban Roads Tamale and the Chemistry Department of the Kwame Nkrumah University of Science and Technology.

Table 1. *Chemical Analysis result of Portland cement & Corn Cob Ash.*

Name of compounds	Cement (%)	CCA (%)
Total Organic Content (Toc)	1.7	0.25
Calcium Oxide (CaO.)	62.32	0.005
Silicate (SiO$_2$) 18.72 0.005	18.72	5.10
Aluminate (Al$_2$O$_3$)	6.2	5.10
Ferrite (Fe$_2$O$_3$)	0.94	2.48
Magnesium Oxide (MgO)	1.62	0.09
Sulphur trioxide (SO$_3$)	1.1	0.12
Sodium oxide (Na$_2$O)	0.34	0.02

Table 2a-e. *Compressive Strength Test Result of Sandcrete block Specimens Containing Various CCA.*

Percentage CCA Content	Age (Days)	Dry Density (x10^{3}kg/m^3)	Compressive Strength at 28 days (N/mm^2)
	7	1.835.30	1.52
0%	14	1.858.40	2.11
(CONTROL)	21	1.784.70	2.41
	28	1.860.70	2.48

Percentage CCA Content	Age (Days)	Dry Density (x10^{3}kg/m^3)	Compressive Strength at 28 days (N/mm^2)
	7	1.819.20	0.70
	14	1.819.20	1.14
10%	21	1.807.70	1.33
	28	1.766.30	1.40

Percentage CCA Content	Age (Days)	Dry Density (x10^{3}kg/m^3)	Compressive Strength at 28 days (N/mm^2)
	7	1.840.00	0.25
	14	1.842.30	0.35
20%	21	1.833.00	0.44
	28	1.853.80	0.51

Percentage CCA Content	Age (Days)	Dry Density ($\times 10^3$ kg/m^3)	Compressive Strength at 28 days (N/mm^2)
30%	7	1.842.30	0.03
	14	1.844.60	0.06
	21	1.837.70	0.11
	28	1.821.50	0.06

Percentage CCA Content	Age (Days)	Dry Density ($\times 10^3$ kg/m^3)	Compressive Strength at 28 days (N/mm^2)
40%	7	1.834.80	1.50
	14	1.859.30	2.10
	21	1.828.35	2.40
	28	1.872.75	2.47

References

[1] Bentur A. Fiber-reinforced cementitious materials. In: Skalny JP, editor. Materials science of concrete. Waterville: The American Ceramic Society; 1989. p. 223–84.

[2] Heinricks H, Berkenkamp R, Lempfer K, Ferchland H-J. Global review of technologies and markets for building materials. In: Moslemi AA, editor. Proceedings of the 7th International Inorganic- Bonded Wood and Fiber Composite Materials Conference. Moscow: University of Idaho; 2000. 12p. [Siempelkamp Handling Systems report].

[3] Harrison PTC, Levy LS, Pratrick G, Pigott GH, Smith LL. Comparative hazards of chrysotile asbestos and its substitutes: a European perspective. Environ Health Perspect 1999; 107(8): 60711.

[4] Giannasi F, Thebaud-Mony A. Occupational exposures to asbestos in Brazil. Int J Occupat Environ Health 1997; 3(2): 15 0–7.

[5] Wood IM. Fibre crops: new opportunities for Australian agriculture. Brisbane: Department of Primary Industries Queensland; 1997. 102p.

[6] John VM, Zordan SE. Research and development methodology for recycling residues as building materials—a proposal. Waste Mgmt 2001; 21: 213–9.

[7] Swamy RN. Design for durability and strength through the use of fly ash and slag in concrete. In: Malhotra VM, editor. Proceedings of the 3rd CANMET/ACI International Conference on Advances in Concrete Technology. Auckland: ACI Publication SP-171-1; 1997. p. 1–72.

[8] Oliveira CTA, John VM, Agopyan V. Pore water composition of activated granulated blast furnace slag cements pastes. In: Proceedings of the 2nd International Conference on Alkaline

[9] Cements and Concretes. Kiev: Kiev State Technical University of Construction and Architecture; 1999. p. 18–20.

[10] Savastano Jr H, Warden PG, Coutts RSP. Brazilian waste fibres as reinforcement for cement based composites. Cem Concr Compos 2000; 22(5): 379–84.

[11] Zhu WH, Tobias BC, Coutts RSP, Langfors G. Air-cured banana-fibre–reinforced cement composites. Cem Concr Compos 1994; 16(1): 3–8.

[12] Higgins HG. Paper physics in Australia. Melbourne: CSIRO Division of Forestry and Forest Products; 1996.

[13] Coutts RSP, Ridikas V. Refined wood fibre-cement products. Appita 1982; 35(5): 395–400.

[14] Eusebio DA, Cabangon RJ, Warden PG, Coutts RSP. The manufacture of wood fibre reinforced cement composites from Eucalyptus pellita and Acacia mangium chemi-thermomechanical pulp. In: Proceedings of the 4th Pacific Rim Bio-Based Composites Symposium. Bogor: Bogor Agricultural University; 1998. p. 428–36.

[15] Coutts RSP, Kightly P. Bonding in wood fibre-cement composites. J Mater Sci 1984; 19:3355–9. Coutts RSP. Wood fibre reinforced cement composites. In: Swamy RN, editor. Natural fibre reinforced cement and concrete. Glasgow: Blackie; 1988. p. 1–62.

[16] Warden PG, Savastano Jr H, Coutts RSP. Fibre-cements fromBrazilian waste materials. In: Evans PD, editor. Proceedings of the 5th Pacific Rim Bio-Based Composites Symposium. Canberra: ANU Forestry; 2000. p. 75–80.

[17] Wang S-D, Pu X-C, Scrivener KL, Pratt PL. Alkali-activated slag cement and concrete: a review of properties and problems. Adv Cem Res 1995; 7(27): 93–102.

[18] Bijen J. Blast furnace slag cement. DM_s-Hertogenbosch: Stichting Beton Prisma; 1996.

[19] Coutts RSP, Warden PG. Effect of compaction on the properties of air-cured wood fibre reinforced cement. Cem Concr Compos 1990; 12: 151–6.

[20] Coutts RSP. Highyield wood pulps as reinforcement for cement products. Appita 1986; 39(1): 31–5.

[21] Savastano H, Mabe I, Devito RA. Fiber cement based composites for civil construction. In: Proceedings of the 2nd International Symposium on Natural Polymers and Composites ISNaPol 98. S~ao Carlos: Unesp/Embrapa/USP; 1998. p. 119–22.

[22] Taylor HFW. Cement chemistry. 2nd ed. London: Thomas Telford; 1997.

[23] Savastano Jr H, Agopyan V. Transition zone studies of vegetable fibre-cement paste composites. Cem Concr Compos 1999; 21(1): 49–57.

[24] Tol^edo RD, Filho K, Scrivener GL, England K. Durability of alkali-sensitive sisal and coconut fibres in cement mortar composites. Cem Concr Compos 2000; 22(2): 127–43.

[25] Olafusi Oladipupo S and Olutoge Festus A. "Strength Properties of Corn Cob Ash Concrete" Journal of Emerging Trends in Engineering and Applied Sciences (JETEAS) 3 (2): 297-301 Scholarlink Research Institute Journals, 2012 (ISSN: 2141-7016) jeteas.scholarlinkresearch.org.

[26] Bentur A, Akers SAS. The microstructure and ageing of cellulose fibre reinforced cement.

[27] Composites cured in a normal environment. Int J Cem Compos Lightweight Concr 1989; 11(2): 99–109.

[28] SARI, 1996. Savanna Agricultural Research Institute. *Annual Report. 1996.*

[29] FAO Statistical Databases. 2008. FAOSTAT: Agriculture Data. Available online: http://faostat.fao.org.

[30] Morris, M. L., Tripp, R. and Dankyi, A. A. 1999. Adoption and Impacts of Improved Maize Production Technology. A Case Study of the Ghana Grains Development Project, Economics Program Paper 99-01. Mexico, D. F., CIMMYT. Available online http://www.cimmyt.org/Research/economics/map/research_results/program_papers/pdf/EPP%2099_01.pdf.

[31] Neville, A. M. (1996). *Properties of Concrete*, 3rd edition, Longman Scientific and Technical Publishing, London. Pp 58–70.

[32] Saraswathy, V. and Ha-Won, S. (2007). Corrosion Performance of Rice Husk Ash Blended Concrete, Construction and Building Materials, Gale Group, Farmington Hills, Michigan. Retrieved on 12/10/2009 fromhttp://www.encyclopedia.com/doc/1G1-163421748.html.

[33] Sima, J. (1974). Portland-Pozzolona Cement: Need For a Better Application". *Indian Concrete J.* 48: 33-34.

[34] Jackson, N. and Dhir, R. K. (1991). *Civil Engineering Materials*, 4th Edition, Macmillan ELBS, Hong Kong, Pp 144 –160.

[35] Grith, D. H. (1974). Rice, 2nd Edition, Longman Limited, London. pp12-18. Adesanya D. A., [26] [34] [35] Raheem A. A. 2009. Development of Corn Cob Ash Blended Cement, Construction and Building Materials, (Vol. 23, pp. 347-352).

[36] Adesanya D. A., Raheem A. A. 2010. A study of the workability and compressive strength characteristics of corn cob ash blended cement concrete, Construction and Building Materials, (Vol. 23, pp. 311-317).

[37] Ogunfolami T. F. 1995. The Effect Of Thermal Conductivity and Chemical Attack on Corn Cob Ash Cement Concrete, Unpublished B. Sc. Project Report, Department of Building, Obafemi Awolowo University, Ile-Ife.

[38] Olutoge F. A., Bhashya V., Bharatkumar B. H., and Sundar Kumar S. 2010. Comparative Studies on Fly Ash and GGBS High Performance Concrete, Proceeding of National Conference on Recent Trend and Advance in Civil Engineering-TRACE 2010.

[39] Rice Husk Ash (2009). Rice Husk Ash. Retrieved on 8/10/2009 from http://www.ricehuskash.com/details.htm.

20

Premature Failure of Apedwa-Bunsu Junction Section of N6 in Ghana: Some Notes for Consideration

Yaw Adubofour Tuffour[1], Nana Kwesi Agyepong[2], Daniel Atuah Obeng[1]

[1]Department of Civil Engineering, Kwame Nkrumah University of Science and Technology, Kumasi, Ghana
[2]Materials Division, Ghana Highway Authority, Ministry of Roads and Highways, Accra, Ghana

Email address:
yat@engineer.com (Y. A. Tuffour), k_agyeponguk@yahoo.co.uk (N. K. Agyepong), obengatuah@yahoo.co.uk (D. A. Obeng)
[*]Corresponding author

Abstract: This study investigated premature and continual failure of the Apedwa-Bunsu Junction section of Route N6 in Ghana despite an earlier maintenance intervention which included geotextile installation and placement of a new wearing course. It involved a condition survey, density, asphalt content, gradation, stiffness modulus and Falling Weight Deflectometer (FWD) tests on the section. The condition survey revealed cracking (alligator, transverse and longitudinal), ravelling, potholes, rutting and shoving as the predominant defects on the road. The density tests on the bituminous layers revealed relative compaction levels which, in most cases, did not meet the minimum required by the technical specifications despite the additional densification by traffic. The poor compaction was corroborated by high pavement deflections from the FWD device. Asphalt cores revealed a friable dense bituminous macadam (DBM) layer although bitumen extraction tests indicated all design asphalt contents were met. Lack of inter-particle cohesion within the DBM layer was suggestive of stripping damage to the asphalt concrete. Some samples of the crushed rock base contained plastic fines and fines content that exceeded specification limits. High stiffness modulus values of the bituminous layers suggested possible premature aging of the asphalt binder which probably accelerated crack development. An earlier intervention in the form of placement of geotextile in the wearing course failed to arrest cracking because the material had been placed at a shallow depth rendering it ineffective. It was concluded that inadequate compaction of the bituminous layers and the use of crushed rock and other pavement materials that did not wholly meet the technical specifications were the root causes of the premature failure of the section.

Keywords: Premature Failure, Compaction, Cracking, Relative Density, Rutting, Shoving

1. Introduction

Route N6 in Ghana is a major transport link between the north and south of the country and also an important trade route linking Ghana to several of her West African neighbours to the north. Safety and uninterrupted flow on N6 are very important as the route forms part of the trans-West Africa trade route. In 2002, the Ghana Highway Authority (GHA) awarded the contract for the construction of the Apedwa-Bunsu Junction Road, which is 23km long, as a new alignment to the Apedwa-Potroase-Bunsu Junction section of N6 which was accident-prone, to improve safety on N6 as a whole. The pavement structure consists of 40mm wearing course, 60mm binder course, 80mm dense bituminous mac-

adam (DBM), 200mm crushed stone base and 200mm natural gravel sub-base. Construction was completed and the road opened to traffic in 2004 but since then, performance has been poor.

Within the first few years of its service life, the section began to experience mainly cracking, ravelling and rutting although there were other minor defects. In 2007, the Ghana Highway Authority (GHA) initiated investigation into the causes of the premature failures the outcome of which led to the removal of the wearing course, installation of geotextile as reinforcement to arrest the cracks and placement of a new layer of wearing course. The intervention notwithstanding, the distresses continued and accelerated. By 2011, deterioration had become so severe at several locations as to prompt another investigation into the causes of the failures.

Premature failures of asphalt overlays within the first few years of in-service life are not uncommon and have been well investigated by several researchers. Himeno and Watnabe [1] have noted that fatigue failure can initiate at the top of a new asphalt concrete layer with low stiffness arising from poor compaction. According to Button and Lytton [2], distresses in the wheel path and rapid deterioration of asphalt overlays may arise if moisture accumulates through evapo-transpiration from beneath or infiltration from the top in hot-mix asphalt susceptible to moisture damage. The accumulated water tends to cause stripping damage and weakens the pavement structure. In cold regions, freeze-thaw cycles may induce thermal cracking and moisture distresses to reduce the capacity of asphalt pavements [3]. Excess asphalt content in bituminous layers, particularly the wearing course, and a change in aggregate gradation may lead to early rutting [4].

Early brittleness, cracking and stripping in an asphalt pavement due to the use of super fine filler have been reported by Horak and Emery [5]. According to Muench and Willoughby [6], construction-related temperature differentials may lead to the placement of cooler mats that may resist adequate compaction and result in localised open-textured surfaces having high air voids. Generally, overlays with high air voids content have a higher risk of moisture damage than those with low air voids content due to the ease of water penetration [7]. Oxidative aging may also accelerate in such overlays leading to binder embrittlement and subsequent cracking and ravelling. Significant rutting due to densification under traffic, especially for thick lifts, may occur in new overlays compacted to voids content higher than the long-term air voids content pertaining to the mix design [8]. Pavements with excessive fines and excessive asphalt content as well as improper aggregate grading are also likely to suffer early shear deformation if they come under heavy loads and high tyre pressures [8]. In some cases, inadequate bonding between the base and intermediate asphalt concrete lifts, arising from inadequate or non-uniform tack coat application, could lead to middle-up cracking and cause unanticipated pavement failure to occur [9]. De-bonding and slippage failure could also occur if tack coat application was non-uniform or the material was removed by construction trucks before placement of the asphalt concrete lift [10].

This paper reports on the outcome of the second investigation into the failures on the Apedwa-Bunsu Junction section of N6. It is expected that the outcome of the study would provide some useful notes for better construction practices that would reduce the incidence of premature distresses in future asphalt pavement constructions in the country.

2. Materials and Methods

2.1. Road Condition Survey

A thorough condition survey was carried out on both the north- and south-bound lanes of the Apedwa-Bunsu Junction section of N6 to note the types of distresses and extent of coverage. The survey also provided opportunity to map out uniform sub-sections and determine locations for sampling.

2.2. Uniform Sectioning and Sampling

The condition survey was used as a basis for dividing the section under study into uniform sub-sections with seemingly pristine conditions to enable samples to be taken for laboratory analysis. In all, a total of 10 uniform sub-sections with 10 sampling locations as detailed in Table 1 were selected.

Table 1. Details of uniform sections and sampling points.

Km	Length (m)	Sampling Point
6+000 – 7+000	1000	6+750
9+300 – 10+900	1600	10+050
10+900 – 11+400	500	10+500
12+800 – 13+600	800	13+250
14+200 – 15+700	1500	14+130
15+700 – 16+200	500	16+000
17+000 – 17+800	800	17+750
17+800 – 18+000	200	18+000
18+000 – 19+000	1000	18+750
19+000 – 21+000	2000	20+750

2.3. Trial Pitting

Layer materials for testing were obtained through trial pitting. Bituminous materials as well as unbound pavement layer materials beneath the asphalt concrete layers were sampled. Of the 10 locations selected for sampling, 5 were sited on the south-bound lane and the other 5 on the north-bound lane. Samples were taken at depths corresponding to the wearing and binder courses, the dense bituminous macadam (DBM) layer, the crushed stone base (CSB) layer, the sub-base layer and the sub-grade.

2.4. Coring of Asphalt Concrete Layers

Asphalt cores were taken at locations adjacent to the trial pits for laboratory testing. In all, 4 cores were taken at each of the 10 locations. Cores for the wearing course, the binder and the DBM layers were separated. Two cores out of the four samples taken at each location were used for density tests and the remaining for indirect tensile stiffness modulus tests.

2.5. Falling Weight Deflectometer Test

At locations where the elastic modulus and surface deflections were measured, the FWD equipment was set up and then a load pulse applied to the pavement through a piston by means of a computerised system attached to the device. Deflections were picked up by seven geophones. The elastic modulus and surface deflection measurements were taken on both the north-bound and south-bound lanes.

2.6. Indirect Tensile Stiffness Modulus Test

Asphalt concrete cores taken from the field were prepared

and tested in accordance with BS DD213 [11]. Deformations during testing were measured by high-speed transducers.

2.7. Other Laboratory Tests

Other laboratory tests conducted on the field samples were:
- Bitumen extraction
- Grading of the residual aggregates after bitumen extraction
- Grading and Atterberg limits of non-bituminous layer materials

3. Results and Discussion

3.1. Distress Types

The major distress identified on the study section and their prevalence in terms of percentage of road length coverage have been summarised in Table 2. Figures 1-5 show graphically how severe some of the distresses were.

Table 2. *Distress types and coverage on study section.*

Distress Type	North-bound Lane		South-bound Lane	
	Length (m)	Coverage (%)	Length (m)	Coverage (%)
Raveling	2280	9.9	1320	5.7
Longitudinal & Transverse Cracks	15070	65.5	15570	67.7
Alligator Cracks	6720	29.2	4340	18.9
Potholes	Localised	-		-
Rutting	10500	45.7	10280	50.3
Shoving	Localised	-		-

Figure 3. *Rut in wearing course.*

Figure 4. *Ravelling with incipient pothole development.*

Figure 1. *Alligator cracks with pothole development.*

Figure 5. *Structural failure and shoving of outer edge of surface course.*

Overall, the following were established from the survey:
- Transverse and longitudinal cracking was very extensive in both travelled lanes and affected a greater length of the section.
- In terms of structural deformation, rutting dominated on both travelled lanes while shear failure and associated shoving tended to be confined to the outer edges of the wearing course.
- Alligator cracking was prevalent on both travelled lanes but affected a longer length of the north-bound lane than

Figure 2. *Extensive alligator cracks in wearing course.*

the south-bound lane.

- Rutting was prevalent on both travelled lanes but affected a slightly longer length of the north-bound lane than the south-bound lane.

At some locations, the presence of ruts and cracks allowed run-off to seep into the pavement structure to saturate the underlying layers which became evident during the trial pitting (see Figure 6). Differences in prevalence of the distresses on the two travelled lanes could not be linked to differences in lane loading as this portion of N6 comes under heavy goods transport in both travelled directions almost equally. Besides, portions of N6 abutting the study section and constructed earlier under a different contract did not exhibit many of the observed defects on the study section, despite coming under the same loading regime.

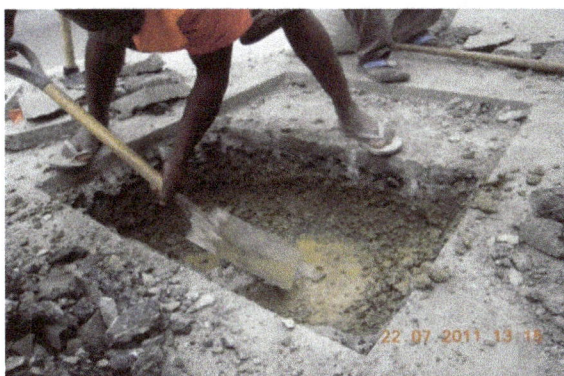

Figure 6. *Trial pit showing water-soaked crushed stone base.*

3.2. Level of Compaction of Bituminous Layers

The bulk densities of the cores taken from the uniform sub-sections have been detailed in Table 3 for the wearing, binder and DBM layers.

Table 3. *Bulk densities of bituminous layers.*

Km	Bulk Density (kg/m³)		
	Wearing Course	Binder Course	DBM Layer
6 + 750	2450	2378	2341
10 + 050	2310	2382	2347
10 + 500	2280	2288	2310
13 + 250	2355	2310	2388
14 + 130	2276	2176	2378
16 + 000	2368	2257	2411
17 + 750	2373	2302	2418
18 + 000	2431	2326	2248
18 + 750	2401	2346	2403
20 + 750	2352	2350	2366

a) Wearing Course

The bulk density values ranged between 2276kg/m³ and 2450kg/m³ with an average value of 2360kg/m³. The reference laboratory bulk density for hot-mix asphalt used for the paving operation was 2490kg/m³. This translates to relative compaction achieved in the field that ranged between 91% and 98%. Cored samples from 5 out of the 10 sample locations had relative compaction values below the minimum of 95% specified by the special technical specifications.

b) Binder Course

The density values ranged between 2176kg/m³ and 2382kg/m³ with an average value of 2312kg/m³. The laboratory bulk density achieved for the paving mix was 2497kg/m³. Based on the densities of the cores taken from the road, the relative compaction achieved ranged between 87% and 96%, with 8 out of 10 sample locations (80%) having relative compaction values that were below the minimum of 95% specified by the special technical specifications.

c) DBM Layer

The density of the DBM layer ranged between 2248kg/m³ and 2418kg/m³ with an average value of 2361kg/m³. The bulk density of the samples taken from the hot-mix plant for the paving operation was 2497kg/m³. The relative compaction achieved for the layer ranged between 90% and 97%. Relative compaction values for 4 locations out of the 10 sampled were below the minimum specified by the special technical specifications.

It was noted that the cored DBM layer tended to be friable with the mix hardly able to hold together. This suggested a lack of cohesion caused by stripping. There was also evidence of pitting on the cored surface suggesting material segregation and poor compaction during placement as well as stripping (see Fig. 7).

Figure 7. *Asphalt cores with pitted surfaces.*

The low relative compaction values associated with the bituminous layers are indicative of inadequate compaction during construction. It is believed that the poor compaction rendered the DBM layer permeable and probably made it easy for water to penetrate the pavement structure to cause stripping damage. Even though Chen [12] has cautioned that density alone may not engender quality construction, nevertheless, dense gradation and high density are important for achieving minimum permeability [2]. It is believed that moisture damage may have led to loss of inter-particle cohesion within the DBM layer causing it to behave essentially like a compacted granular layer.

3.3. Bitumen Content of Bituminous Layers

Table 4 contains the results of the bitumen content for the bituminous samples taken from the homogenous sections without distinction between the two travelled lanes.

Table 4. Bitumen content of bituminous layers.

Km	Bitumen Content (%)		
	Wearing Course	Binder Course	DBM
6+750	5.1	4.8	3.9
10+050	5.1	5.2	4.2
10+500	5.1	4.7	4.3
13+250	5.0	4.5	4.1
14+130	5.0	4.9	4.2
16+000	5.2	4.9	4.0
17+750	5.2	4.8	4.2
18+000	5.1	4.8	3.9
18+750	5.1	4.9	4.1
20+750	5.1	4.9	4.2

Table 5. Indirect Tensile Stiffness Modulus.

Km	Stiffness Modulus (MPa)		
	Wearing Course	Binder Course	BDM
6 + 750	30362	21056	16269
10 + 050	25803	12504	10557
10 + 500	15052	20472	6846
13 + 250	11959	16255	18820
14 + 130	23747	20999	16505
16 + 000	13530	19549	17714
17 + 750	21980	27351	20251
18 + 000	15294	27531	13483
18 + 750	21894	14598	15255
20 + 750	11345	18028	17488

a) Wearing and Binder Courses

The asphalt content of the wearing course ranged between 5.0% and 5.2% with a mean of 5.1% which was the same as the mix design value. In the case of the binder course, it ranged between 4.5% and 4.9% with a mean value of 4.8%. The mean was the same as the mix design value.

b) Dense bituminous macadam

The asphalt content values ranged between 3.9% and 4.3% with a mean of 4.1%. This differed only marginally from the mix design value of 4.0% but was within tolerance limits.

3.4. Elastic Moduli of Bituminous Layers

3.4.1. Indirect Tensile Stiffness Modulus

Table 5 contains the Indirect Tensile Stiffness Modulus as measured in the Indirect Tensile Stiffness test. In comparison, results from similar tests conducted at the GHA Materials Lab on new un-aged asphalt concrete briquettes yielded values in the range of 4,000MPa-8,000MPa.

In general, high stiffness indicates brittle material and, hence, high cracking potential. This suggests that the bituminous layers had undergone premature aging which must have contributed to crack development.

3.4.2. Moduli from FWD Device

Table 6 details the elastic moduli obtained from the FWD device. The average value for the binder and wearing course considered as a composite layer is about 4,200MPa whilst that for the DBM layer is about 430MPa, a tenfold difference. The average value 350MPa for the crushed stone base was similar to that of the DBM layer. The low modulus of the DBM layer corroborates the assertion made earlier that the poor compaction and the lack of cohesion caused by stripping within the DBM layer made the layer behave much like a granular layer and not a bound layer.

Table 6. Elastic Moduli of Pavement Layers from FWD Tests.

Km	Elastic Modulus (MPa)									
	Wearing & Binder Course		DBM		CSB		Sub-base		Sub-grade	
	NBL	SBL	NBL	SBL	NBL	SBL	NBL	SBL	NBL	SBL
6 + 750	5586	2867	591	384	489	307	270	162	134	81
10 + 050	7554	8997	400	623	331	498	183	264	141	142
10 + 500	1977	2967	486	509	402	407	222	216	230	232
13 + 250	3666	2488	447	328	369	262	204	139	151	132
14 + 130	3348	4753	400	448	330	359	182	190	179	99
16 + 000	2633	2510	344	320	284	256	157	136	96	77
17 + 750	2786	7198	299	511	248	409	137	217	80	80
18 + 000	2072	4679	429	545	355	437	196	231	62	78
18 + 750	4943	4716	349	425	288	341	159	180	40	45
20 + 750	3796	-	419	393	347	314	192	166	74	79

NBL=north-bound lane, SBL=south-bound lane

Table 7. Surface deflections from FWD tests.

Km	Maximum Deflection (microns)	
	North-bound Lane	South-bound Lane
6 + 750	318	444
10 + 050	385	255
10 + 500	309	366
13 + 250	364	469
14 + 130	411	410
16 + 000	485	474
17 + 750	519	338
18 + 000	421	347
18 + 750	476	477
20 + 750	425	435

3.5. Surface Deflections

The deflections on the sections obtained from the FWD device have been shown in Table 7. The values refer to the maximum deflections measured of the deflection bowl and ranged between 300 to 520 microns for the north-bound lane and 200 to 480 microns for the south-bound lane.

Typical deflection values obtained by the Ghana Highway Authority (GHA) on some roads with similar age and pavement structure were in the range of 198-210 microns. The high values recorded in the current study reflect a weak composite pavement structure.

3.6. Properties of Unbound Granular Materials

3.6.1. Atterberg Limits

Table 8 contains data on the Atterberg limits (LL and PI) of the unbound pavement materials. It is seen from the table that the CSB material had a PI of the order of 7%-8% even though the specification required non plastic material. The values suggest the presence of some amount of clayey material. In the case of the sub-base material, the PI value was essentially the same as the maximum specified and therefore, the material may be considered as being of marginal quality with respect to Atterberg limits. The subgrade material, on the other hand, met the specification requirement.

Table 8. *Atterberg limits of unbound pavement materials.*

	CSB		Sub-base		Sub-grade	
Km	PI	LL	PI	LL	PI	LL
6 + 750	-	-	-	-	-	-
10 + 050	7.5	-	11.9	33.6	21.3	44.7
10 + 500	7.5	-	10.3	28	-	-
13 + 250	7.6	-	13.7	35.2	-	-
14 + 130	7.4	-	12.4	30.5	21.2	45.8
16 + 000	7.3	-	11	28.2	-	-
17 + 750	7.6	-	-	-	23.9	51
18 + 000	7.7	-	12.9	30	-	-
18 + 750	7.3	-	-	-	-	-
20 + 750	7	-	10	28.7	-	-

3.6.2. Particle Size Distribution

a) Wearing and Binder Courses

Figures 8 and 9 show the particle size distribution curves for samples of the wearing course and binder course materials, respectively, after bitumen extraction. The curves have been superimposed on the corresponding gradation envelope per the Ministry of Roads and Highways Specification (MRH) [13]. It is seen that all the curves essentially fell within the gradation envelope, although in the case of the binder course material, there was slight violation for some of the samples in the fine fraction content. In addition, most of the sample curves tended to gravitate a bit more toward the upper limits of the specification size ranges.

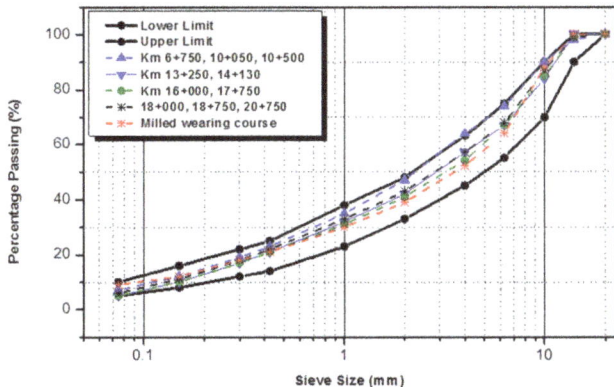

Figure 8. *Particle size distribution of wearing course aggregate.*

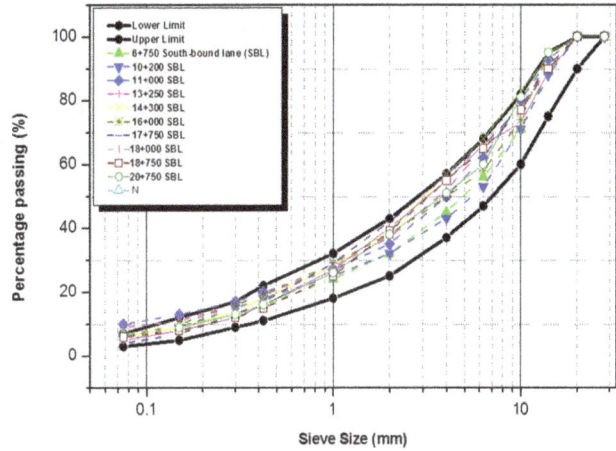

Figure 9. *Particle size distribution of binder course aggregate.*

b) DBM Layer

Figure 10 shows the particle size distribution of the residual aggregates of the DBM layer after bitumen extraction. While most of the samples had the particle size falling within the specification envelope, a few fell outside. Also, most of the samples with size distribution falling within the envelope tended to gravitate toward the upper limit of the size ranges.

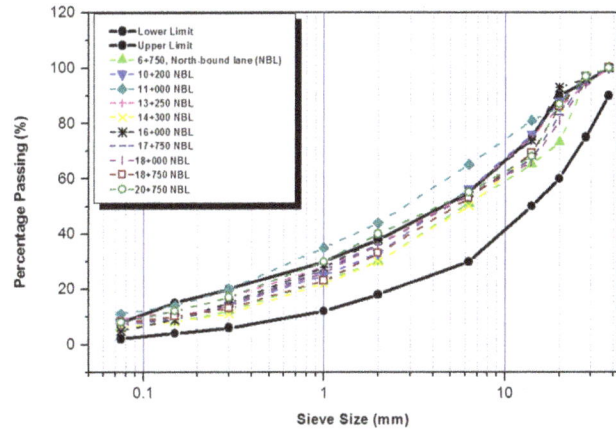

Figure 10. *Particle size distribution of DBM aggregate.*

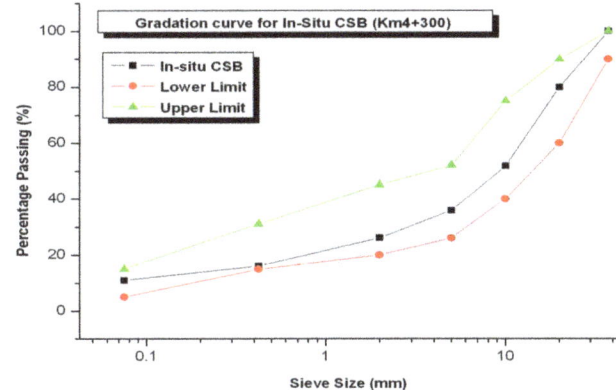

Figure 11. *Particle size distribution of CSB material.*

Hence, on the whole, compliance of the DBM material with the particle size distribution requirement of the MRH Specification was not total.

c) Sub-base and Crushed Stone Base Layers

Figures 11 and 12 show the particle size distribution curves of the CSB material and the sub-base, respectively. It is seen from the curves that whereas compliance with the MRH specification for the base material was total, it was not so for the sub-base material. The fines fraction of some of the samples went above the upper limits of the specification.

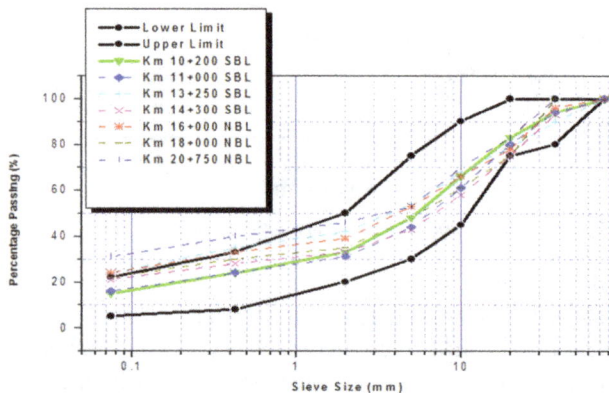

Figure 12. Particle size distribution of sub-base material.

3.7. Geotextile Placement

In the course of the fieldwork, a geotextile material placed in an earlier intervention work was established to have been placed at a depth with only about 15-20mm asphalt concrete cover (see Fig. 13). Literature on geotextile placement in asphalt overlays recommends a minimum compacted cover of 38mm (1.5inches) as first lift [2] to make it effective. This suggests that the placement of the geotextile did not meet this minimum requirement and probably partly explains why the material had not been effective in arresting cracks.

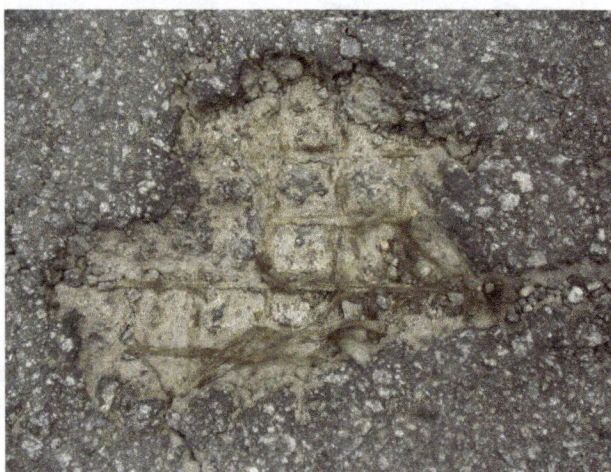

Figure 13. Exposed geotextile placed in an earlier maintenance intervention with thin cover.

4. Conclusions

Field investigations and laboratory tests conducted in this study were aimed at investigating the premature failure of the Apedwa-Bunsu Junction section of Route N6. Properties of pavement materials and construction quality at several loca-

tions along the section were investigated. It was established that poor construction quality and, in some cases, the use of sub-standard pavement materials were the major causes of the many defects on the road. In particular, densities of the bituminous layers (wearing course, binder course and DBM layer), in most cases, did not meet requirements of the technical specifications. The low compaction resulted in a permeable overlay that facilitated moisture penetration from below and above to cause moisture-induced damage. The DBM layer did not benefit from the asphalt binder within the matrix nor exhibit the characteristics of a reinforcing layer because of moisture damage effects. High stiffness modulus values of the bituminous layers suggested possible premature aging of the asphalt binder which contributed to the many cracks observed on the road. The presence of plastic fines in the crushed rock base at some locations was suggestive of the presence of clayey material, with the fines content in most locations exceeding specification limits. A geotextile material placed earlier as an intervention measure to arrest cracks was ineffective because it had been placed within the wearing course with insufficient asphalt concrete cover.

References

[1]   Himeno, K., and Watnabe, T. (1987). Design of Asphalt Pavements. Sixth Int. Conf. on Structural Design of Asphalt Pavements. Ann-Abor, Michigan.

[2]   Button, J. and Lytton, R. (2003). Guidelines for using geosnythetics with HMA overlays to reduce reflective cracking; Transportation Research Institute, Report No. 1777-P2, Texas A&M University. Available at http://www.utexas.edu/research/tppc/pubs/button_geosynthetics.pdf

[3]   Si, W., Ma, B., Li, N., Ren. J. and Wang, H. (2014). Reliability-based assessment of deteriorating performance to asphalt pavement under freeze-thaw cycles in cold regions. *Construction and Building Materials*, 68 (2014): 572-579.

[4]   Chen, D., Bilyeu, J., Scullion, T., Lin, D. and Zhou, F. (2003). "Forensic Evaluation of Premature Failures of Texas Specific Pavement Study–1 Sections." *J. Perform. Constr. Facil.*, 10.1061/(ASCE)0887-3828(2003)17: 2 (67), 67-74.

[5]   Horak, E. and Emery, S. J. (2010). Forensic Investigation to determine the reasons for premature failure in asphalt surface layer. *Road Materials and Pavement Design*, 11(3): 511-527.

[6]   Muench, S. and Willoughby, K. (2006). Preventing pavement failure caused by hot-mix asphalt pavement temperature differentials: Washington State's systematic approach. TR News 246, September-October 2006, pp 26-28.

[7]   Gubler, R., Partl, M. N., Canestrari, F. and Grilli, A. (2005). Influence of water and temperature on mechanical properties of selected asphalt pavements. *Materials and Structures*, 38(5): 523-532.

[8]   Anochie-Boateng, J. K, Mataka, M. O., Malisa, J. T. and Komba, J. J. (2015). Forensic study into the causes of premature failures in asphalt pavements in Tanzania. Road pavements of the XXVth World Road Congress in Seoul, Seoul, South Korea, November 2015.

[9] Vrtis, M. and Timm, D. (2015). Case Study on Premature Pavement Failure and Successful Reconstruction of a High RAP Section at the NCAT Test Track. Airfield and Highway Pavements 2015: pp. 260-271. doi: 10.1061/9780784479216.024

[10] Mohammad, L. N., Elseifi, M. A., Bae, A., Patel, N., Button, J. N. and Scherocman, J. A. (2012). Optimization of tack coat for HMA placement. National Cooperative Highway Research Program, Report No. 712. Available at http://onlinepubs.trb.org/onlinepubs/nchrp/nchro_rpt_712.pdf

[11] BS DD213 (1993). Method for determination of the indirect tensile stiffness modulus of bituminous mixtures. British Standard Institutions, London.

[12] Chen, D. (2009). "Investigation of a Pavement Premature Failure on a Weak and Moisture Susceptible Base." *Journal of Performance of Constructed Facilitie*s, 23 (5): 309-313

[13] MRH (2007). Specifications for Road Works and Bridges, Ministry of Roads and Highways, Accra, Ghana.

21

Experimental Study on the Quality of Concrete Strengthened by the Means of Infrared Thermal Imager

Fuchun Song, Jie Zhao, Mengchen Li

School of Traffic Engineering, Shenyang Jianzhu University, Shenyang, China

Email address:

Songfch@163.com (Fuchun Song), 530779841@qq.com (Jie Zhao), 891677574@qq.com (Mengchen Li)

Abstract: Testing of concrete quality of bonded steel reinforcement by the means of infrared thermal imager, study of sticky steel reinforcement concrete component under the irradiation of an external heat source surface temperature with time, spatial distribution and variation analysis internal hollowing defect thickness, size and thermal infrared imager vertical shooting angle and emission rate and other factors on the effect of infrared thermal imaging. The test results show that, the greater the thickness of internal defects, hollowing the greater area is easy to be detected. In the presence of internal defects, contrast the thickness of the site with no defect parts of the surface temperature, defect site temperature is significantly lower.

Keywords: Bonded Steel Plate, Concrete, Infrared Thermal Imaging Technology, Nondestructive Testing

1. Introduction

In foreign countries, many scholars have studied the infrared thermal image method in the concrete quality control of bonded steel plate reinforcement [1-2], and have achieved good results in practical engineering applications [3-5]. In China, the research of this aspect is relatively few [6-7], and the case of engineering application is not much [8-9]. From the existing research results, the infrared thermal imaging detection technology in the paste steel plate reinforced concrete structure quality judgment, mainly in the qualitative analysis stage, still lacks quantitative analysis model. In order to better promote the application of the technology in the construction, especially the steel reinforcement and detection of the bridge engineering in China with the rapid increase of highway mileage, bridge maintenance and maintenance tasks are becoming more and more difficult.

Infrared Thermal Imaging Nondestructive Testing (IRNDT) has the advantages of non-contact, fast, large measurement area, high temperature measurement resolution and no interference to the surface temperature field [10]. It has been widely used in the detection of the electrical equipment and machine equipment, the fault diagnosis of petrochemical equipment, fire detection, material internal defect detection. Single side heating method is applied to nondestructive testing of metal, alloy, plastic, ceramic and composite materials. The working principle is to measure the surface of the object by heat flux, then the surface temperature of the object is recorded by infrared thermal imaging system and infrared detector. Because of the difference in the geometrical structure and thermal physical property of the object, results in the difference of the surface temperature, so make a judgement for defects.

2. Detection Principle of Infrared Thermal Imager

Infrared radiation is a common phenomenon in nature. It is based on the microscopic view that any object can generate its own molecular and atomic motion without any object at room temperature. The more intense of the movement of molecules and atoms, the greater of the radiation energy is and vice versa. The surface temperature of the object is closely related to the thermal infrared energy emitted by the object. Temperature in the absolute zero of the object, will be due to their own molecular movement and radiation from the infrared. Object radiated power signal is converted into an electrical signal through the infrared detector, the output signal of the imaging device can fully faithful corresponding simulation scan the surface temperature of the space distribution, the electronic system, transmitted to the display, and surface heat distribution corresponding the program is obtained. Using this method, we can achieve the goal of remote thermal state imaging and temperature measurement, at the same time, the state of the object to be measured. The working principle is shown in Figure 1 [11].

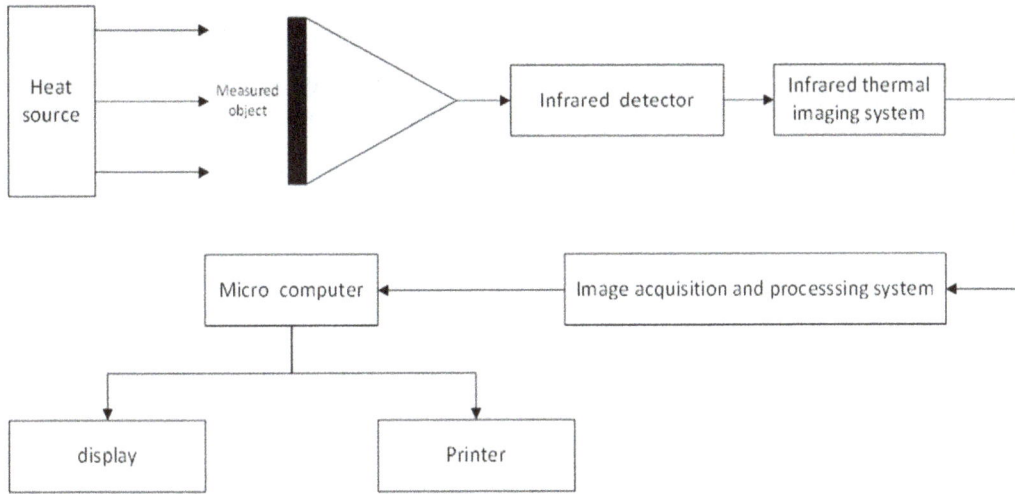

Figure 1. Sketch of infrared thermal imaging system.

Infrared thermal imaging system shows that the thermal image can reflect the thermal distribution and temperature difference of the surface of the detected object. When the internal part of the bonded steel plate is defective, the thermal conductivity of the defect site can be different with the other position due to the external heat source. Description: the thickness of the defect is 1mm.

3. The Test Situation

3.1. Shape and Size of Specimen

In this experiment, 2 groups of specimens were designed, with a total of 12 test pieces. The size of the specimen is as follows: L×W×H=600mm×400mm×300mm, as shown in Figure 2.

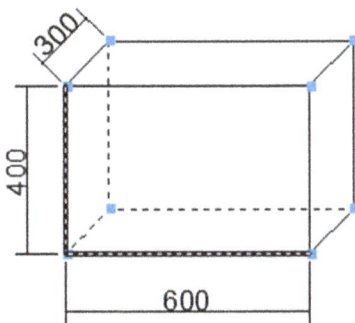

Figure 2. The specimen size.

3.2. Specimen Processing and Design Defects

Grouping of specimen, specimen size and defects were showed in table 1.

After the completion of the concrete pouring, places ahead of prefabricated density for 18 Styrene board in the concrete surface the requirements of benzene plate and concrete surface flush to reserved defects of concrete normal curing 28 days after, dismantling, removing the hard surface of Styrene board in. The production of the defects were showed in Figure 3.

Table 1. Grouping design.

Group	NO.		Size (H×W×L mm)	Defect type	Defect size (L or R mm)	Defect thickness (mm)
1	TIS1-A TIS1-B	2	300×400×600	Square	15/10/6/4/2	1
	TIS2-A TIS2-B	2	300×400×600	Square	15/10/6/4/2	2
	TIS3-A TIS3-B	2	300×400×600	Square	15/10/6/4/2	3
2	TIS4-A TIS4-B	2	300×400×600	Round	15/10/6/4/2	1
	TIS5-A TIS5-B	2	300×400×600	Round	15/10/6/4/2	2
	TIS6-A TIS6-B	2	300×400×600	Round	15/10/6/4/2	3

Figure 3. Specimen making.

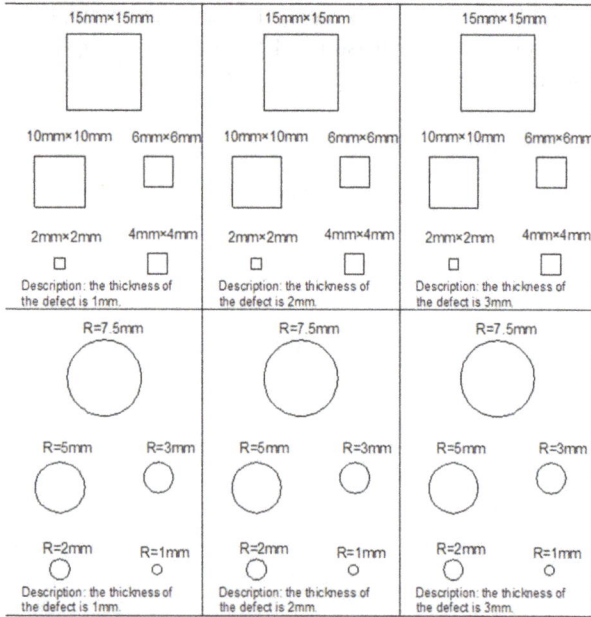

Figure 4. *Sketch of hollowing distribution (mm).*

air, so density hollowing material selection and heat transfer of air performance similar to 8 styrene foam board. The distribution of the defects in each group is showed in Figure 4.

3.3. Strengthening by Gluing Steels

Test using SKO epoxy adhesive, test the steel steps:

(1) the test was carried out by using 6mm steel plate. Steel plate bonding surface with a flat grinding wheel burnish derusting, until metal luster, and then the structural adhesive coated on the plate teach the bonding surface and the plate central position smear glue of a thickness of 3 mm, both sides can be thinne r [12-15]. (2) with blade closely, uniformly were applied in surface cleaning concrete joint surface and plate combined with surface to fully–immerse in the glue on the surface. (3) in order to prevent the plate from sliding on the concrete surface, some fixed measures should be taken, and a certain load is applied to the steel plate to make the structural adhesive on the steel plate and the concrete surface.

Due to the practical engineering, hollowing media mainly is

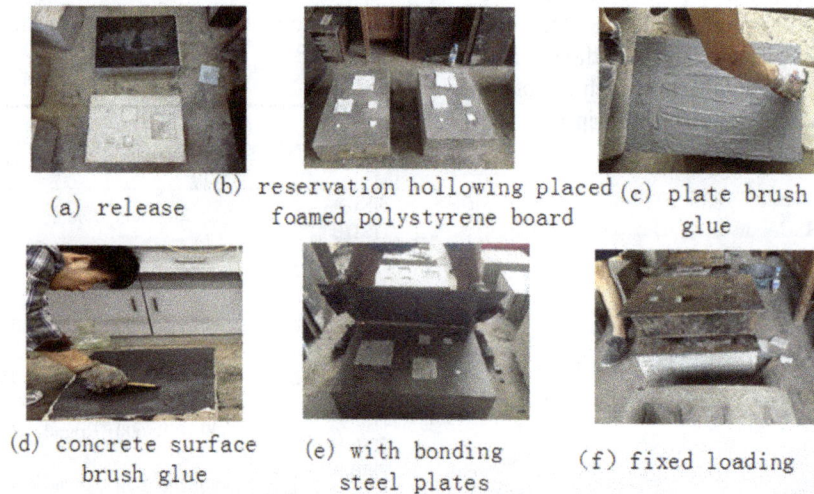

(a) release
(b) reservation hollowing placed foamed polystyrene board
(c) plate brush glue
(d) concrete surface brush glue
(e) with bonding steel plates
(f) fixed loading

Figure 5. *Reinforcement test piece defect production process.*

Test pieces of the production process: (a) release, (b) reservation hollowing placed foamed polystyrene board, (c) plate brush glue, (d) concrete surface brush glue, (E) with bonding steel plates, (f) fixed loading. As shown in Figure 5.

3.4. Experimental Apparatus and Test Method

Experiment instrument for R1 FlukeTi infrared thermal image instrument for measuring temperature range: - 20°C to 250°C, precision: ±0.1°C) handheld in detected paste plate specimen before, adjust the distance, angle and focal length, to the imaging effect best as the standard, while the use of thermometer measuring accident surface temperature. On the specimen surface of continuous observation, every 10 minutes to paste plate filming an infrared thermography and record temperature inspecting instrument on the corresponding readings; indoor heating device for electric

heating fan (power of 1200W to 2000W). Test apparatus and equipment are as shown in Figure 6.

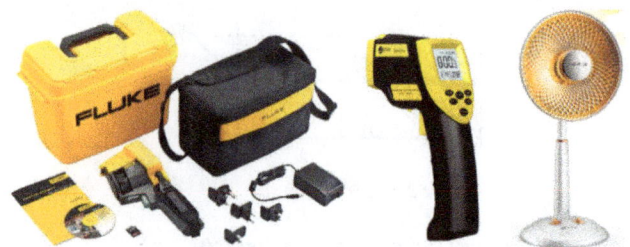

(a) Thethermalimager (b) Thermodetector (c) The heater

Figure 6. *Test equipment.*

Because of the structure of the glued steel specimen is not hot, it is required to apply the active heating to the surface of the steel plate. At present, some scholars have carried out a special research on the method. Usually can be used for high-power infrared led to an external heating source, there are also some scholars by surface cooling method, this principle and the heating is the same, by heat transfer from the plate surface temperature difference is formed [16-19]. Whether the method of heating or cooling, to ensure that the uniformity of heat transfer is essential, this experiment in the indoor heating, so that the heating of the test piece surface of the test piece is heated, the test piece is heated to a certain temperature, the use of cooling method to test pieces of testing.

4. Experimental Results and Analysis

4.1. Thermal Imaging Results

At normal temperature, the TIS1-A is tested, and the results were showed in Figure 7. It can be seen from the figure, normal temperature test edge higher temperature on the other region, the main reason is due to the specimen edges paste without full that hollowing and due to the specimen edges with the surrounding air contact heating temperature is slightly higher. Thus it can be seen that the measured is the ambient temperature which also has certain influence on the specimen, but the main effect of specimen edge position.

The test was conducted in September, the indoor temperature is about 20°C , the test uses the indoor cooling method, first of all,the test pieces of heating, when the temperature of the surface of the specimen is about 80°C , the time is about 50 minutes, and then stop external heating; the test results are not obvious in the 80°C to 60°C range of the test results, when the temperature drops to about 50°C to30°C, the temperature difference between the steel defect area and the non-defect area is beginning to appear.

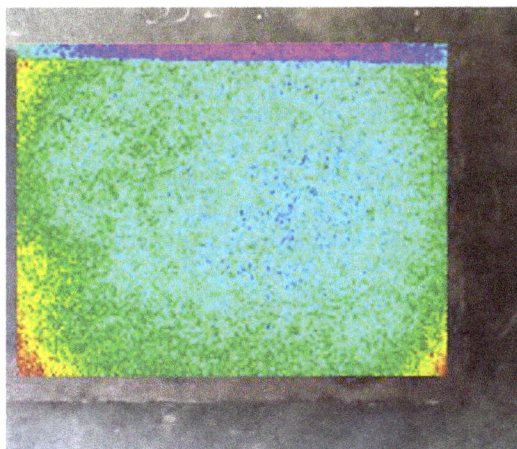

Figure 7. Steel plate surface temperature infrared thermal image map pasting.

In this experiment, the cooling method used to test the temperature is not easy to control, and the operation process is complicated. In order to measure the effect of temperature, the temperature of the sample is measured. The average temperature of the specimen TIS1-A is shown in Figure 8.

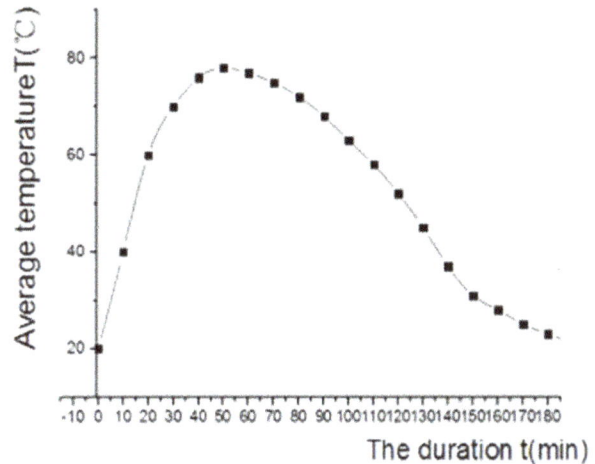

Figure 8. Variation of surface mean temperature field with time.

4.2. Analysis of Factors Affecting Thermal Images

4.2.1. The Internal Defect Size

In this experiment, 5 defects were detected by thermal imaging, which can be detected by thermal imaging, which can be detected in 4. From a large number of thermal imaging analysis results we can see that the same size of the internal defects, the greater the thickness of the surface temperature is higher. This is mainly due to the greater the heat capacity of large area and thickness of hollowing and the cold zone effect more obvious; when the ambient temperature tends to be constant and paste steel plate and concrete specimen reached thermal equilibrium, hollowing part of surface temperature and no hollowing part of the surface temperature did not differ. At the same time cannot detected by infrared thermography to judge hollowing exists. In this test, the test pieces are uniformly reduced to 50°C to 30°C (need to define the optimum temperature range of the test environment). The test results are good, and the infrared thermal image is shown in Figure 9. Thus, the larger the area, the greater the thickness of the internal defects can be detected.

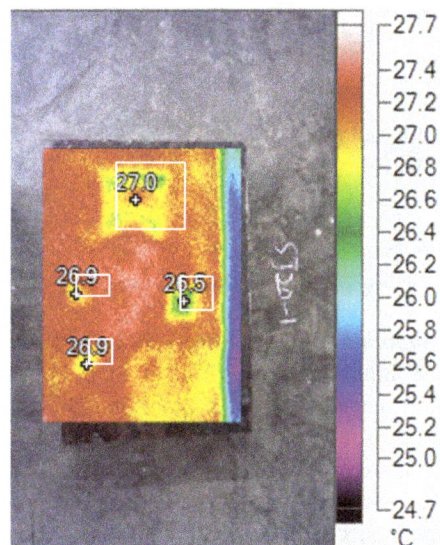

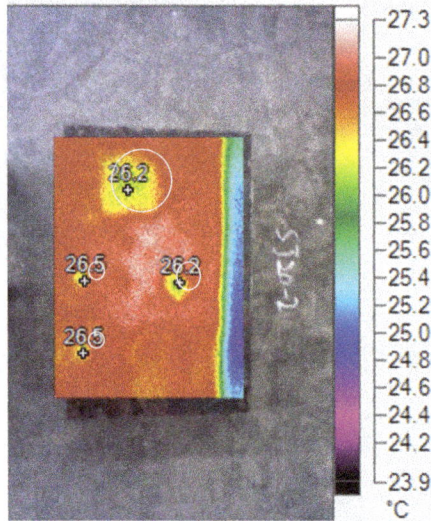

Figure 9. *The infrared thermal image.*

4.2.2. The Size and Thickness of Defects

The infrared thermal images of different sizes and thicknesses were analyzed, and the surface temperature was extracted. Analysis shows that the hollowing part of the infrared image surface is low temperature zone; instead, no hollowing part of the surface is relatively high temperature region, which is mainly due to the large defects of the dissipation of heat faster, and small defects of the local heat accumulation and dissipation slower. According to the imaging image, the position, shape and size of the defect can be estimated, and the approximate area of the defect can be estimated by using the ratio of the differential calculus. When the thickness of the defect is 3mm, the imaging effect is better than that of 1mm and 2mm. At the same time, the effect of thermal imaging is not obvious when the defect area is small (less than 4mm2). Thus, infrared thermography on the detection effect of the internal defects of the larger and thicker hollowing is more obvious.

Figure 10. *Comparison of complete specimens before and after damaged.*

4.2.3. Vertical Shooting Angle of Infrared Thermal Imager

Due to space limitations, the vertical thermal infrared imager in this test the shooting angle of 45° and 30° angle of depression and head up (0°). According to the experimental results, the average temperature of the surface of the sample is three times of the same temperature. By comparison of temperature, the surface temperature distribution of inner defects in the same time is the same, and the surface

temperature decreases with the increase of the angle. In actual engineering, the influence of the vertical angle of the infrared thermal imager on the inner defects of the inner defects in the 45° range is small.

4.2.4. Infrared Emissivity

The emissivity is the ratio of the energy of the body to the radiation of the body surface and the energy of the body at the same temperature. The surface temperature variation of the same sample with the same temperature and the surface temperature of the sample is different. By the graph, the surface temperature increases with the decrease of the emission rate, and the surface temperature at the center of the defect increases with the increase of the speed of the other position. Therefore, in the actual engineering detection, it can be based on the nature of the finish layer material to choose the right infrared thermal imager emissivity.

4.2.5. The Comparison of Test Results

The sticking steel concrete parts in the use of infrared thermography detection, the percussion method and the local damage approach to judge the hollowing finishes parts were carried out to confirm the detection. The results show that the infrared thermography detection result is correct. To further verify the feasibility of the experimental results, after the test was carried out to observe the damage, found that the defect position and the actual location of the defects are basically consistent, and the test pieces are shown in Figure 10.

5. Conclusion

At present, although there have been many studies on the infrared thermal imaging technology, there are great progress in engineering application, but overall speaking, it still need to be improved in many aspects, the infrared thermal imaging method to detect and paste steel plate reinforced concrete quality test research can provide reference for the relevant research and practical engineering application. In this paper, the normal construction process of the 12 groups of 2 interface in the presence of debonding of concrete sticky steel reinforcement members, application of infrared thermal imaging testing technology for the 2 groups of test pieces of non-destructive-testing,and the main influencing factors were compared with the experimental study, the main conclusions are as follows:

(1) it is feasible to detect the quality of concrete structure by infrared thermal imaging method. The method can check the reinforcement of concrete structure surface quality rapid, non-contact, large area scanning, more hollowing layer thicker, hollowing the position range is easier to tell;

(2) the surface temperature increases with the decrease of the vertical angle and the emissivity of the infrared thermal imager, but the effect of the vertical angle of the infrared thermal imager on the detection results can be neglected;

(3) the accuracy of the infrared thermal imager is greatly influenced by the external conditions, so it is suitable

to choose the appropriate time and the suitable environment condition. Therefore, it is needed to study the influence factors and the method to reduce the influence of the thermal imager. At the same time, other methods (such as tapping method) were used to test the results of infrared thermal imaging method.

References

[1]   Datcu S, Ibos L, Candau Y, et al. Improvement of building wall surface temperature measurements by infrared thermography. Infrared Physics & Technology, 2005, 46(6): 451-467.

[2]   Youcef M H A L, Mazioud A, Bremond P, et al. A nondestructive method for diagnostic of insulated building walls using infrared thermography//Proceedings of the SPIE, 2007.

[3]   Titman D J. Applications of thermography in non-destructive testing of structures. NDT&E Intertnational, 2001, 34(2): 149-154.

[4]   Edis E, Flores-Colen I, Brito J. Passive thermographic inspection of adhered cera mic claddings: Iimitation and conditioning factors. Journal of Performance of Constr ucted Facilities, 2012, 27 (6): 737-747.

[5]   ZhuBin, Shen Shaijun. Inspection on adhesion defect of external-wall coating brick with infrared thermography technology Construction Technology, 2009, 38 (SI): 449-151.

[6]   AM Birk, MarkH Cunningham. Thermographic Inspection of Rail-Car Thermal Insulation. Transactions of ASME, 2000, 122(11): 494 -501.

[7]   Maldague X P V. Theory and practice oI infrared technology for nondestructive testing. A Wiley Inter science Publication, 2001: 495.

[8]   Takahide Sakagami, ShiroKubo. Development of a New Non-destructive Testing Technique for Quantitative Evaluations of Delamination Defects in Concrete Structures Based on Phase Delay Measurement Using Lock-in Thermography. Infrared Physics & Technology, 2002, 43: 311-316.

[9]   Sakagami T, Izumi Y, Kubo S. Application of infrared thermography to structural integrity evaluation of steel bridges. Journal of Modern 0ptics, 2010, 57 (18): 1738-1746.

[10] Uhosh K K, Karbhari V M. Use of infrared thermography for quantitative non-destructive evaluation in FRP strengthened bridge systems. Materials and Structures, 2011, 44(1): graphy of cladding debond in solid rockets. Journal of Mechanical Engineering, 2011, 47(2): 9-15.

[11] Feng Liqiang, Wang Huanxiang, Yan Dawei et al. Experimental study on internal defects detection of exterior wallfinish coat by infrared thermography. Journal of Civil Engineering, 2014, 47(6): 51-56.

[12] Huang Pei, Xie Huicai, Yuan Xin. Concrete bonded steel reinforcement method of infrared thermal image quality. Laser & infrared, 2004. 34 (5): 50-353.

[13] Huang Wenhao et al. New method of testing the quality of steel bonded reinforcement structure steel sheets. Construction technology, 2006. 37 (6): 465-467.

[14] Li Xiaogang, Fu Dongmei. Infrared thermal-imaging diagnostic technique. Beijing: China Electric Power Press2006.

[15] JG/T 269-2010 Building infrared thermography detection requirements. Bei jing：Standards Press of China, 2010.

[16] Sun Li, Huan Kewei. The effect of distance on temperature measurement accuracy of the infrared thermal image instrument and correction method. Journal of Changchun University of Science and Technology, 2008, 25 (3): 33-35.

[17] Wang Kai Yan Bogang Zhang Xiyuan et al. Infrared thermograph nondestructive evaluation technique and its application in construction engineering. Concrete, 2015, 5: 154-160.

[18] Li Jiawei, Chen Jimao. Nondestructive testing handbook. Beijing: China Machine PRESS, 2002.

[19] Wu Jiaye, An Xuehui, Tian Beiping. The status and progress of Concrete nondestructive detection technology. Sichuan University of Science & Engineering (natural sicence edition), 2009, 22 (4): 4-7.

22

Evaluating methods for 3D CFD Models in sediment transport computations

Hamid Reza Madihi[*], Sina Bani Amerian

Graduate Faculty of Environment, Tehran University, Tehran, Iran

Email address:
hrmadihi@ut.ac.ir (H.R. Madihi), Sina_Bani Amerian@ut.ac.ir (S. Amerian)

Abstract: Usual uncertainties in computational fluid dynamics (CFD) results include numerical errors, modeling errors, program bugs, mistakes in input parameters and boundary conditions. The errors can be assessed using itemized results from the CFD program together with its documentation. Each uncertainty can be assessed by evaluating the variables against each other, parameter responsiveness examinations and testing for simplified items. The role is made easier if the program is as transparent as possible. This means that the user can see the values of all the pertinent variables and mediatory results for the whole computational domain. Together with an extensive collection of documents of the computer program that includes all the formulas used, the user can be able to find the causes of suspect results, including an estimation of possible bugs. An important tool in the testing of a CFD program is using simplified cases, generally channels with uniform one-dimensional flow.

Keywords: Sediment, Models, CFD, Transport

1. Introduction

Engineering investigations of sediment transport problems have until recent years been done thorough physical model studies. Laboratory modeling have the disadvantage of several scaling problems, for example modeling the interaction between bed load and suspended load, and modeling bed forms. The scaling problems have often necessitated very large laboratory models, which are costly and time-consuming to build. An alternative approach that has evolved with the increased speed of computers in recent years is to calculate bed elevation changes by using of computational fluid dynamics (CFD).

The Navier-Stokes equations are solved in three dimensions together with equations for sediment transport. Empirical formulas are used for critical bed shear stress, roughness estimations, bed form effects etc. The approach complicated and it takes several years to make a computer program capable of carrying out the task. The number of people who have made such programs is therefore limited, and most of engineers involved in sediment transport studies will only use programs others have made. It is technically possible to use the programs without knowing the details of the numerical algorithms. Generating the input data is not always too complicated, and the models will then produce a result, also for an inexperienced user. The main problem is to assess how accurate and reliable the results will be. Some method is needed to address this problem and ensure sufficient quality of the results. Knowledge about numerical modeling theory is essential in this solution. Establishments of best practice guidelines would be very useful.

2. Data and Material

It is often not too complicated to produce results using commercial CFD programs. The main problem is to find out how accurate the results are, and if they are good enough for practical engineering purposes. An organization that has worked on this topic is ERCOFTAC. ERCOFTAC is an acronym for European Research Community on Flow, Turbulence And Combustion [1]. The organization has produced a set of best practice guidelines for CFD. Included is a list of errors and uncertainties in CFD. The list includes the following topics:

- Errors in numerical approximations
- Modeling errors
- Errors in input data
- Programming errors

Best practice would involve an assessment of these errors and uncertainties. Errors in the numerical approximations include false diffusion, arising from a grid that is too coarse and/or inaccuracies in the discretisation schemes. The false diffusion can be investigated by varying the grid size or using different discretization methods. The difference between the results with the varying parameters is an estimation of the errors due to false diffusion. A CFD program should make it easy for the user to vary the grid size and make parameter tests with different grid resolutions. The program should also include multiple discretization schemes with varying accuracy. Modeling bed elevation changes over time involves a time-dependent computation. The total time is each time step. Because iterative solvers are used and computational time may be long, sometimes the iterations are not done until complete convergence for each time step. Then an error will be introduced [1, 2]

Assessment of this error can be done by looking at the residual values over time or the average residual for each time step. The CFD program should therefore produce such information to the user. For most practical engineering cases, the needed input data for the CFD model is not available. Typical data needed for the model is a digital terrain model and spatial variation in bed grain size distribution at the bed. Also, a time series of inflow of sediments together with its grain size distribution can be difficult to find. To assess the errors involved in possible uncertainties in the input data, a parameter sensitivity test can be used. Errors in the computer program are also possible.

Abbreviations such as IEEE, SI, MKS, CGS, sc, dc, and a 3D CFD program with sediment transport may have over 100 000 lines of code, and bugs are therefore likely. Bugs are often difficult to find, but they may show up during a detailed analysis of the computed variables or in a thorough parameter sensitivity test. For example, it is possible to see if the empirical formulas for sediment parameters are coded correctly by viewing all the variables in a cell and computing the results by hand.

The CFD program should therefore provide the user with parameters like the water depth, bed shear stress, grain size distribution etc. Bugs may also show up by giving unphysical results for a parameter. Examples are: erosion at a section with non-erodible bed, negative sediment concentrations, high fractions of fine sediments at the bed in areas of erosion, sediment deposition at the outside of a bend instead of at the inside etc. Possible bugs can be investigated using simplified cases, as described in more detail later.

Modeling errors include the turbulence model, but also all the empirical functions related to sediment transport and bed roughness. The best way of assessing the error is to look at the results with varying parameters in the formulas. This is a standard parameter sensitivity test. There are a number of empirical parameters in the formulas, giving a number of output variables to assess.

Most CFD computations today are done in mechanical engineering, use only one fluid, smooth walls and a fixed geometry. In such a situation, the main processes causing inaccurate results are false diffusion if the grid is too coarse and inaccuracies in the turbulence model. Research on CFD in mechanical and aerospace engineering therefore have focused considerably on the turbulent processes. In using CFD for sediment transport modeling, the main uncertainty is the physics of the sediment processes. The large errors involved in the sediment computations often make small inaccuracies in the turbulence model insignificant.

3. Research Methodology

3.1. Assessing Sediment Transport Results from CFD Models

The goal of a sediment transport computation is often to predict bed changes in a fluvial system. The changes are mostly in the vertical direction, but also lateral erosion can be computed.

The basis of a vertical bed elevation change in a computational cell is the continuity of sediments flowing in and out of the cell. The sediment transport flux is a function of an empirical formula for sediment transport capacity or sediment concentration. This formula is again a function of the particle size on the bed, the bed grain size distribution and the effective bed shear stress. The bed shear stress is a function of the water velocity close to the bed, the water velocity gradient, the turbulent kinetic energy, the bed form sizes and the roughness. The roughness and the bed form height is a function of the grain size distribution, the water depth and the sediment fluxes [1, 2, 10].

The water velocity and the turbulence is a function of the roughness, the geometry of the case, the water level, discharge, the grid and the turbulence model. Another process taking part is secondary currents. Experienced and knowledgeable scientists may also use a CFD program to compute sediment transport and not get the results they see in the field or in the laboratory. If the program is a black box, it is impossible for the scientist to know why the results are incorrect. The present paper proposes features of the program, which allows for transparency of the algorithms used with intermediate results. The user is thereby given the possibility to follow the computations and see where the problem arises. Together with information about the algorithms and empirical formulas used by the program, the user can assess the need to make improvements in the program.

Most CFD programs will provide an output of the computed water velocity field and turbulence parameters. A program computing bed elevation changes must of course also show this variable graphically. A transparent CFD program will also show all other variables that are used in the computation of the bed elevation. In relation to computing the water velocity, the following variables are useful:

- Bed shear stress
- Roughness
- Water depth
- Froude number

The bed shear stress is the most important parameter for the sediment transport computations, so this parameter needs to be shown graphically. Comparing the water velocity field close to the bed with the bed shear stress should show a similar pattern. The shear stress is also a function of the bed roughness, so it is important to both parameters into account in the assessment. The water depth is also an important parameter. Using a structured grid, the aspect ratio of the cells will be unfavorable in shallow areas and can lead to convergence problems. For unstructured grids with a varying number of cells in the vertical direction, this will not be a problem. The number of cells in the vertical direction can also be shown, to check if the values correspond well to the water depths, and that there are not too many cells in shallow areas. The Froude number is important for free surface algorithms. Areas of Froude number above unity indicate possible existence of hydraulic jumps, which often cause problems for these algorithms. Instabilities may be identified by looking at the water continuity [13]. Using the SIMPLE method, the residual for the pressure-correction equation is the water continuity defect for each time step [10]. Looking at this variable in a graphics plot can identify problematic areas of instabilities. Watching the velocity vectors as they, change over time can also give useful information.

For sediment computations, important parameters are:
- Sediment concentration for all size fractions
- Bed grain size distribution
- Bed form height
- Sediment transport capacity
- Thickness of sediment layer

The sediment transport concentration close to the bed should be strongly related to the bed shear stress. It is possible to use hand computations with the empirical formulas to control if the computer program computes the correct values. The bed elevation changes are again related to the sediment concentration in the cells near the bed. A high sediment concentration at the bed will lead to erosion, and a decreasing concentration gradient along the direction of the bed velocity vectors will lead to deposition. If the bed shear stress is high and the concentration is not, then the reason can be that the thickness of the available sediment layer may not be sufficiently high. This can then be checked by looking at this parameter.   Using a non-uniform sediment grain size distribution in a CFD program is most often done by modeling multiple sediment sizes. Then sediment concentrations for each size could be checked. If there is low concentration of one size, this may be caused by the limited availability of this particular size in the bed material in the particular cell. This can be checked by looking at graphics of the grain size distribution. The bed form height affects the sediment transport capacity by increasing the roughness and thereby the bed shears stress. Note the shear stress partition between the friction due to the sediment grains and the bed forms will also affect the effective bed shear stress used in the computation of the sediment concentration.

The sediment transport will be affected by the secondary currents that exist in bends and around obstacles. The secondary current can be shown graphically by displaying both the velocity vectors at the bed and at the water surface at the same time (Fig. 2). The angle between the vectors will be a function of the strength of the secondary current. This angle may also be displayed as a variable in a contour map (Fig. 1). The figure with the bed and surface vectors can also be used to check that the computed velocity field is in the correct direction in bends.

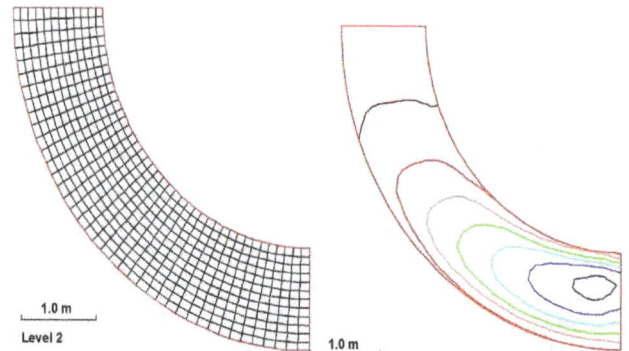

Figure 1. *Plan view of the grid (left) for the bend. The water is flowing in from above and out to the right.The right figure shows a contour map of the secondary current angle. The dark blue value is -14 degrees. The dark red value is – 2 degrees.*

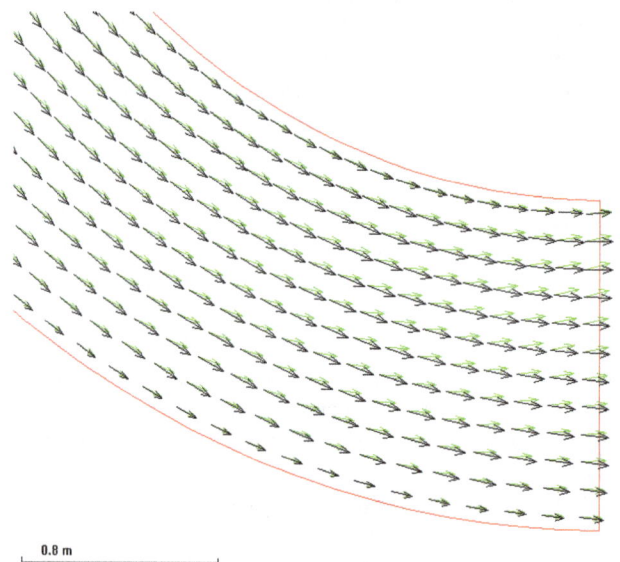

Figure 2. *Velocity vectors at the surface and the bed for flow in a bend, showing the secondary current*

3.2. Testing for Simplified Situations

The case should be designed so that focus is given on one effect, and other processes do not affect the results. An example is if the user wants to find out if the program computes the correct bed shear stress. A straight flume can be modeled with zero gradient boundary conditions on the sides, ensuring a 1D situation. The resulting pressure gradient can then be compared with hand calculations using for example Manning's formula. The same approach can be used to compute the vertical velocity profile and compare it with the logarithmic formula of Schlichting [4]. Similarly, the vertical profiles of turbulence parameters or sediment concentrations can be tested. The 1D case can also be used to test other parameters against empirical formulas, for example the sediment transport capacity, bed form height etc. Simplified geometries are also used for test cases when validating computer programs and numerical algorithms.

Examples are flow over a backward facing step or cavity flow to test the turbulence model. The length of the recirculation zone is then the main testing parameter. A collapsing water column or a wave can be used for testing free surface algorithms Simplified cases do not always have to involve a simplified geometry [3].

A CFD program computing sediment transport will also include algorithms for turbulence, free surface location, roughness effects etc. It is possible to test the accuracy of these algorithms without computing the sediment transport. An example is given in Fig: 1, where the flow in a bend is computed. The water surface algorithm will then give a transverse slope of the channel. This can be compared with simple analytical formulas for the centrifugal acceleration related to gravity.

4. Result and Analysis

Some CFD programs use a primitive turbulence model that contains calibration parameters. An example is the use of a constant-eddy viscosity model for the horizontal directions in some programs solving the shallow-water equations. However, most general-purpose CFD programs will use an advanced turbulence model that does not need calibration, for example the k-epsilon model.

The main calibration parameter for solving the Navier-Stokes equations will then be the bed roughness. This calibration parameter is physical, and is also used in laboratory model studies. Modeling bed elevation changes in rivers, it is possible to use empirical formulas that relate the sediment grain size distribution and the bed form height to the bed roughness. If these formulas were exactly correct, the CFD model would in principle not need any calibration. However, most of the empirical functions describing the sediment flow have large uncertainties and low accuracy. This applies for example for the formulas for critical bed shear stress for multiple grain size mixtures, effect of cohesive forces, bed form height, roughness as a function of grain size distribution and bed form height, and the formula for the sediment transport capacity itself. It is a well accepted fact that

The sediment transport capacity formula will have less accuracy than 30-40 % for many cases [8, 9, 11]. The large number of formulas giving highly different predictions is an evidence for this. When empirical formulas with so large inaccuracies are used in a CFD model, the results will also have similarly large errors, even if the velocity field and turbulence would be theoretically exactly computed [5,6,7]. Then calibration of the empirical parameters may be the only option of ensuring some reliability in the CFD results.

In the future, we can hope that more accurate empirical formulas will be developed. Then it would be possible to compute bed elevation changes in rivers without calibration. This would be a great advantage for cases assessing the impact

of future man-made impacts in rivers, where calibration data do not exist.

5. Discussion and Conclusions

Three-dimensional numerical models for sediment transport should be designed as transparent as possible, providing the user with graphical output of all relevant parameters. Also, a thorough documentation of all equations and empirical formulas used by the program should be given, enabling the user to assess the computational procedures of the program to see possible bugs.

When testing numerical model results against observed bed elevation changes, a procedure should be used where the user look at the values from all the parameters and checks the empirical formulas used in the program. Starting from the velocity field to the bed elevations changes, the parameter distribution in the geometry should be investigated to see if it is reasonable. How each variable changes over time can also give very useful information.

When using 3D CFD model to predict bed elevation changes for a situation where no calibration data exist, best practice should be to investigate all uncertain parameters and formulas in parameter sensitivity tests. A large number of parameters should be tested, from grid size, discretization schemes, turbulence models, values for critical bed shear stress parameters, empirical coefficients in the sediment transport capacity formula, parameters in bed form and roughness prediction formulas etc. The test will give the total accuracy of the results and expose influence of each parameter on the final result.

If the user is in doubt whether the CFD program computes the correct result, a testing should be done for a simplified case, focusing on the potential problem. A typical simplification is to model a straight flume with uniform, steady flow using a very coarse grid. Such a case can expose errors in the computation of the shear stress on the bed or the sediment transport formula. If the problem is believed to be in the free surface or water flow part of the program, then the simplified case could be run without sediments.

The main uncertainty in today's prediction of bed elevation changes using 3D CFD models is the empirical formulas for the sediment transport processes. Future research should therefore focus on improving the accuracy of these formulas. This is an area where physical model studies will be needed for many years to come.

References

[1]   Baranya, S. and Jozsa, J.) "Morphological modeling of a sand-bed reach in the Hungarian Danube", Proceedings of the 33rd Congress of the International Association of Hydraulic Engineering and Research, Vancouver, Canada, 2009.

[2]   T. Fischer-Antze, N. R. B. Olsen. and D.Gutknecht, "Three-dimensional CFD modeling of morphological bed changes in the Danube River", Water Resources Research, 44, W09422,doi:10.1029/2007WR006402,2008.

[3]   T.Fischer-Antze,N., Ruether,.,N.R.B Olsen, and D .Gutknecht,. "3D modeling of non-uniform sediment transport in a channel bend with unsteady flow", Journal of Hydraulic Engineering and Research, Vol. 47, No. 5, pp. 670-675, 2009.

[4]   H. Schlichting, "Boundary layer theory", McGraw-Hill, 1979.

[5]   M.Abboltt, An introduction to the Method of characteristics, Thames and Hudsun, 1966.

[6]   M.Abotte, Computational Hydraulics, Pitman, 1979.

[7]   F.Henderson,Open Channel Flow, Macmillan,1996.

[8]   W.Gray and G.. Pinder, On the relationship between the FEand FD method, Int.j.Num. Methods Engng, Vol.12, No.9, 1976.

[9]   S.Nakamura, Computational   Method Engineering and Science, Wiley,1997.

[10] A.Raudkivi and R Callander, Advanced Fluid Mechanics, Arnold, 1975.

[11] R.H Gallagher(ED).Finite Element Techniques for Fluid Flow, Newnes Bruteerworths,1976.

[12] R. Hamming, Introduction to Applied Numerical Analysis, McGraw-Hill,1971.

[13] E.Isaaacson and H. Keller, Analysis of Numerical Methods, Willy, 1966.

23

Effects on Environment and Health by Garments Factory Waste in Narayanganj City, Dhaka

Md. Masud Alom

Department of Civil & Environmental Engineering, Uttara University, Dhaka, Bangladesh

Email address:

masud_ruet50@yahoo.com

Abstract: Bangladesh has more reputation for Readymade Garments in the world and which plays a significant role in the economical enhancement as well. Most of the garments factories in Bangladesh are located at Gazipur and Narayanganj industrial area, Dhaka. But in terms of pollution, the garments factory has been disgraced as being one of the world's most perpetrators. Now a day's environmental problems are completely anchored in our ways of life. The main goal of this study is to find out the environmental and social problems which arise from garments waste of Narayanganj city and propose some ways of mitigation measures. This is done by analyzing and observing numerous data acquired from field survey. The field surveys were conducted among 100 general people near the study area, some VIP's and some NGO's. Findings from this study are that, the waste management systems are progressing very softly. So the garments factory need to improve the waste management system immediately and NGO side to help to raise the environmental condition.

Keywords: Garment Waste, Environmental Effects, Social Problems, Waste Management, Mitigation Measures

1. Introduction

Bangladesh is a developing country. Between 2004 and 2014, the averaged GDP growth rate is 6%. Economic development of this country is depend on firstly on agriculture and secondly on industry. Although Bangladesh is not rich in industry, it has been enriched in Garment industries in the recent past years, which is a promising step. At present Bangladesh is the third largest garment manufacturer and exporter country in the world [8]. Narayanganj is a city in central Bangladesh. It is located in the Narayanganj District, near the capital city of Dhaka and has a population of 29, 48,217 [2]. The city area of Narayanganj is 760 km^2. The city is on the bank of the Shitalakshya River. The river port of Narayanganj is one of the oldest in Bangladesh. It is also a center of business and industry, especially the jute trade and processing plants, and the textile sector of the country. As most of the garments industry are situated here and the proper waste management system are not followed by the garments factory, the environmental condition of this area are going to be worst situation day by day. Environment management system and

policy impersonation are still far away from being workable. Policies of environment have often been lack ofcongruence among environmental acts, antithetical interests at several levels of the propulsion and resources available to environmental institutions to carry out their responsibilities. Solid waste disposal possesses a greater problem because it leads to land pollution when openly dumped, water pollution when dumped in low land and air pollution when burnt [1].

2. Objectives

- Find out the waste generation source of Garments industry in Narayanganj city.
- To investigate the environmental effects and health hazard of the people who are living near the study area.
- Provide some recommendations for solving the existing problems.

3. Background of the Study

The numbers of garments are increasing significantly in Narayanganj city due to heavy inrush of immigrants from rural area. In general, environmental scenario of this area is too hazardous for garments waste. At present there are about 6500 garment industries in Bangladesh and 75% of them are in Dhaka. The rest are in Chittagong and Khulna. These Industries have employed 5000000 of people and 85 percent of them are letter less rural women. About 76 percent of our export earning comes from this sector.

According to Jaspal Singh, textile processing are very important and major stages which are from raw material fiber to yarn and then fabric until at the end of ready products [5]. In order to understand the process of textile production it is necessary to get familiar with key process of entire textile making. There are some phases like carding, spinning, warping and weaving these four steps are mostly important [3]. In general, in order to improve entire process of textile wastes, under the technology prospective managing and utilization of whole process is necessary for efficient [6]. Reserve logistics are the terms of activities which are included in the field of waste management. It is the methods of implementing, systematizing the efficiency, making cost-effective flow of raw materials, updating and upbringing information within logistics activities, which are connected to each other such as process inventory, finished goods and other information from the point of origin to the point of expenditure and other modes of disposal. Waste management is important for the processing of logistics comes into the category of green logistics; the process which manages all the activities at least cost is termed as green logistics.

In the 1980's the environmental debate gained momentum as methods such as the precautionary principle were being developed and implemented against a rising tide of litigation against organizations possible for environmental degradation [4]. Environmental management practices are often promoted as a cost-cutting opportunity but for many managers this is not a reality. the pressure of implementing green centered around structures such as environmental auditing, impact assessment and accreditation, which has high [10].

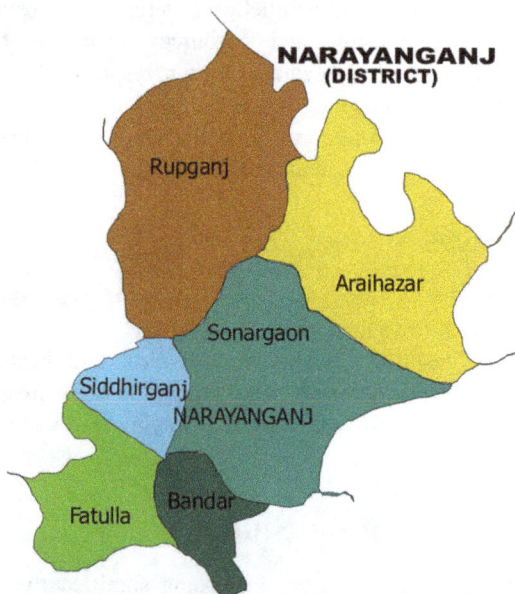

Figure 1. Narayanganj District Map.

4. Methodology

A methodology is the systematic, theoretical analysis of the methods applied to a field of study, or the theoretical analysis of the methods and principles associated with a branch of knowledge. Methodology cannot provide solutions but offers the theoretical underpinning which method can be applied to a specific case.

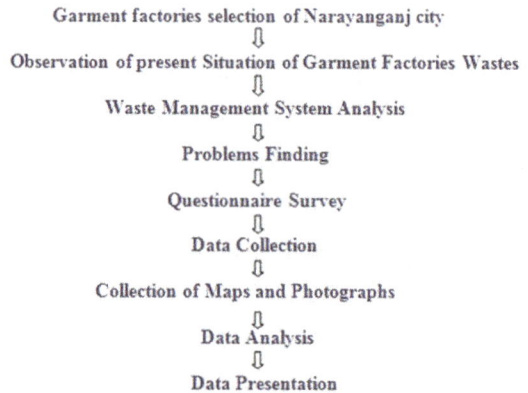

Figure 2. Flow Chart of Methodology.

4.1. Selection of Garment Factory

In Narayanganj city significantly the numbers of garments are increasing everyday due to heavy influx of migrants from rural area. At present there are about 6500 garment industries in Bangladesh and 75 percent of them are in Dhaka. The rest are in Chittagong and Khulna. In this study 10 effective Garments factories are selected from Narayanganj city, which are: DNV Clothing, NR Knit Composite, Pioneer Sweaters & Knit wears Unit-1, Creative Wool wear, Sharoms Samsons Winter Wear, Shore to Shore Textiles, Young 4 Ever, Space Sweater, Aboni Knitwear, Shasha Garments.

4.2. Present Situation of Garment Factories Wastes

Predominantly, Narayanganj city is called an industrial city. Most of the textile waste originates during yarn and fabric propagation, processes of garment-making and from the retail industry. According to Jing, waste is one kind of discarded materials which have to discarded, in order to make new raw material for reuse for new products [7]. The extensive amount of industrial waste such as, polyethylene, cloth and papers are generated here daily. Narayanganj Pourashava authority is the only responsible organization for waste management in this area. Everyday 120 to 125 tons waste are generated in this area [9].

Table 1. Type of Garments Waste.

Types of waste	Total waste (%)
Cutting	59%
Dyeing	21%
Knitting	13%
Sewing	3%
Others	4%

(Source: field survey)

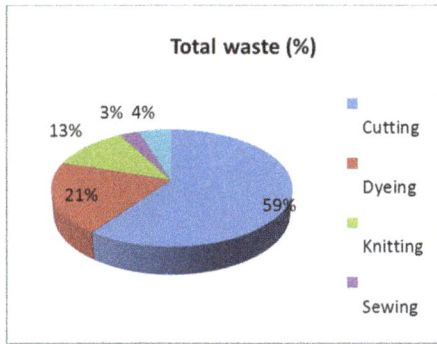

Figure 3. *Status of Garments Waste.*

4.3. Waste Management System

The compositions of various wastes have varied over time and location, with industrial development and which is directly linked to waste materials. Some of the components of waste have beneficial value and can be recycled once correctly recovered. Proper management of waste can be reduced the negative impacts on environment and society. Waste management system can be divided into five key components which are: i) Generation, ii) Storage, iii) Collection, iv) Transportation and v) Disposal of waste. In European textile waste management, there are important activities which are included in the field of waste management. thus the term is called reserve logistics, reverse logistics is the process of implementing, controlling the efficiency, making cost effective flow of raw materials, updating and keeping information within logistics activities which are interlinked to each others such as, process inventory, finished goods and other information from the point of origin to the point of consumption and other process of disposal. It is important the applications of logistics in textile waste management, when the process of logistics comes into the category of green logistics; "green logistic" is the process which manages all the activities at minimum cost of process.

4.4. Problems Finding

Various problems are found by this study which is created by the garments waste. Some kinds of problems are environmental and some are human health related. From the observation of questioner survey it is found that, by improper garments waste management, it affects: i) human health ii) Socio-economic conditions iii) Coastal and marine environment iv) Rivers and also v) Affects soil by land filling. The opinions of the people whose are facing different types of problems around the study area, obtained by field survey are listed below.

Table 2. *Types of problems around the study area.*

Types of Problem	Opinion of total people by (%)
Health Risk	60%
Insects	11%
Toxic substances	19%
Odour	10%

(Source: field survey)

4.5. Questionnaire Survey and Informal Interview

A field survey as questionnaire survey and open discussion has been conducted with the authorities of different concerned organizations, experts and people living near the study area for evaluating the waste management system of garment factories which affect on Environment and health of general people. The questionnaire was designed in such a way that it would track down the problems of environment and general people which help us to sort out some recommendations. About 100 respondents were interviewed in different zones near the garment factories. Informal interview of official experts of different development organizations and some VIP's were also done in order to know their views of the waste management system which affects on environment of Narayanganj city.

4.6. Data Collection and Other Secondary Data

To fulfill the objectives of this study both primary and secondary data were needed. All the necessary data has been collected from various sources. We have collected these data's through 3 different processes, i) Questionnaire survey with general people, ii) Informal interview with different organization and iii) Informal interview with different garment's VIP persons. The data's are collected from the organization's who works for the welfare of Environment and from the VIP persons who are experts of their fields and can give a valuable opinion about this environmental study.

4.7. Collection of Photographs

Figure 4. *Garments factory waste generation.*

Lots of photographs were collected to illustrate the waste management system of the garment factories in Narayanganj city. Some of the photographs have been collected directly from the field survey and some others from daily newspapers as well as from internet website.

4.8. Data Analysis and Presentation

All the collected data from field survey were putted in a spread sheet and analyzed separately according to the respondent opinion and finally all the analyzed data (%) have been integrated and presented as tables and graphs and putted in the report.

5. Recommendations

As a conscious citizen of any country, everyone have some common dreams like neat, clean and poverty free society. Government cannot bear this responsibility alone. For fulfilling these dreams, we have to take some initiatives together and share these obligations each other. For keeping the locality clean, City Corporations already start to organize their own waste collection services. Following steps can help the authorities for extensive management of Garments waste and reduce the enduring of the city dwellers from physical, social and environmental point of view

- Government and owner of garments factories should conscious about environmental Act and try to follow the rules and regulations strictly.
- Effluent treatment plant (ETP) installation can be mandatory for all garments industry for decreasing the toxicity of the produce waste.
- Where the amount of waste is more, the number of bins should be increased and placed at proper place of garments side.
- For quick transport of waste the vehicles such as Container carrier, Compactor, Arm roller, Van etc. should be on proper operation and if necessary, more vehicles need to be included.
- Can be organized effective training programe about health & hygiene as well as overall environment among the employees of the factory.
- Properly follow the systematic procedure of waste disposal and ensure while handling the wastes, workers must use their safety equipment like mask, hand gloves, boot etc.
- The NGOs- Government partnership should be exhibited and promote the citizen monitoring mechanism to proper evaluation, efficiency and effectiveness of national as well as foreign aid program in the sector of environment.
- The industries like Textile and others harmful factories can be transferred outside the City and proper drainage system should be ensured by the government.

6. Conclusion

In industrial areas, the local environment and peoples are influenced by both chemical and organic pollutants, because of textile dyeing industries. From this study it is found that, large quantities of effluent are discharging by the textile dyeing industries in Narayanganj City. These can be highly toxic and composed of various physicochemical pollutants at significant upper level than standard value of department of environment. Some of the industries have authentic effluent treatment plant (ETP) for treating waste water. As most of the factories specially dyeing industries are throwing their waste directly to Shitalakkhya River without treatment, the water of this river is getting highly polluted. Most of the people living near this area are facing different health related problems. The increasing number of textile dyeing industries in this area, the concentration of these pollutants is increasing in an alarming rate. So it is necessary to take initiatives immediately to minimize the pollution to a significant extent and reduce health hazard problems. That is the way to increase environmental benefits for future generation.

References

[1] Akter NR, Acott E, Sattar MG, Chowdhury SA (1997). Environmental Investigation of Medical W aste Disposal at BRAC Health Centre's. BRAC, Research and Evaluation Division, 75 Mohakhali, Dhaka 1212, Bangladesh. pp. 16-18.

[2] BBS (2011). Bangladesh Bureau of Statistics Bangladesh. Community Report: Nayarayanganj, Dhaka, Bangladesh.

[3] Bhushan (2009, 165). The process and production of textile. Viewed 02.02.2015

[4] Glasson, J. Therivel, R. and Chadwick, A.(1999) Introduction to Environmental Impact Assessment: Principles and procedures, process, practice and prospects, 2nd edition, UCL Press, London

[5] Jaspal Singh (2009) Textile processing. Viewed 02.02.2015.

[6] Jing, Z. (2012, 7) an analysis of textile waste management. HAMK University of applied sciences. Supply Chain Management Degree program. Bachelor's Thesis.

[7] Jing, Z. (2012, 9) an analysis of textile waste management. HAMK University of applied sciences. Supply Chain Management Degree program. Bachelor's Thesis.

[8] Kakuli A. and Risberg V., 2012. A lost Revolution? Enpowered but trapped in poverty. Women in the garment industry in Bangladesh want more, Swedwatch report #47.

[9] Narayanganj Pourashava. (2009). Rough estimate of Solid Waste Collection Database.

[10] Tzschentke, N., Kirk, D. & Lynch, A. (2008) "Going green: Decisional factors in small hospitality operators" International Journal of Hospitality Management, Volume 27, p. 126-133.

24

Reviewing the FRP Strengthening Systems

Seyyed Mohammad Banijamali[1, *], Mohammad Reza Esfahani[1], Shoeib Nosratollahi[2], Mohammad Reza Sohrabi[2], Seyyed Roohollah Mousavi[2]

[1]Dept. of Civil Engineering, Ferdowsi University of Mashhad (FUM), Mashhad, Iran
[2]Dept. of Civil Engineering, University of Sistan and Baluchestan (USB), Zahedan, Iran

Email addresses:

Banijamali.sm@gmail.com (S. M. Banijamali), Esfahani@um.ac.ir (M. R. Esfahani), Shoeib.nosratollahi@yahoo.com (S. Nosratollahi), Sohrabi@hamoon.usb.ac.ir (R. Sohrabi), S.R.Mousavi@eng.usb.ac.ir (S. R. Mousavi)

Abstract: Several methods have been invented for flexural strengthening of RC beams using the FRP materials in recent years. These techniques are categorized into two main groups: Externally Bonded Reinforcement (EBR) techniques and Near Surface Mounted (NSM) techniques. The EBR family contains EBR (with conventional surface preparation), EBROG, EBRIG, MF-EBR, HOLING methods. The NSM family contains NSM-FRP Rods, NSM-FRP Sheets, NSM-MMFRP Rods methods. The EBR family techniques Despite the ease of implementation have weaknesses such as vulnerability against sever environmental conditions. Although the NSM techniques, have longer installation time than the EBR family techniques but in these methods the strengthening materials are greatly protected against the environmental effects. In this paper, the various techniques for flexural strengthening of RC members from both the EBR and NSM families have been fully described and advantages and disadvantages of each technique have been discussed.

Keywords: Externally Bonded Reinforcement, Near Surface Mounted, Fiber Reinforced Polymers

1. Introduction

Replacing the existing structures with the new structures is often not economically cost effective, therefore finding an appropriate solution for repairing and strengthening structures has a great importance. In comparison to building a new structure, strengthening an existing structure is often more complicated, since the structural conditions are already set. In addition, it is not always easy to reach the areas that need to be strengthened. The improper choice of an inappropriate repair or strengthening method can even worsen structure`s function. Traditional methods have been used as strengthening techniques for concrete structures, such as: different kinds of reinforced overlays, shotcrete or post-tensioned cables placed on the outside of the structure which normally need much space [1, 2].

The use of Fiber Reinforced Polymers (FRP) in the last few years in various engineering applications, forums and configurations offers an alternative design approach for the construction of new concrete structures [3] and rehabilitation of existing ones [4]. The key properties that make FRP materials suitable for structural strengthening are their non-corrodible nature and high strength-to-weight ratio. Hence extensive researches on the development and use of these materials have been or are being conducted. The result of these researches is the emergence of different techniques, such as: EBR1, EBROG2, EBRIG3, MF-EBR4, NSM5-FRP rods, NSM-FRP strips, NSM-MMFRP 6rods and HOLING methods for strengthening concrete structures. In this paper each of these techniques will be introduced and their advantages and disadvantages will be discussed.

2. The EBR method

2.1. General features

Externally Bonded Reinforcement (EBR) technique is the

---

1 Externally Bonded Reinforcement
2 Externally Bonded Reinforcement On Grooves
3 Externally Bonded Reinforcement In Grooves
4 Mechanically Fastened and Externally Bonded Reinforcement
5 Near Surface Mounted
6 Manually Made Fiber Reinforced Polymers

most common method to strengthen reinforced concrete structures. In this method, after surface preparation, FRP sheet is adhesively bonded to the tension face of the concrete member. The purpose of surface preparation is to remove contamination and weak surface layers and polish the concrete surface to promote its adherence capacity. However, the main deficiencies in the performance of EBR technique are high possibility of brittle failure mode that is mostly due to premature debonding of FRP sheet from the concrete substrate and vulnerability of FRP materials against the environmental conditions [5, 6]. Provision of adequate end anchorages at the ends of the plates and at critical sections along the span such as: U, L and X shaped wrappings and various surface preparation processes such as: surface roughening by sandblast, water jet and air jet, can slightly postpone the debonding phenomenon [7].

"Fig.1" shows a strengthened cross-section with EBR method.

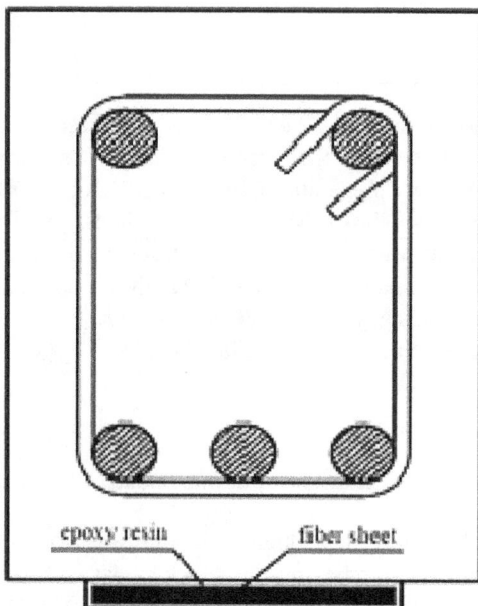

Figure 1. A strengthened cross section with EBR method

2.2. Advantages

a) Quick and easy installation
b) Low performing costs
c) Immediate use of strengthened structures
d) No need to specific labor skills

2.3. Disadvantages

a) Brittle failure mode due to premature debonding of FRP sheet from the concrete substrate
b) Vulnerability of FRP materials against the environmental conditions such as: freeze/thaw cycles, abrasion, mechanical impacts, acidic and alkaline environments, fire, vandalism and UV radiations.
c) Changes caused in the appearance of the structure

3. The EBROG method

3.1. General features

Mostofinejad and Mahmoudabadi [8] invented the grooving method (GM) to postpone or even eliminate the debonding of FRP sheets. This technique was later renamed to Externally Bonded Reinforcement On Grooves (EBROG) method. In this method first grooves are cut onto the tension side of the concrete member. These grooves are then cleaned by the air jet and filled with the epoxy resin. FRP sheets are later installed on the concrete surface saturated with the epoxy resin and the resin in excess is removed. Based on their experimental studies, longitudinal grooves are effective than transverse and diagonal grooves, and can be used as an alternative to conventional surface preparation methods [8].

"Fig.2" shows a strengthened cross section with EBROG method.

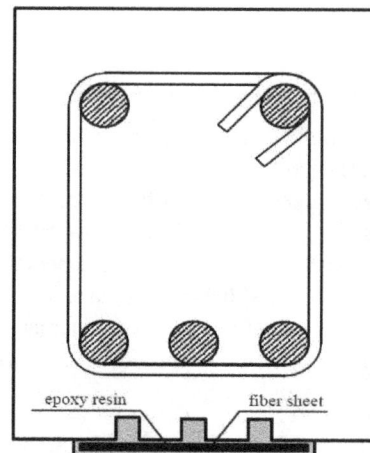

Figure 2. A strengthened cross section with EBROG method

3.2. Advantages

a) Increase in the flexural capacity up to 80 % more than reference specimen
b) Achieving the higher ultimate strain
c) Postponing or even eliminating the debonding phenomenon
d) Immediate use of strengthened structure

3.3. Disadvantages

a) Time-consuming installation process
b) Costly performing
c) Vulnerability of FRP materials against the environmental conditions
d) Environmental pollution caused by the grooving process
e) Change caused in the appearance of the structure

4. The EBRIG Method

4.1. General Features

Mostofinejad and Shameli [9] invented an improved grooving method by penetrating the FRP sheets used in the

EBROG method into the grooves. This technique was named Externally Bonded Reinforcement In Grooves (EBRIG) method. Due to the fact that EBRIG technique provides larger contact area between FRP and the underlying concrete layer compared to EBROG method, this technique significantly promote the structural performance, modify crack initiation and propagation, and increase the failure limits. Based on their result the EBRIG method is effective than EBROG and EBR method when multilayer FRP sheet are used [9].

"Fig.3" shows a strengthened cross section with EBRIG method.

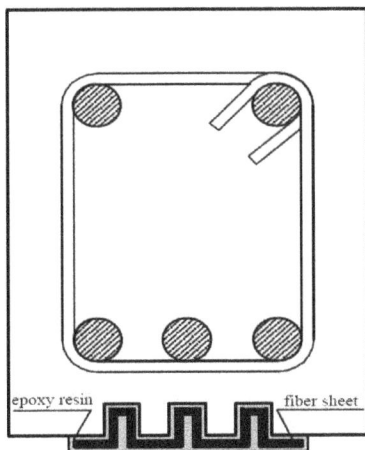

Figure 3. A strengthened cross section with EBRIG method

4.2. Advantages

a) Achieving the ultimate loads higher than the previous methods (up to 142% increase in ultimate load compared to the reference specimen)
b) Achieving the higher ultimate strain
c) Completely eliminating the debonding phenomenon in specimens strengthened with only one layer of FRP sheets.
d) Immediate use of strengthened structures

4.3. Disadvantages

a) Time-consuming installation process
b) Costly performing process
c) Vulnerability of FRP materials against the environmental conditions
d) Environmental pollution caused by the grooving process
e) Changes caused in the appearance of the structure

5. The MF-EBR method

5.1. General Features

Sena-Cruz et al [10] proposed the Mechanically Fastened and Externally Bonded Reinforcement (MF-EBR) method. This technique is based on the MF-FRP method and combines the mechanical Fastening of MF-FRP method with the external adhesion in EBR method. Moreover, all of the anchors are also pre-tensioned [10, 11]. The strengthening process in-volves the following main steps:

a) Holes of specific diameter depth are made in the soffit of the beam. The holes are cleaned using compressed-air and a steel brush.
b) The holes are filled with the chemical adhesive, and the fasteners are then inserted up to a specific depth.
c) A rough concrete surface is assured using a rotary hammer with a needle adapter. Compressed-air is used to clean the final surface.
d) A transparent acrylic strip is used to mark the fasteners position and, then, the holes in the laminates are executed. The laminates are cleaned with acetone.
e) Epoxy adhesive is applied on the treated area in the concrete surface and on the laminate surface that will be in contact.
f) The laminate is placed on the concrete surface and pressed against it to create a uniform thickness of 1–2 mm.
g) The adhesive in excess is removed and the fasteners are cleaned from any dirt attached.
h) h. The pre-defined pre-stress level is applied after the curing time of the epoxy adhesive.

"Fig.4" shows a strengthened cross section with MF-EBR method.

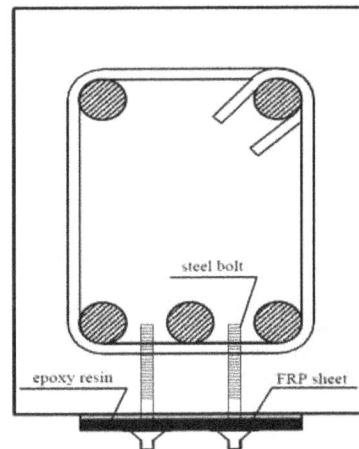

Figure 4. A strengthened cross section with MF- EBR method

5.2. Advantages

a) No need for special labor skills
b) No need for surface preparation
c) Increase in the load carrying capacity up to 87% compared to that of reference specimen
d) Increase in the ductility index compared to that of EBR methid
e) Immediate use of strengthened structures

5.3. Disadvantages

a) The larger initial cracks due to fasteners in high strength concretes
b) Low efficient stress transferability between the FRP and the concrete due to the separated connection points
c) Vulnerability of FRP materials against the environmental

conditions

d) Changes caused in the appearance of the structure

6. The HOLING Method

6.1. General Features

Eftekhar and Yaqubi [12] suggested a new technique as an alternative method for conventional surface preparations. This technique was named HOLING method. In this method first holes of specified depth, diameter and distance from each other are drilled onto the tension side of the concrete member. These holes are then cleaned by air jet and filled with the epoxy resin. FRP sheets are later installed on the concrete surface saturated with the epoxy resin and the resin in excess is removed. This technique provides larger contact area between the concrete and epoxy resin as well as cylindrical anchors at the interface of FRP sheet and concrete. These cylindrical anchors transfer the interfacial stresses between FRP and concrete surface to the strong underlying concrete layers. This technique can postpone the debonding phenomenon and increase the ultimate load and strain [12].

"Fig.5", shows the holes drilled on the tension side of a beam in HOLING method.

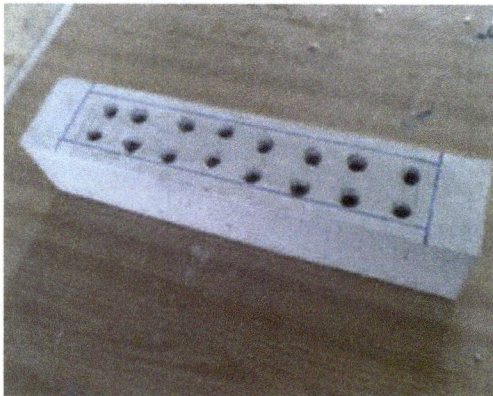

Figure 5. Holes drilled on the tension face in HOLING method

6.2. Advantages

a) Increase in the flexural capacity up to 30 % more than EBR method

b) Increase in the ductility index and energy absorption compared to the EBR method

c) No need for special labor skills

d) Immediate use of strengthened structures

6.3. Disadvantages

a) Premature debonding of FRP sheet from concrete surface

b) Time-consuming drilling process

c) Costly performing process

d) Vulnerability of FRP materials against the environmental conditions

e) Environmental pollution caused by the grooving process

f) Changes caused in the appearance of the structure

7. The NSM method

7.1. General Features

As mentioned in the section 2, the EBR technique is the most common method to strengthen reinforced concrete structures. In this method, due to premature debondng of FRP sheet, the tensile capacity of FRP materials is not completely used. To overcome this defect, modifications were carried on this technique that led to invention of more efficient methods such as EBROG, EBRIG, MF-EBR and HOLING techniques. Although these method postpone or even in some cases completely eliminate the debonding phenomenon but all of them have an important problem. The problem is that the strengthening materials in these techniques are laid on the external surface of the concrete member and are vulnerable against sever environmental conditions. To overcome the mentioned weaknesses, several efforts have been done and one of the most successful methods is the near surface mounted (NSM) method [13].

The NSM method consists of the following steps:

a) Cutting grooves of specified dimensions at designed locations using a diamond blade cutter.

b) Cleaning the grooves using compressed air to remove debris and dust to ensure proper bonding between the epoxy adhesive and concrete.

c) Placing the epoxy paste into the grooves to fill 2/3 of the groove depth.

d) Inserting the strengthening materials (rods or strips) in the groove and lightly pressing it to displace the adhesive, ensuring that the space between the rod or strips and the sides of the groove is completely filled with epoxy without any voids.

e) Placing additional epoxy paste to ensure that the grooves are completely filled and then levelling the surface by removing excessive adhesive.

f) Curing the members at room temperature for at least 5 days [14].

"Figs.6, 7" show two strengthened cross sections with NSM method using FRP rods and FRP strips.

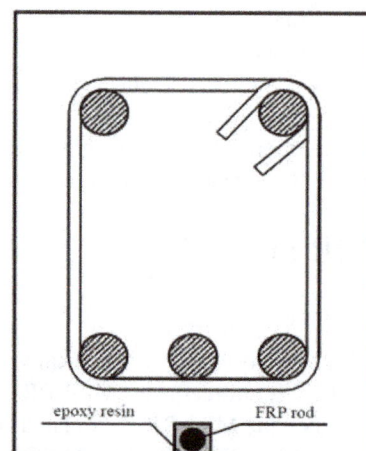

Figure 6. Application of FRP rod in NSM method

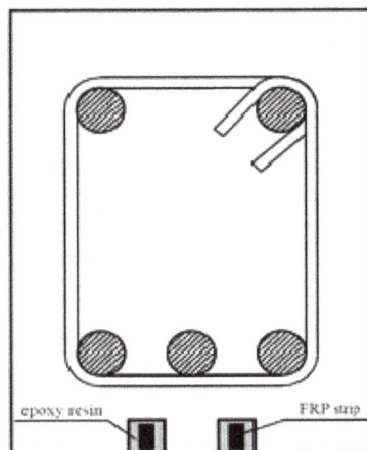

Figure 7. Application of FRP strips in NSM method

7.2. Advantages

a) Reduction in the strengthening operation because of no need for the surface preparation
b) Postponing or even eliminating the debonding phenomenon
c) Possibility of using this technique in the negative moment region of flexural frames because of preserving FRP materials against abrasion
d) Allowing easier pre-stressing strengthening materials
e) Preserving strengthening materials against sever environmental conditions such as: mechanical impacts, abrasion, fire, freeze/thaw cycles, vandalism and UV radiations
f) Possibility of using various types of FRP materials such as: FRP rods, FRP strips and hand-made FRP rods

7.3. Disadvantages

a) Costly and time-consuming installation process
b) Environmental pollution caused by the grooving process

8. Application of MMFRP rods in the NSM method

8.1. General features

The Manually Made Fiber Reinforced Polymers (MMFRP) rods are made from FRP sheets wrapping around a wooden rod. They provide a larger perimeter with respect to conventional FRP bars for the same amount of fiber as the core of the MMFRP rod consists of a low strength wooden rod, providing larger bond circumference and thus potentially higher bond strength. The circular shape of MMFRP rod is not only convenient for production but also suitable for NSM shear strengthening as noted by Rizzo and De Lorenzis [15]. The key advantage of the MMFRP rod, however, is that it allows the incorporation of a novel anchor system that can be used to improve the performance of NSM MMFRP reinforcement for shear strengthening of RC beams with low FRP percentage. This cannot be easily achieved using conventional pultruded

FRP bars or strips [16].

"Figs.8, 9, 10" show a strengthened cross section with NSM-MMFRP rod, the manufacturing procedure of MMFRP rod and it's anchorage system respectively.

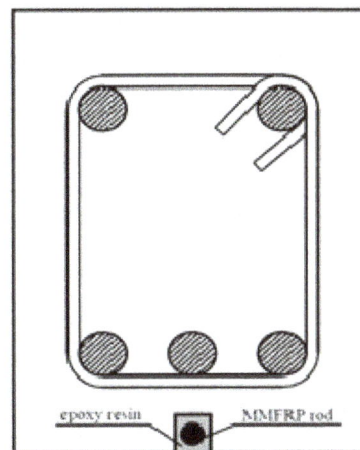

Figure 8. Application of MM FRP strips in NSM method

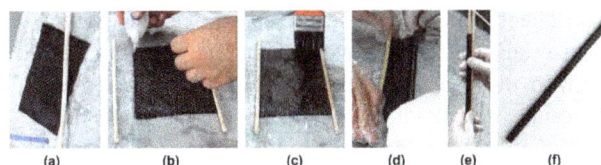

Figure 9. Manufacturing procedure of MMFRP rod [16]

Figure 10. Fabrication procedure of MMFRP end anchorage [16]

8.2. Advantages

a) Possibility of mounting anchorage system at the ends of the rod in order to improve the performance of these rods
b) Possibility of producing rods of various diameters
c) Possibility of using the flexural capacity of core in the cases of steel cores
d) lower cost compared to conventional FRP rods

8.3. Disadvantages

a) Time-consuming manufacturing process
b) Lower quality than that produced in the factories
c) Possibility of core corrosion or core decay.
d) Environmental pollution caused by the grooving process

9. Discussion and Conclusion

The EBR method despite the easy and quick installation process, is faced with serious challenges such as premature debonding of the FRP sheet from the concrete surface and vulnerability against the sever environmental conditions.

The conventional surface preparation processes such as surface roughening by sandblast, air jet and water jet can slightly postpone the debonding phenomenon but cannot eliminate it.

EBROG and EBRIG techniques can significantly increase the load carrying capacity of concrete members. Although these techniques, postpone or even in some cases eliminate the debonding phenomenon but are faced with problems such as vulnerability against sever environmental conditions and environmental pollution caused during the grooving process.

The MF-EBR method which combines the mechanical fastening of MF-FRP method with the external adhesion in EBR method, increases the load carrying capacity and ductility index of strengthened member compared to the EBR method. However, in this method the stress is transferred with low efficiency because of separated connection points. Also in this method the FRP materials are vulnerable against sever environmental conditions.

The HOLING method can increase the load carrying capacity of strengthened members but is faced with problems such as: premature debonding, environmental pollution causedduring the holing process and time-consuming installation process.

The NSM family methods postpone or even eliminate the debonding phenomenon. Also in these methods the strengthening materials are interred in the concrete cover and are confined with the concrete sides of the grooves. Therefore these materials are preserved against sever environmental conditions. Various types of FRP materials such as FRP rods, FRP strips and MMFRP rods can be used in the NSM method. Environmental pollution caused during the grooving process is an important problem in this method.

The key advantage of the MMFRP rod, is that it allows the incorporation of an anchor system that can be used to improve the performance of NSM method.

References

[1] A. Carolin, "Carbon Fiber Reinforced Polymers for Strengthening of Structural Elements," Doctoral Thesis, Lulea University of Technology, Sweden, p. 247, 2003.

[2] H. Nordin, " Fiber Reinforced Polymers in Civil Engineering," PhD. Thesis, Lulea University of Technology, Sweden, p. 57, 2003.

[3] S. M. Soliman, "Flexural Behaviour of Reinforced Concrete Beams Strengthened with Near Surface Mounted FRP Bars," Ph.D Thesis, University of Sherbrooke, Sherbrooke, Quebec, Canada, p. 12, 2008.

[4] M. Arduini, A. Nanni, "Behavior of Precracked RC Beams Strengthened with Carbon FRP Sheets," Journal of Composites for Construction, Vol. 1, No. 2, pp. 63–70, 1997.

[5] D. M. Nguyen, T. K. Chan,H. K. Cheong, "Brittle Failure and Bond Development Length of CFRP-Concrete Beams, " ASCE, Journal of Composites for Construction, Vol. 1, No. 5, pp. 7-12, 2001.

[6] P. Mukhopadhyaya, N. Swamy, "Interface Shear Stress: A New Design Criterion for Plate Debonding," ASCE, Journal of Composites for Construction, Vol. 1, No. 5, pp. 35-43, 2001.

[7] N. F. Grace, G. A. Sayed, A. K. Soliman, K. R. Saleh, "Strengthening of Reinforced Concrete Beams Using Fiber Reinforced Polymer (FRP) Laminates," Journal of Structures, ACI, Vol. 5, No. 96, pp. 865-74, 1999.

[8] D. Mostofinejad, E. Mahmoudabadi, "Grooving as an Alternative Method of Surface Preparation to Postpone Debonding of FRP Laminates in Concrete Beams," Journal of Composites for Construction, ASCE, V. 6, No. 14, pp. 804-11, 2010.

[9] D. Mostofinejad, S. M. Shameli, "Externally Bonded Reinforcement In Grooves (EBRIG) Technique to Postpone Debonding of FRP Sheets in Strengthened Concrete Beams," Journal of Construction and Building Materials, Vol. 38, pp. 751-758, 2013.

[10] F. Micelli, A. Rizzo, D. Galati, "Anchorage of Composite Laminates in RC Flexural," Journal of Composites for Construction,Vol. 3, No. 11, pp. 117-26, 2010.

[11] J. M. Sena-Cruz, J. A. O. Barros, M. R. F. Coelho, L. F. F. T. Silva, "Efficiency of Different Techniques in Flexural Strengthening of RC Beams under Monotonic and Fatigue Loading," Construction and Building Materials, No. 29, pp. 175-182, 2012.

[12] M. R. Eftekhar, M. Yaqubi, "Holing method to postpone debonding in strengthened RC beams with FRP," 4th annual national conference of concrete, Iran, 2012.

[13] A. Hajihashemi, D. Mostofinejad, M. Azhari, "Investigation of RC Beams Strengthened with Prestressed NSM CFRP Laminates," ASCE, Journal of Composites for Construction, Vol. 6, No. 15, pp. 887-95, 2011.

[14] L. De Lorenzis, A. Nanni, "Shear Strengthening of Reinforced Concrete Beams with Near-Surface Mounted FRP Rods," ACI Structural Journal, Vol. 98, No. 1, pp. 60-68, 2001.

[15] A. Rizzo, L. De Lorenzis, "Behavior and capacity of RC beams strengthened in shear with NSM FRP reinforcement," Journal of Composites for Construction, Vol. 5, No. 2, p. 114, 2001.

[16] M. Jalali, M. K. Sharbatdar, J. Fei-Chen, F. Jandaghi Alaee, "Shear Strengthening of RC Beams Using Innovative Manually Made NSM FRP Bars," Construction and Building Materials Journal, Vol. 36, pp. 990-1000, 2011.

25

Seismic Performance Analysis of RCC Multi-Storied Buildings with Plan Irregularity

Mohaiminul Haque[1], Sourav Ray[1, *], Amit Chakraborty[2], Mohammad Elias[1], Iftekharul Alam[1]

[1]Department of Civil and Environmental Engineering, Shahjalal University of Science and Technology (SUST), Sylhet, Bangladesh
[2]Department of Civil Engineering, Leading University, Sylhet, Bangladesh

Email address:
sourav.ceesust@gmail.com (S. Ray), sourav-cee@sust.edu (S. Ray), pallab87.sust@gmail.com (M. Haque)
[*]Corresponding author

Abstract: Bangladesh is one of the most earthquake prone areas in South-Asia and Sylhet is the most seismic vulnerable region in Bangladesh. Seismic performance analysis is highly recommended to ensure safe and sound building structures for this region. To get better performance from reinforced concrete (RCC) structure, new seismic design provisions require structural engineers to perform both static and dynamic analysis for the design of structures. The objective of the this study is to carry out static and dynamic analysis i.e. equivalent static analysis, response spectrum analysis (RSA) and time history analysis (THA) over different regular and irregular shaped RCC building frame considering the equal span of each frame as per Bangladesh National Building Code (BNBC)- 2006. In this study, four different shaped (W-shape, L-shape, Rectangle, Square) ten storied RCC building frames are analysed using ETABS v9.7.1 and SAP 2000 v14.0.0 for seismic zone 3 (Sylhet) in Bangladesh. Comparative study on the maximum displacement of different shaped buildings due to static loading and dynamic response spectrum has been explored. From the analyzed results it has been found that, for static load analysis, effects of earthquake force approximately same to all models except model-1(W-shape).W-shape has been found most vulnerable for earthquake load case. It is also found from the response spectrum analysis that the displacements for irregular shaped building frames are more than that of regular shaped building. The overall performance of regular structures is found better than irregular structures.

Keywords: Equivalent Static Analysis, Time History Analysis, Response Spectrum Analysis,
Regular and Irregular Shape Building, Displacement, Seismic Evaluation

1. Introduction

Bangladesh is one of the most densely populated country of the world. Due to the large population and small per capita area, the construction of mid to high-rise buildings is becoming quite familiar in the country [1]. As Bangladesh is located in one of the most active seismic region of the world, consideration of earthquake loads in structural design has become a significant issue [2]. The behaviour of a building during an earthquake depends on several factors, stiffness, adequate lateral strength and ductility, simple and regular configurations [3]. At the time of any disaster like earthquake, cyclone or tornado, failure of structure starts at points of weakness. This weakness arises due to discontinuity in mass, stiffness and geometry of structure [4, 5]. The structures having this discontinuity are termed as irregular structures. Irregularities are one of the major reasons of failures of structures during earthquakes [4]. Among all the factors configuration of a building is an important feature which has huge influence on the damage during the earthquake shaking [6, 7]. The feature of the regularity and symmetry in the overall shape of the building both in plan and elevation enormously affects the response of the building under static and dynamic loading [8]. But nowadays the need and demand of the modern era and growing population has made the architects or engineers forced towards planning of irregular structures [9] which needs additional careful structural analysis so that acceptable behaviour of the structures can be ensured throughout a devastating earthquake [10]. So seismic analysis must be done for regular and irregular medium to high-rise buildings.

In a study, Bagheri et. al. (2012) compared the damage assessment of an irregular building based on static and dynamic analysis [11] and found greater displacement by static analysis compared to dynamic analysis. In a study of Ravikumar et. al. (2012), performance of various irregular building was observed for the hard rock region in India [3]. In Sharma's (2013) study, the effects of various vertical irregularities on seismic response of a structure has been discussed [4]. Response spectrum analysis and time history analysis was done to observe the seicmic response but equivalent static analysis was not considered in that study. He observed that the geometrically irregular shaped buildings experienced higher displacements than regular shaped building.

This study aims is to evaluate the impact of shape on earthquake response of RCC multi-storied building frames in according to BNBC-2006. The storey displacements have been obtained by using equivalent static, time history and response spectrum analysis. The results obtained are compared to determine the structural performance.

2. Materials and Methods

2.1. Methods of Seismic Analysis

Seismic analysis is a major tool in earthquake engineering which is used to understand the response of buildings due to seismic excitations in a simpler manner. In the past the buildings were designed just for gravity loads and seismic analysis is a recent development [4]. It is a part of structural analysis and a part of structural design where earthquake is prevalent. Different types of earthquake load analysis methods are used in this study those are given below.

2.1.1. The Equivalent Static Analysis

Dynamic nature of the load must be considered to analyse all the structures under seismic load. However, in most codes equivalent linear static methods is permitted to analyse regular, low to medium-rise buildings. It can be done with an estimation of base shear load and its distribution on each storey calculated by using formulas given in the code [12]. Then the displacement demand of model must be checked with code limitation [8]. According to BNBC-2006 displacement limitation is,

i) $\Delta < 0.04h/R < 0.005h$ for $T < 0.7$ seconds
ii) $\Delta < 0.03h/R < 0.004h$ for $T \geq 0.7$ seconds

where, h = height of building or structure, R = Response modification co-efficient and T = fundamental period of vibration in seconds, of the structure.

In this study equivalent static analysis carried out by ETABS v9.7.1. The data used for this study are given in the Table 1.

Table 1. *Necessary data for static load analysis.*

Seismic zone coefficient.	Z=0.25(zone 3)
Soil profile type	S=1.5
Structural importance factor	I=1
Response modification factor for RCC frame	R=12

2.1.2. Time History Analysis

Time history analysis is a powerful tool for the study of structural seismic response [13]. It is an analysis of the dynamic response of the structure at each increment of time, when its base is subjected to a specific ground motion time history. Recorded ground motion from past natural earthquakes can be used for time history analysis [11, 14]. An earthquake ground motion records during the earthquake at 'Loma Prieta' in 1989 in Loss Angel's area has been selected for this purpose as there is no recorded data available for past earthquakes in Bangladesh region [15]. The magnitude of earthquake was 7 with ground motion having 2% probability of being exceeded in 50 years. Minimum and maximum PGA is 0.39g and 0.41g respectively. In this study SAP 2000 software is used for time history analysis. Details of this earthquake data are given in the Table 2.

Table 2. *Data for time history analysis [16].*

Parameters	Numerical value
PGA	409.95 cm/sec^2
Duration of Earthquake	25 seconds
Time interval	0.01
Distance from epicenter	3.5 km

2.1.3. Response Spectrum Analysis

As per BNBC-2006, a site specific response spectra is required based on the geologic, tectonic, seismologic and soil characteristics associated with the specific site. In absence of a site specific response spectrum, the normalized response spectra for damping ratio 5% shall be used in the dynamic analysis [8]. BNBC response spectrum curve which has been used in this study is given in the Fig. 1. SAP-2000 is used for response spectrum analysis.

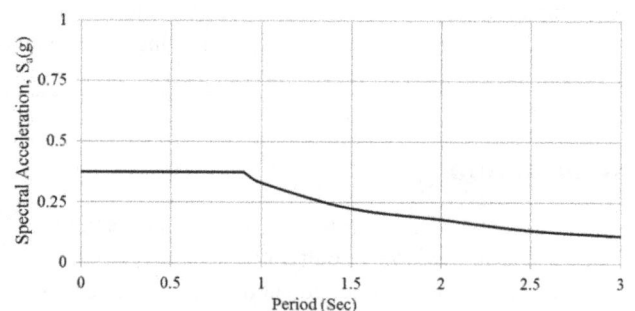

Fig. 1. *BNBC response spectrum curve for 5% damping ratio.*

2.2. Details of Models

Among different regular structures, square and rectangular shaped structures, which are the most common shape of regular structures in Sylhet have been chosen in this study. Irregular structures, those are very common in Sylhet, have been selected for this study. However, this paper is focused only with plan irregularity. 2D views of Model-1 (W-shape), Model-2 (L-shape), Model-3 (Rectangular shape) and Model-4 (Square-shape) are shown below in Fig. 2.

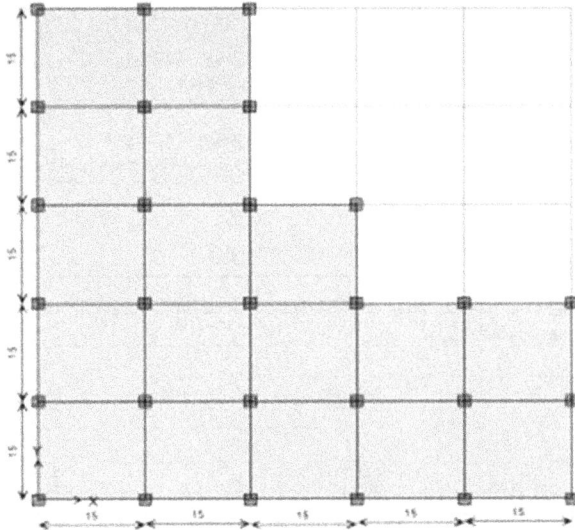

a) Model-1 (W-shape)

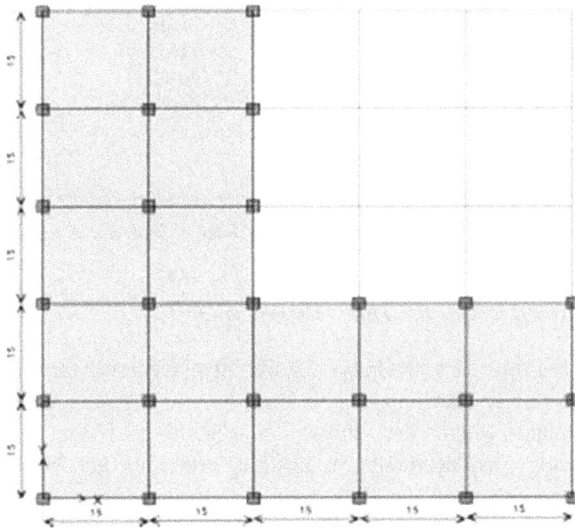

b) Model-2 (L-shape)

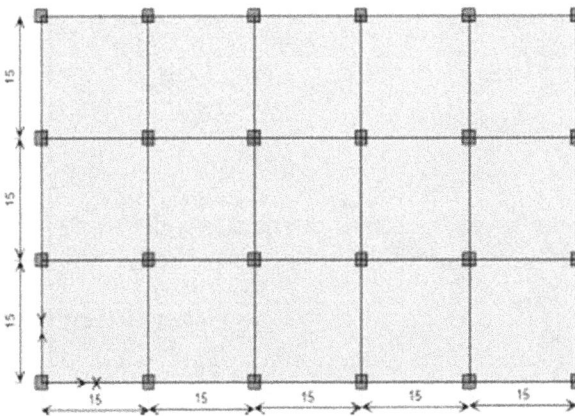

c) Model-3 (Rectangular shape)

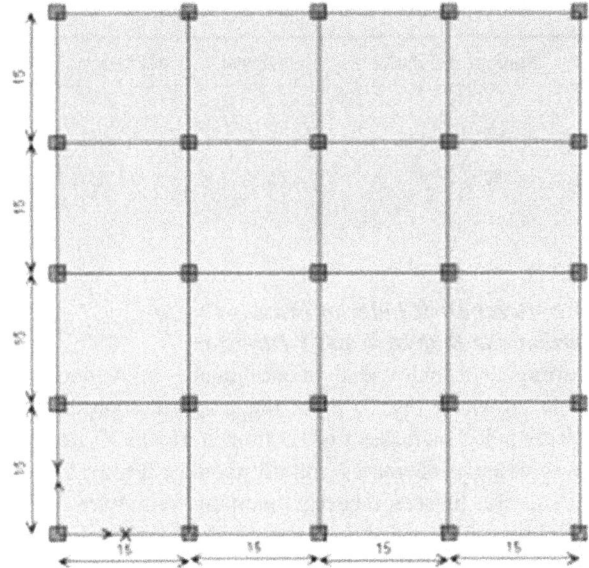

d) Model-4 (Square shape)

Fig. 2. *2D view of Models.*

Size of columns, beams and slabs are taken approximately for 10 storied buildings considered in this study. The area of the structures are kept close to each other to keep their influence similar. The span lengths are kept constant both in X and Y direction of different shapes. Storey height is taken as 10 ft. Loads applied on structures include dead load, live load, and earthquake load according to BNBC-2006. Structural dimension details and material properties are presented in Table 3 and Table 4 respectively.

Here the specifications are in a tabular form:

Table 3. *Structural Dimensions.*

Component	Model-1	Model-2	Model-3	Model-4
Area (sft)	3825	3600	3375	3600
Beam size	18 in x 12 in			
Column size	20 in x 20 in			
Slab thickness	5 in			
Height of storey	10 ft			
Span	15 ft			

Table 4. *Materials properties.*

Component	Values (unit)
Compressive strength of concrete	4000 psi
Modulus of elasticity of concrete	3600 ksi
Shear modulus of concrete	1500 ksi
Unit Weight of concrete	150 pcf
Yield stress of steel	60 ksi

3. Analysis Results and Discussion

3.1. Analyzed Result for Equivalent Static Analysis

Base shear found from ETABS and SAP 2000 was compared with manual calculation which is shown in Table 5. From this table it can be seen that the result obtained from ETABS and SAP 2000 varies from manual calculation only by 2.03% and 0.245% respectively.

Table 5. *Comparison of base shear of Models.*

Model	Manual calculation	SAP2000	ETABS
1			
2			
3	4.08% of W	4.09 % of W	3.997 % of W
4			

3.1.1. Inter-storey Drift Index of Frames Due to Earthquake Both in X and Y Direction

Inter-storey drift index due to earthquake in X and Y direction is shown in Fig. 3. From the graph it is observed that the drifts index increases from bottom storey to 3rd storey and then gradually decreases for all models. From these figures it can also be seen that maximum and minimum drift index is observed in Model-1 and Model-4 respectively. Maximum drifts in Model-1 0.002919 whereas in Model-4 it is found 0.001604. Model-2 and Model-3 shows almost similar values of Model-4. So Model-1 shows poor performance in terms of drift index compared to other models.

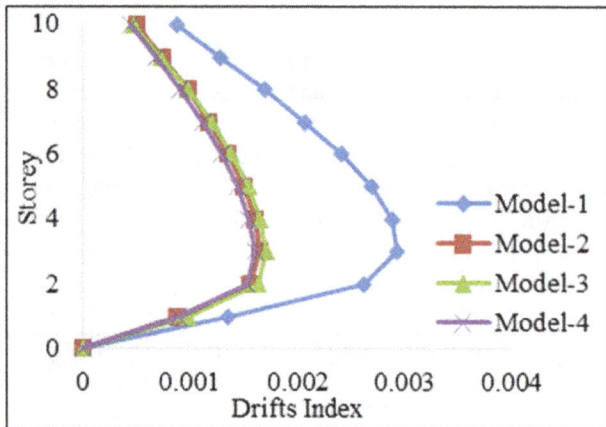

Fig. 3. *Storey vs drift index for Earthquake load in X &Y direction.*

3.1.2. Comparison of Displacements for Static Load Analysis

From equivalent static analysis, storey displacements for all structures due to Earthquake load both in X and Y directions are obtained which are shown in Fig. 4 and Fig. 5 respectively. Storey displacement is found almost similar for all the structures except Model-3 due to its change of moment of inertia in X and Y direction. It is observed that maximum displacement yields for Model-1 at every storey compared to other models whereas minimum displacement yields for Model-4. Maximum storey displacement of Model-1 is 2.4906 inch which is 1.8 times of Model-4. So Model-1 will experience more damage during earthquake. Allowable deflection is calculated for all the structures. It is found that deflections of all structures lie within allowable limit. But deflections observed in Model-1 were very close to allowable limit. So the performance of Model-1 is not satisfactory.

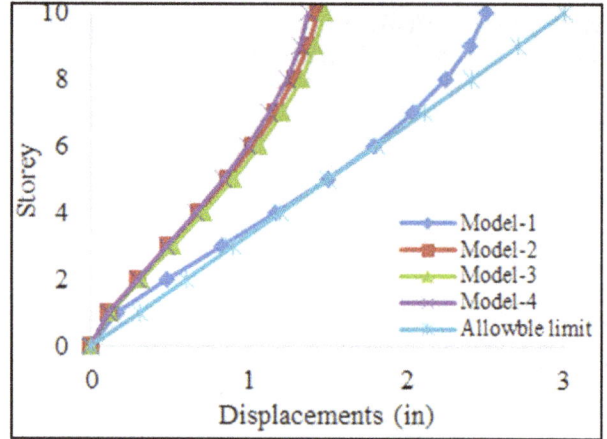

Fig. 4. *Storey vs displacement for earthquake load in X direction using equivalent static method.*

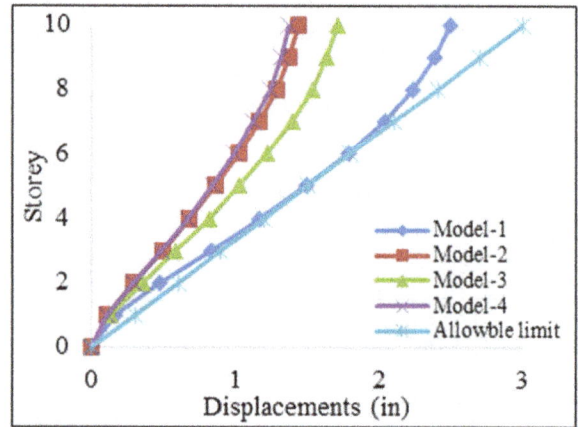

Fig. 5. *Storey vs displacement for earthquake load in Y direction using equivalent static method.*

3.2. Analyzed Result for Time History Analysis

From the time history analysis storey displacements for all structures due to Earthquake load both in X and Y direction are obtained which are shown in Fig. 6 and Fig. 7 respectively. Displacements in all the structures are very close to each other. In this case, all structures exceed displacements criteria for dynamic analysis.

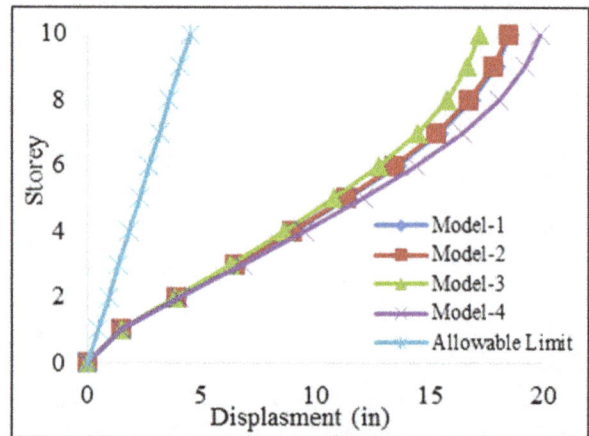

Fig. 6. *Storey vs displacement in X direction using time history analysis.*

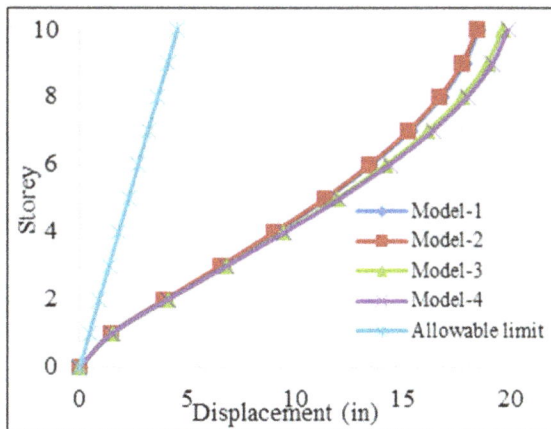

Fig. 7. Storey vs displacement in Y direction using time history analysis.

3.3. Analyzed Result for Response Spectrum Analysis

From the response spectrum analysis storey displacement for all structures are obtained both for X and Y direction which are shown in Fig. 8 and Fig. 9 respectively. From the graphs it is found that all the structures exceed displacements criteria given in BNBC-2006. Greater displacements are found for Model-1 and Model-2 compared to Model-3 and Model-4.

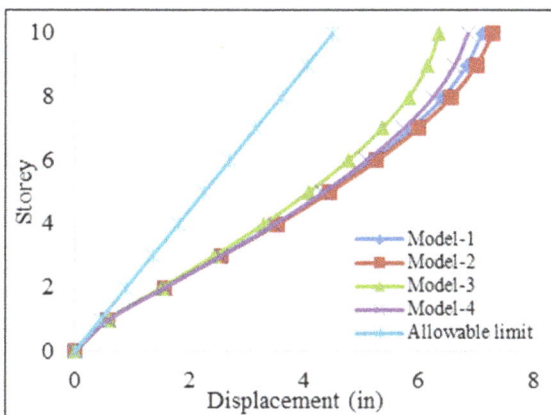

Fig. 8. Storey vs displacement in X direction using response spectrum analysis.

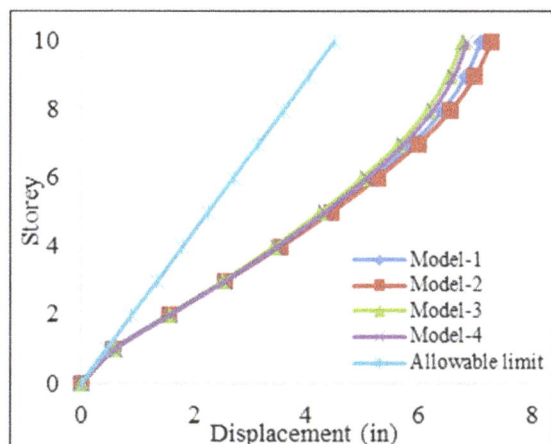

Fig. 9. Storey vs displacement in Y direction using response spectrum analysis.

4. Conclusion

From the analysis of various shaped multi-storied buildings it is found that all structures fulfil displacement criteria for equivalent static analysis though Model-1 just touches allowable limit curve. Deflection in Model-1 has been found more than 80% from Model-4. Storey drift indexes increase with the storey height upto 3rd storey reaching to maximum that start to decrease for all four models. Displacements obtained from the time history analysis are much higher than the allowable limit for all the Models. The difference of displacement values among all the models is insignificant since the weights of the structures are similar. From the response spectrum analysis it is also found that maximum displacement for all the structures exceed allowable limit. However, these values are much lesser than the values obtained from time history analysis. The difference of displacement values among all four shapes is insignificant in lower stories but it increased in higher stories and reached peak at top stories. Irregular shaped structures (Model-1 and Model-2) shows greater displacement than Regular shaped structures (Model-3 and Model-4). So from the overall analysis it can be conclude that performance of buildings irregular in plan is more susceptible to earthquake load than regular shaped buildings.

References

[1] Bari, M. S. and T. Das, *A Comparative Study on Seismic Analysis of Bangladesh National Building Code (BNBC) with Other Building Codes.* Journal of The Institution of Engineers (India): Series A, 2013. 94(3): p. 131-137.

[2] Barua, K., S. M. Hasanur Rahman, and S. Das, *Performance Based Analysis of Seismic Capacity of Mid Rise Building.* International Journal of Emerging Technology and Advanced Engineering, 2013. 3(11): p. 44-52.

[3] Ravikumar, C., et al., *Effect of irregular configurations on seismic vulnerability of RC buildings.* Architecture Research, 2012. 2(3): p. 20-26.

[4] Sharma, A., *Seismic Analysis and Design of Vertically Irregular RC Building Frames.* 2013, National Institute of Technology Rourkela.

[5] Kumar, M. and V. G. Babu, *Comparative Study of Seismic Performance of Building Having Mass Vertical Irregularity at Different Floor Levels.* International Journal of Science and Research, 2016. 5(1): p. 895-899.

[6] Kabir, R., D. Sen, and M. Islam, *Response of multi-storey regular and irregular buildings of identical weight under static and dynamic loading in context of Bangladesh.* International Journal of Civil & Structural Engineering, 2015. 5(3): p. 252-260.

[7] Islam, S. and M. M. Islam, *Analysis on the structural systems for drift control of tall buildings due to wind load: critical investigation on building heights.* The AUST Journal of Science and Technology, 2014. 5(1).

[8] BNBC, *Bangladesh National Building Code.* 2006: Housing and Building Research Institute Dhaka, Bangladesh.

[9]   Valmundsson, E. V. and J. M. Nau, *Seismic response of building frames with vertical structural irregularities.* Journal of Structural Engineering, 1997. 123(1): p. 30-41.

[10] Herrera, R. G. and C. G. Soberon. *Influence of plan irregularity of buildings.* in *The 14th World Conference on Earthquake Engineering.* 2008.

[11] Bagheri, B., E. S. Firoozabad, and M. Yahyaei, *Comparative Study of the Static and Dynamic Analysis of Multi-Storey Irregular Building.* World Academy of Science, Engineering and Technology, International Journal of Civil, Environmental, Structural, Construction and Architectural Engineering, 2012. 6(11): p. 1045-1049.

[12] Bagheri, B., K. A. Nivedita, and E. S. Firoozabad, *Comparative damage assessment of irregular building based on static and dynamic analysis.* International Journal of Civil and Structural Engineering, 2013. 3(3): p. 505.

[13] Mwafy, A. and A. Elnashai, *Static pushover versus dynamic collapse analysis of RC buildings.* Engineering structures, 2001. 23(5): p. 407-424.

[14] Arvindreddy and F. R. J., *Seismic analysis of RC regular and irregular frame structures.* International Research Journal of Engineering and Technology (IRJET), 2015. 2(5): p. 44-47.

[15] Ansary, M. and M. Sharfuddin, *Proposal for a new seismic zoning map for Bangladesh.* J Civ Eng, 2002. 30(2): p. 77-89.

[16] *Pacific Earthquake Engineering Research Center (PEER).* NGA database, http://nisee.berkeley.edu/.

26

Study on the Influence of Chloride Ions Content on the Sea Sand Concrete Performance

Wu Sun, Junzhe Liu*, **Yanhua Dai, Jiali Yan**

School of Civil Engineering, Ningbo University, Ningbo, Zhejiang, China

Email address:

491480031@qq.com (Wu Sun), junzheliu@163.com (Junzhe Liu), 963122649@qq.com (Jiali Yan), 759782287@qq.com (Yanhua Dai)

*Corresponding author

Abstract: The influence of chloride ions content on the sea sand concrete performance was investigated through testing the sea sand mortar strength in this paper. The concrete strength rule was analyzed to discover the early strength for the sea sand concrete by the presence of chloride. In addition, the XRD microscopic analysis and TG/DTA were observed the composition of concrete to research the influence on hydration processes of concrete caused by sea sand, and from two aspects to find out the differences of the internal microcosmic structure and the chemical composition respectively. And Micro-technique was used to determine the water-soluble chloride ion concentration of different types of sea sand concrete which were maintained 28 days. The results show that the chloride ion can improve concrete strength value and the concentration of free chloride increases with the rise of extraction temperature.

Keywords: Sea Sand Mortar, Water Soluble Chloride Ions, Micro Structures

1. Introduction

Sea sand concrete is a kind of concrete in which mixed sea sand as fine aggregate, which is large-scale application in the coastal areas and gets more and more attention of the researchers in recent years [1], especially in Ningbo area. The sea sand solves the problem of river sand shortage [2-6]. Therefore, scientific, standardized and reasonable use of local resources of sea sand concrete is of great practical significance to the local economic development. [7]

Generally, there are three types of chloride ions in the sea sand concrete: free chloride ions, physically bound and chemically bound chloride ions. Only the first type is responsible for the corrosion of rebars. Even though the pH of concrete is more than 12, the free chloride ions accumulated on the surface ofrebars can cause or aggravate the corrosion [8-9].

The history of using sea sand in China is very short, so the durability of the sea sand concrete need to be tested by many projects in the future. The chloride ions in sea sand is the one of main reasons which cause the concrete reinforcement corrosion. So in order to get a fine durability concrete, this paper clarified the actionmechanism of chlorine ions in sea

sand concrete, as well as the effects of internal microstructure of concrete and material compositions.

2. Experimental Procedure

2.1. Materials and Mix Proportions

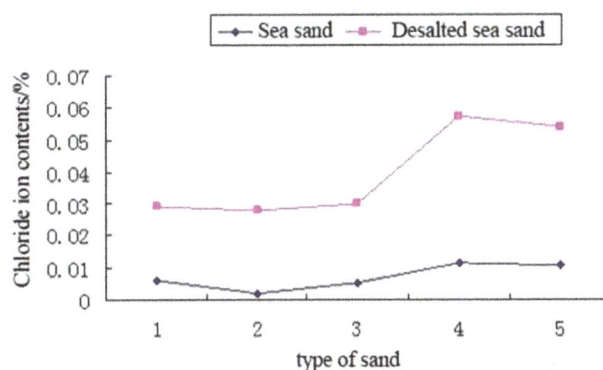

Figure 1. Chloride ion contents in sea sand (W%).

Grade 42.5 ordinary Portland cement produced by Ningboshunjiang Cement Plant was used in this study; River sand (RS) was used as fine aggregate, and its fineness modulus is 2.27. Regular sea sand (SS) and desalted sea sand (DSS) came from five plants and its chloride ion contents are shown in Figure 1.

Mixture proportions are shown in table 1. The mortar cube specimens of 70.7mm×70.7mm×70.7mm were prepared and tested after being cured in standard curing room for 28 days.

Table 1. Mix proportions of the mortar.

Sample	Sand type	Cl⁻ contents in sand (%)	Mix proportion (C:S:W)
HCA	RS	0.01	1:2.5:0.45
HCB	RS	0.06	1:2.5:0.45
D	DSS	0.01	1:2.5:0.45
W	SS	0.054	1:2.5:0.45

* RS—River sand

2.2. Test Method

2.2.1. Test of Sea Sand Mortar Strength

Different types of sea sand mortar were prepared according to JTJ 270-1998 testing code of concrete for port and waterwog engineering.

2.2.2. Test of Product Analysis and Microstructure of Mortar

The proportions of three samples HCA, HCB, W, Dfor TG/DTA and XRD tests were shown in Table 1. The mortar used for the SEM test was mixed with regular sea sand. The scanning electron microscope was produced by Japan Hitachi Company. The polycrystalline X-ray diffraction and TG/DTA thermal analyzer were produced by German Brueck and America Perkin Amelmer Company. The inner products, microstructure and its content in mortars can be determined by use of these analyzers.

2.2.3. Test of Total Chloride Content in Regular Sea Sand and Desalted Sea Sand

The sand was quartered to 1500g and dried in an oven with 105°C. Then the sand was cooled to room temperature.500g sand was put in a reagent bottle with rubber stopper then 500ml distilled water was added. For the complete extraction of chloride in sand, the bottle was vibrated once in 24h and thereafter three times every five minutes. After a certain time, the clear solution in the bottle was filtered and let the filtrate flow into a glass beaker. 50ml filtrate was transferred into a triangular flask by pipette. Then the filtrate was titrated by 0.01mol/L standard silver nitrate solution with 5% potassium chromate as indicator until the solution become red and the red color can be maintained for 5-10s. In the course of titration, the amount of standard silver nitrate solution consumed was recorded and the total chloride content in sea sand can be calculated according to the amount.

2.2.4. Test of Free Chloride Ion Content in Mortar

The motar cubes were cured at the standard condition for 28 days. The mole method was adopted to test the free chloride ion content. The pH of the solution under the test must be near neutral according to the requirement of mole method, however, the solution of mortar is alkaline. The dilute sulfuric acid was used for the neutralization in this study.

To investigate the effect of extraction temperature on the total or free chloride concentration of mortar, the tests of chloride ion content were carried out at 15°C and 65°C, respectively.

3. Experimental Results and Discussion

3.1. The Influence of Chlorine Salt Content on the Sea Sand Concrete Strength

Figure 2 and 3 shows no obvious regularity of between chlorine salt content and sea sand concrete strength. But simply fromgoup D and W , it can be foundthat higher sea sand concrete strength in higher chlorine salt content.It may be for two reasons, one reason is that the chloride ion can improve concrete strength value, another isthat the grading of not desalinate sea sand is slightly better than the grading of desalination sea sand.

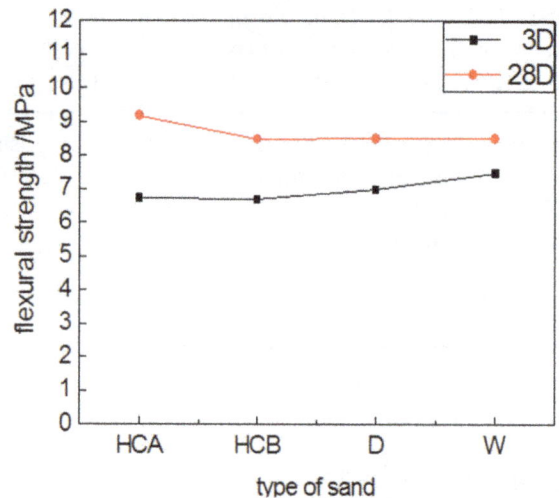

Figure 2. Sea sand mortar flexuralstrength (MPa).

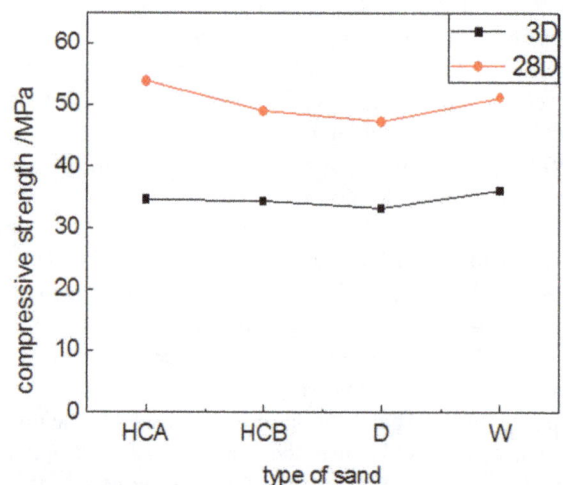

Figure 3. Sea sand mortar compressive strength (MPa).

3.2. Free Chloride Ion Content in Mortar

3.2.1. Total Chloride Ion Content in Regular Sea Sand and Desalted Sea Sand

Figure 1 shows chloride ion contents of samples derived from the five plants are different.The chloride ion content in desalted sea sand is significantly lower than regular sea sand. It also shows that the the higher the original chloride ion content, the higher the residual content is after the sea sand is desalted.

3.2.2. Free Chloride Ion Content in Sea Sand Mortar

Figure 4 shows the soluble and bonded chloride ion contents at 15°C and 65°C in mortars. It can be seen that there is a close correlation between free chloride concentration and extraction temperatures. At 15°C, free chloride concentration content varies from 22%-34% and at 65°C, it varies from 52%~65%. This indicates that the free chloride concentration of mortars at extraction temperature 65°C is about twice as mortars at 15°C. This may be due to that the extraction rate of physically bound chloride ion increase with an increase in temperature. Therefore, taking the differential between total chloride ion content and free chloride ion content as the amount of chemically bound chloride is somewhat inaccurate.

There are physically and chemically bound chloride ions in mortars. Chemical bonding generally results from the reaction between chlorides and C3A to produce Friedel's salt or the reaction with C4AF to produce a Friedel's salt analogue. Physical binding is owing to the adsorption of chloride ion to the C-S-H surfaces.

Majority of the chloride ion is in the form of physically bound to ion exchange sites of C-S-H gel and there exists a significant degree of reversibility. In fact, for a physical adsorption, an increased temperature accelerates the thermal vibration of absorbates, bring about more free chloride.

In addition, Figure 5 shows the free chloride extraction rate of the mortar with 20% fly ash is higher than that of the cement mortar without fly ash at extraction temperature15°C, however, the free chloride extraction rate of the mortars with or without fly ash is almost the same at extraction temperature 65°C. This may be due to less chemical bound and more physical bound chloride ions at lower extraction temperature.

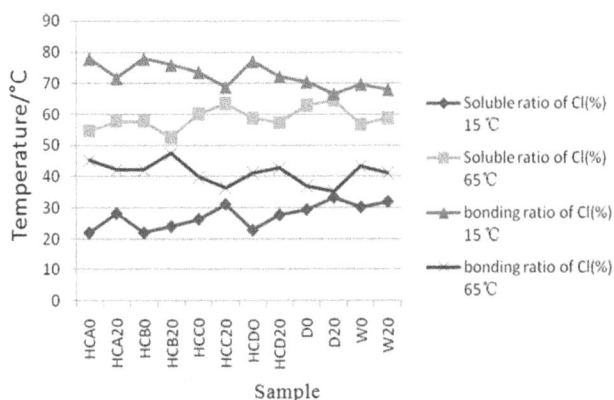

Figure 4. Existing status of chloride ions at different temperatures.

3.3. The Characteristic of Friedel's Salt in Sea Sand Mortar

As chloride enters the cementitious material, it may be converted to Friedel's salt due to chemical binding. Chloride ion can be introduced into concrete by two methods: (1) mixingas an additional agent (internal chloride); (2) penetrating from outside (external chloride). In the literature review, the chloride was frequently dissolved in the mixing water and then entered the mixture. At the same time, the amount of chloride introduced varies 2%~10% by weight of the cementitious material. Many researchers investigated the Friedel's salt with the above mentioned condition. However, because the chloride contents in the regular sea sand and desalted sea sandare much lower than that of abovementioned researches, it has not been reported if the Friedel's salt exsits in regular sea sand or desalted sea sand mortars and concretes and if it can be observed [10].

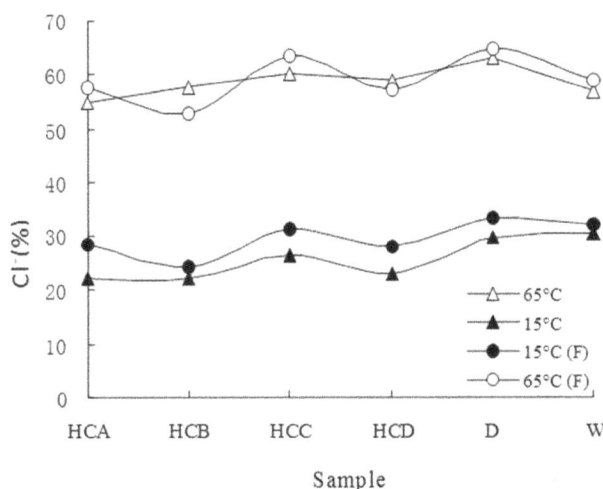

Figure 5. Relationship between the temperature and soluble chlorideion concentrations.

Generally, Friedel's salt yields an endothermal effect at about 360°C [11, 12]. Figure 6 shows the TG/DTA curves of mortars with river sand (sample HCD0), regular sea sand (sample W0) and desalted sea sand (sample D0), and the chloride content of these sands were 0.1%, 0.054% and 0.01%, respectively. Unfortunately, no clear endothermal peaks of Friedel's salt appear for the three TG/DTA curves of mortars.

To further clarify weather Friedel's salt exists in regular sea sand and desalted sea sand mortar, the XRD tests were carried out for the grounded mortars samples at 28 days of age and the results are shown in Figure 7. From Figure 7, regardless of the mortars with regular sea sand or desalted sea sand, several intensity peaks of Friedel's salt come ourat the corresponding situation. However, the intensity peak of the Friedel's salt is quite lower than the other compositions (for example Ca(OH)$_2$) of the mortar. This indicates that the chloride ions introduced by regular sea sand or desalted sea sand still form some Friedel's salt in the mortars.

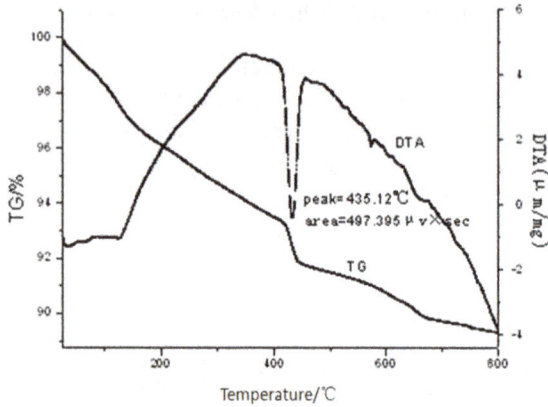

(a) TG/DTA curves of D

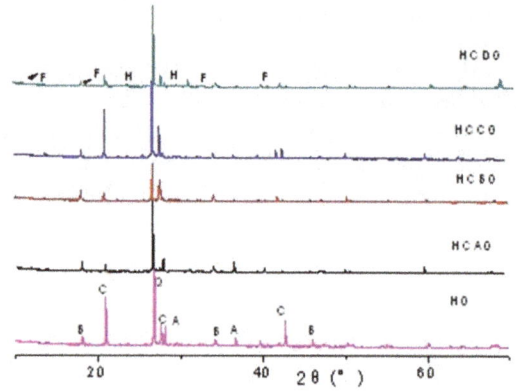

(a) XRD patterns of mortars with river sand

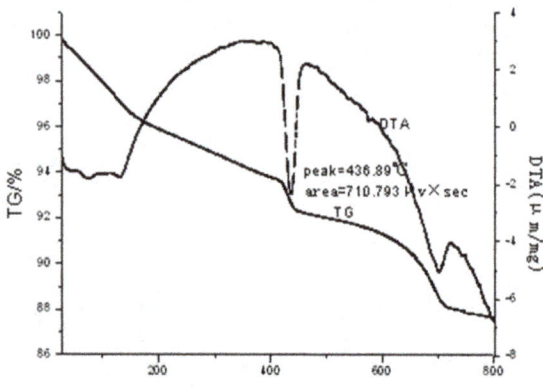

(b) TG/DTA curves of D

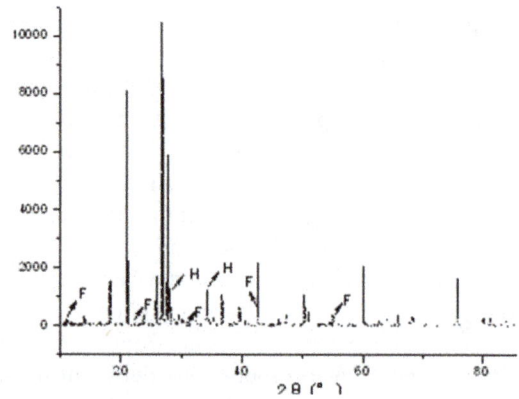

(b) XRD patterns of mortars with desalted sea sand

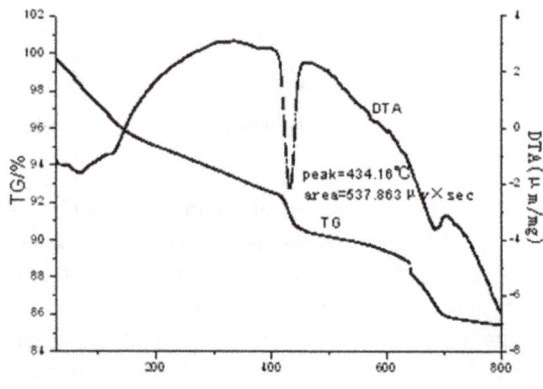

(c) TG/DTA curves of W

Figure 6. *TG/DTA curves of mortar.*

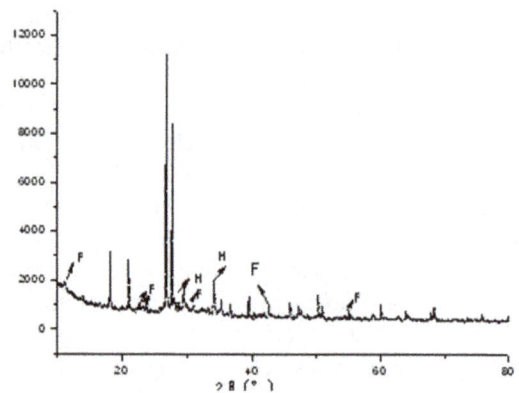

(c) XRD patterns of mortars with sea sand

Figure 7. *XRD patterns of mortars.*

Because the total chloride content in regular sea sand or desalted sea sand is relatively low, the intensity peak of Friedel's salt is very low. Moreover, the abovementioned TG/DTA curves of mortars do not show the end other mal peak of Friedel's salt. This is because the small amount of chloride ions introduced by regular sea sand or desalted sea sand formed a very small amount of unstable Friedel's salt. In this paper, the intensity peak of Friedel's salt can be clearly observed in the mortars with river sand containing more than 0.03% chloride content, and not in the mortars with river sand containing less than 0.03% chloride content.

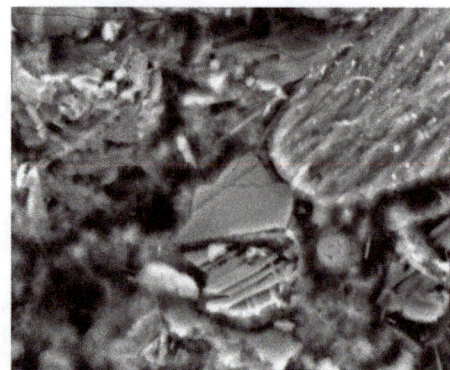

Figure 8. *SEM micrographs of Friedel's salt.*

The SEM micrographs of Friedel's salt in regular sea sand mortars is showed by Figure 8. It is clear to see that Friedel's salt's morphology is hexagonal slice in size of about 2~3µm. Base on the investigation of XRD and SEM, we can conclude that Friedel's salt exisits in sea sand mortars.

4. Conclusions

(1) The chloride ion can improve concrete strength value.

(2) Regardless of the mortars with river sand, regular sea sand or desalted sea sand, the free chloride content correlates closely to the extraction temperature. The concentration of free chloride increases with the rise of extraction temperature, and the free chloride concentration increases by about two times when the extraction temperature varies from 15°C to 65°C.

(3) The free chloride concentration of the mortar with fly ash is higher than that of the mortar without fly ash at 15extraction temperature. This may be due to the chemically bound chloride ion content is low and the physically bound chloride ion content is relatively high.

(4) Small quantity of unstable Friedel's salts exists in the mortars. It explains why the TG/DTA curves of regular sea sand and desalted sea sand mortars do not show end other mal peaks. Thus, it is unreasonable to determine the existence of Friedel's salt only using TG/DTA curves, and the invest igation of XRD and SEM should be carried out at same time.

Acknowledgements

This work has been supported by K. C. Wong Magna Fund in Ningbo University, the National Natural Science Foundation of China (No. 51278255, 51478227) and Ningbo Municipal Science and Technology Project (No. 2013C51006).

References

[1]   QI Gui-hai, WANG Yu-lin, LI Shuo, WANG Zhang-li. Review on sea sand concrete research in China. Concrete, 2013, 557-61.

[2]   S. Pack, M. Jung, H. Song, S. Kim, K. Ann. Prediction of time dependent chloride transport in concrete structures exposed to a marine environment. Cement and Concrete Research, 2010, 40(2) 302-312.

[3]   J. J SHI, W. SUN. Recent research on steel corrosion in concrete Chinese Ceramic Society, 2010, 38(9) 1753-1764.

[4]   Q. Yuan, C. J. Shi, G. D. Schutter, K. Audenaert, D. H. Deng. Chloride binding of cement-based materials subjected to external chloride environment-A review. Construction and Building Materials, 2009, 23(1) 1-13.

[5]   K. V. Subramaniam, M. D. Bi. Investigation of steel corrosion in cracked concrete: Evaluation of macrocell and microcell rates using Tafel polarization response. Corrosion Science, 2010, 52(8) 2725-2735.

[6]   Jianbin Chen,Junzhe Liu, Guoliang Zhang, Zhimin He. Study on the strength of sea sand concrete introduced by chloride ions, 2011 Second International Conference on Mechanic Automation and Control Engineering, 2011, 250-253 262-265.

[7]   S. HE. The application of desalination sea sand in commercial concrete. Fujian Building Materials, 2013, 11 45-46.

[8]   Z. Q. JIN, W. SUN, T. J. ZHAO, Q. Y. LI. Chloride binding in concrete exposed to corrosive solutions. Chinese Ceramic Society, 2009, 37(7) 1068-1072.

[9]   S. M. Abd El Haleem, S. Abd El Wanees .Environmental factors affecting the corrosion behavior of reinforcing steel. IV. Variation in the pitting corrosion current in relation to the concentration of the aggressive and the inhibitive anions. Corrosion Science, 2010, 52(5) 1675-1683.

[10] J. Z. LIU, F. XING, Z. M. HE, Z. DING. Study on critical value of $n(NO_2^-)/n(Cl^-)$ in reinforced concrete. Chinese Ceramic Society, 2010, 38(4) 68-73.

[11] C. Abate, B. E. Scheetz. Aqueous phase equilibria in the system $CaO–Al_2O_3–CaCl_2–H_2O$: The significance and stability of Friedel's salt. Journal of American Ceramic Society, 1995, 78(4) 939-944.

[12] Rafael Talero. Synergic effect of Friedel's salt from pozzolan and from OPC co-precipitating in a chloride solution. Construction and Building Materials, 2012, 33(8) 164-180.

Permissions

List of Contributors

Atik Sarraz, Md. Khorshed Ali and Debesh Chandra Das
Department of Civil Engineering, University of Information Technology and Sciences (UITS), Chittagong, Bangladesh

Ahmat-Charfadine Mahamat, Mahamat Barka, Abakar Mahamat Tahir and Malloum Soultan
Laboratoire des Energies Renouvelables et des Matériaux Locaux de Faculté des Sciences Exactes et Appliquées de l'Université de N'Djaména, Tchad

Salif Gaye and Aboubakar Cheikh Beye
Laboratoire de Matériaux, Mécanique et Hydraulique de la Faculté des Sciences et Techniques de l'Université de Thiès, Sénégal

Najm Obaid Salim Alghazali
Corresponding author, Asst. Prof. Doctor, Civil Engineering Department, Babylon University, Iraq

Hala Kathem Taeh Alnealy
M. Sc. Student, Civil Engineering Department, Babylon University, Iraq

Nabil I. El-Sawalhi
Civil Engineering Department, The Islamic University, Gaza, Palestinian Territories

Ahmed El-Riyati
United Nations Development Program, Gaza, Palestinian Territories

Hala Kathem Taeh Alnealy and Najm Obaid Salim Alghazali
Civil Engineering Department, Babylon University, Babylon, Iraq

Md. Serazul Islam
School of Agriculture and Rural Development, Bangladesh Open University, Gazipur-1705, Bangladesh

Tetsuro Tsujimoto
Department of Civil Engineering, Nagoya University, Nagoya, Japan

Lekariap Edwin Mararo and Abiero Gariy
Department of Civil, Construction and Environmental Engineering, Jomo Kenyatta University of Agriculture and Technology, Nairobi, Kenya

Mwatelah Josphat
Department of Geomatic Engineering and Geospatial Information Systems, Jomo Kenyatta University of Agriculture and Technology, Nairobi, Kenya

Abdulkadhim A. Hasan
Department of Electronics and Communications Engineering, Kufa University, Al-Najaf, Iraq

Aram Mohammed Raheem
Civil Engineering Department, University of Kirkuk, Kirkuk, Iraq

Mohammed Abdulsalam Abdulkarem
Geotechnical Engineer, Ministry of Construction and Housing, Kirkuk, Iraq

Hala Kathem Taeh Alnealy
Civil Engineering Department, Babylon University, Iraq

Hesham Abdel Khalik, Shafik Khoury and Remon Aziz
Structural Engineering Department, Alexandria University, Alexandria, Egypt

Mohamed Abdel Hakam
Construction Engineering and Management Department, Pharos University in Alexandria, Alexandria, Egypt

Hesham Abd El Khalek
Construction Engineering and Management, Faculty of Engineering, Alexandria University, Alexandria, Egypt

Remon Fayek Aziz
Construction Engineering and Management, Faculty of Engineering, Alexandria University, Egypt

Hamada Kamel
Faculty of Engineering, Alexandria University, Alexandria, Egypt

Dipu Sutradhar, Mintu Miah, Golam Jilany Chowdhury and Mohd. Abdus Sobhan
Department of Civil Engineering, Rajshahi University of Engineering & Technology, Rajshahi, Bangladesh

Mohammed Naguib, Fikry A. Salem and Khloud El-Bayoumi
Civil Engineering Dep., Faculty of Engineering, Mansoura University, Egypt

Mohammad Taghipour
Industrial Engineering, Science & Research Branch of Islamic Azad University, Tehran, Iran

Hesamoldin Yazdi
Civil engineering, non-profit institution of higher education, Aba - Abyek, Qazvin, Iran

Msafiri Atibu Seboru
Department of Education and External Studies, University of Nairobi, Nairobi, Kenya

Mark Adom-Asamoah
Department of Civil Engineering, Kwame Nkrumah University of Science and Technology, Kumasi, Ghana

Nobel Obeng Ankamah
Department of Civil Engineering, Sunyani Polytechnic, Sunyani, Ghana

Mohammad Naderi Pour and Adel Asakereh
Department of Civil Engineering, University of Hormozgan, Bandar Abbas, Iran

Peng Qu
College of Civil Engineering, Chongqing Jiaotong University, Chongqing, China
China Merchants Chongqing Communications Technology Research & Design Institute Co., Ltd., Chongqing, China

Cong Li and Xueming Jia
College of Civil Engineering, Chongqing Jiaotong University, Chongqing, China

China Merchants Chongqing Communications Technology Research & Design Institute Co., Ltd., Chongqing, China

Abdul-Manan Dauda
Tamale Polytechnic, Department of Building Technology, Tamale, Ghana

Yaw Adubofour Tuffour and Daniel Atuah Obeng
Department of Civil Engineering, Kwame Nkrumah University of Science and Technology, Kumasi, Ghana

Nana Kwesi Agyepong
Materials Division, Ghana Highway Authority, Ministry of Roads and Highways, Accra, Ghana

Fuchun Song, Jie Zhao and Mengchen Li
School of Traffic Engineering, Shenyang Jianzhu University, Shenyang, China

Hamid Reza Madihi and Sina Bani Amerian
Graduate Faculty of Environment, Tehran University, Tehran, Iran

Md. Masud Alom
Department of Civil & Environmental Engineering, Uttara University, Dhaka, Bangladesh

Seyyed Mohammad Banijamali and Mohammad Reza Esfahani
Dept. of Civil Engineering, Ferdowsi University of Mashhad (FUM), Mashhad, Iran

Shoeib Nosratollahi, Mohammad Reza Sohrabi and Seyyed Roohollah Mousavi
Dept. of Civil Engineering, University of Sistan and Baluchestan (USB), Zahedan, Iran

Mohaiminul Haque, Sourav Ray, Mohammad Elias and Iftekharul Alam
Department of Civil and Environmental Engineering, Shahjalal University of Science and Technology (SUST), Sylhet, Bangladesh

Amit Chakraborty
Department of Civil Engineering, Leading University, Sylhet, Bangladesh

Wu Sun, Junzhe Liu, Yanhua Dai and Jiali Yan
School of Civil Engineering, Ningbo University, Ningbo, Zhejiang, China

Index